T0250277

Artificial Intelligence for Health 4.0: Challenges and Applications

RIVER PUBLISHERS SERIES IN BIOTECHNOLOGY AND MEDICAL RESEARCH

Series Editors

PAOLO DI NARDO
University of Rome Tor Vergata,
Italy

PRANELA RAMESHWAR
Rutgers University,
USA

ALAIN VERTES
London Business School,
UK and NxR Biotechnologies,
Switzerland

Aiming primarily at providing detailed snapshots of critical issues in biotechnology and medicine that are reaching a tipping point in financial investment or industrial deployment, the scope of the series encompasses various specialty areas including pharmaceutical sciences and healthcare, industrial biotechnology, and biomaterials. Areas of primary interest comprise immunology, virology, microbiology, molecular biology, stem cells, hematopoiesis, oncology, regenerative medicine, biologics, polymer science, formulation and drug delivery, renewable chemicals, manufacturing, and biorefineries.

Each volume presents comprehensive review and opinion articles covering all fundamental aspect of the focus topic. The editors/authors of each volume are experts in their respective fields and publications are peer-reviewed.

For a list of other books in this series, visit www.riverpublishers.com

Artificial Intelligence for Health 4.0: Challenges and Applications

Editors

Rishabha Malviya
Galgotias University, India

Naveen Chilamkurti
La Trobe University, Australia

Sonali Sundram
Galgotias University, India

Rajesh Kumar Dhanaraj
Galgotias University, India

Balamurugan Balusamy
Galgotias University, India

River Publishers

Routledge
Taylor & Francis Group

LONDON AND NEW YORK

Published 2022 by River Publishers

River Publishers

Alsbjergvej 10, 9260 Gistrup, Denmark

www.riverpublishers.com

Distributed exclusively by Routledge

4 Park Square, Milton Park, Abingdon, Oxon OX14 4RN

605 Third Avenue, New York, NY 10017, USA

Artificial Intelligence for Health 4.0: Challenges and Applications / Rishabha Malviya, Naveen Chilamkurti, Sonali Sundram, Rajesh Kumar Dhanaraj and Balamurugan Balusamy.

Routledge is an imprint of the Taylor & Francis Group, an informa business

ISBN 978-87-7022-784-1 (print)

ISBN 978-10-008-4701-7 (online)

ISBN 978-1-003-37358-2 (ebook master)

While every effort is made to provide dependable information, the publisher, authors, and editors cannot be held responsible for any errors or omissions.

Contents

2 Data Imaging, Clinical Studies, and Disease Diagnosis using Artificial Intelligence in Healthcare   19

Vandana Tyagi, Neelam Dhankher, Bhavna Tyagi,
Iidiko Csoka, and Amrish Chandra

6   Artificial-Intelligence-Based Cloud Computing Techniques for Patient Data Management   **149**

Akanksha Sharma, Ashish Verma, Rishabha Malviya, and Mahendran Sekar

**8 Artificial Intelligence and Machine Learning Approach for
Development and Discovery of Drug**   **211**

*Shweta Kumari, Akhalesh Kumar, Pawan Upadhyay, Ruchi Singh,
Sudhanshu Mishra, Smriti Ojha, and Neeraj Kumar*

Preface

The healthcare industry is ripe for some major changes. From chronic diseases and cancer to radiology and risk assessment, there are nearly endless opportunities to leverage technology to deploy more precise, efficient, and impactful interventions at exactly the right moment in a patient's care. As payment structures evolve, patients demand more from their providers, and the volume of available data continues to increase at a staggering rate. Artificial intelligence (AI) is poised to be the engine that drives improvements across the care continuum. AI offers a number of advantages over traditional analytics and clinical decision-making techniques. Learning algorithms can become more precise and accurate as they interact with training data, allowing humans to gain unprecedented insights into diagnostics, care processes, treatment variability, and patient outcomes. Major disease areas that use AI tools include cancer, neurology, and cardiology. Along with this, AI is at the forefront of scientific research and its progress also complements the growth of cloud computing. The effort to resolve many of the most challenging pharmaceutical aspects, with positive implications to pharmaceutical companies, researchers, patients, regulators, and payers, is enabled by advanced technology, i.e., artificial intelligence. This book focuses on how artificial intelligence brings innovations to the future of healthcare. This book gives an overview on how AI can accelerate the process of clinical studies, which can reduce clinical trial cycle times while improving the costs of productivity and outcomes of clinical development.

This book will help many pharmaceutical companies, CROs, IT companies, and researchers to explore how AI can be effectively utilized to revolutionize the healthcare. This book covers basic- to advanced-level topics related to the applications of artificial intelligence in healthcare. It can be offered as an elective course for graduate and postgraduate students of medical and IT sectors. This book gives compiled information about introduction to AI in different fields of health sector or pharmaceutical industry with their application in various disease managements, cancer diagnoses, and treatments. This book contains 12 chapters that are written by profound researchers from many parts of the world. The book is profusely referenced and copiously illustrated. It should be noted that all chapters were deliberately reviewed and all were suitably revised one or two times. So, the content presented in this book is of greatest value and meets the highest standard of publication.

This book should be of immense interest and usefulness to researchers and industrialists working in clinical research, disease management, pharmacist, formulation scientist working in R&D, remote healthcare management, health analysts, and researchers from pharmaceutical industry.

Finally, here comes the best part. We thank everyone who helped to make this book possible. First and foremost, we express our heartfelt gratitude to the authors for their contribution, dedication, participation, and willingness to share their significant research experience in the form of written testimonials, which would not have been possible without them. Lastly, we are feeling fortunate to express our gratitude to River Publishers for their unwavering support.

Acknowledgment

Having an idea and turning it into a book is as hard as it sounds. The experience is both internally challenging and rewarding. At the very outset, we fail to find adequate words, with limited vocabulary to our command, to express our emotion to almighty, whose eternal blessing, divine presence, and masterly guidelines help us to fulfill all our goals.

When emotions are profound, words sometimes are not sufficient to express our thanks and gratitude. We especially want to thank the individuals who helped make this happen. Without the experiences and support from my peers and team, this book would not exist.

No words can describe the immense contribution of our parents and friends, without whose support this work would not have been possible.

Last but not least, we would like to thank our publisher for their support, innovative suggestions, and guidance in bringing out this edition.

List of Contributors

Akanksha Pandey, *Research Scholar, Department of Pharmacy, School of Medical and Allied Sciences, Galgotias University, India*

Akanksha Sharma, *Assistant Professor, Department of Pharmacy, Monad College of Pharmacy, Monad University, India*

Akash Chauhan, *Research Scholar, Department of Pharmacy, School of Medical and Allied Sciences, Galgotias University, India*

Akhalesh Kumar, *Assistant Professor, Department of Pharmaceutical Science, Maharishi University of Information Technology, India*

Amrish Chandra, *Associate Professor, Amity Institute of Pharmacy, Amity University, India*

Amrita Shukla, *Assistant Professor, Department of Pharmacy, Dr MC Saxena College of Pharmacy, India*

Ankita Moharana, *Assistant Professor, Department of Pharmacy, Jain University, India*

Ashish Verma, *Research Scholar, Department of Pharmacy, Monad University, India*

Bhavna Tyagi, *Department of Oral Maxillofacial Pathology and Microbiology, Sharda University, India*

Bhuneshwar Dutta Tripathi, *Department of Pharmacy, Narayan Institute of Pharmacy, India*

Brojendra Nath Saren, *National Institute of Pharmaceutical Education and Research, India*

Deepika Bairagee, *Pacific College of Pharmacy, Pacific University, Udaipur Rajasthan, India*

Deepshikha Bhargava, *Computer Science Engineering, Amity University, Greater Noida, UP, India*

Dilip Kumar Pal, *Associate Professor, Department of Pharmacy, Guru Ghasidas Vishwavidalaya, Bilaspur, India*

xxiii

Iidiko Csoka, *Faculty of Pharmacy, Institute of Pharmaceutical Technology & Regulatory Affairs, University of Szeged, Hungary*

Indrani Maji, *National Institute of Pharmaceutical Education and Research, India*

Javed Ahamad, *Assistant Professor, Department of Pharmacognosy, Faculty of Pharmacy, Tishk International University, Iraq*

Jitender Madan, *Department of Pharmaceutics, National Institute of Pharmaceutical Education and Research (NIPER), India*

Khushboo Raj, *Research Scholar, Department of Pharmacy, Guru Ghasidas Vishwavidalaya, India*

Mahendran Sekar, *Associate Professor, Department of Pharmaceutical Chemistry, Faculty of Pharmacy and Health Sciences, Royal College of Medicine Perak, University Kuala Lumpur, Malaysia*

Mayur Aalhate, *National Institute of Pharmaceutical Education and Research, India*

Mohamed Yousuff, *Research Scholar, School of Computer Science and Engineering, Vellore Institute of Technology (VIT University), India*

Neelam Dhanker, *Associate Professor, School of Pharmaceutical Sciences, Starex University, India*

Neelam Jain, *Associate Professor, Oriental College of Pharmacy and Research, Oriental University, India*

Neeraj Kumar, *Associate Professor, Faculty of Pharmacy, AIMST University, Malaysia*

Neetesh Kumar Jain, *Dean, Oriental College of Pharmacy and Research, Oriental University, India*

Pankaj Kumar Singh, *National Institute of Pharmaceutical Education and Research, India*

Pawan Upadhyay, *Assistant Professor, Department of Pharmaceutical Science, Maharishi University of Information Technology, India*

Poojashree Verma, *Assistant Professor, Department of Pharmacy, Oriental University, India*

Rajasekhara Babu, *Professor, School of Computer Science and Engineering, Vellore Institute of Technology (VIT University), India*

Rishabha Malviya, *Associate Professor, Department of Pharmacy, School of Medical and Allied Science, Galgotias University, India*

Ruchi Singh, *Assistant Professor, Department of Pharmacy, Narayan Institute of Pharmacy, Gopal Narayan Singh University, India*

Samiya Khan, *Research Fellow, Faculty of Science & Engineering University of Wolverhampton, United Kingdom*

Shahid Rja, *Lecturer, Department of Pharmacy, Dr. M. C. Saxena College of Pharmacy, India*

Shilpa Rawat, *Research Scholar, Department of Pharmacy, School of Medical and Allied Sciences, Galgotias University, India*

Shilpa Singh, *Department of Pharmacy, School of Medical and Allied Sciences, Galgotias University, India*

Shivkanya Fuloria, *Associate Professor, Faculty of Pharmacy, AIMST University, Malaysia*

Shweta Kumari, *Assistant Professor, Department of Pharmacy, Narayan Institute of Pharmacy, Gopal Narayan Singh University, India*

Simran Ludhiani, *Research Scholar, Shri Govindram Sakseria Institute of Technology and Science, India*

Smriti Ojha, *Assistant Professor, Department of Pharmaceutical Science & Technology, Madan Mohan Malaviya University of Technology, India*

Sonali Sundram, *Assistant Professor, Department of Pharmacy, School of Medical and Allied Science, Galgotias University, India*

Sonali Vyas, *Assistant Professor, School of Computer Science, University of Petroleum and Energy Studies, India*

Subasini Uthirapathy, *Associate Professor, Department of Pharmacognosy, Tishk International University, Iraq*

Subham Appasaheb Awaghad, *National Institute of Pharmaceutical Education and Research, India*

Sudhanshu Mishra, *Assistant Professor, Department of Pharmaceutical Science & Technology, Madan Mohan Malaviya University of Technology, India*

Sumeet Dwivedi, *Principal, Oriental College of Pharmacy and Research, Oriental University, India*

Sunita Dahiya, *Assistant Professor, Department of Pharmaceutical Sciences, School of Pharmacy, University of Pureto Rico, Medical Sciences Campus, USA*

Swati Verma, *Assistant Professor, Department of Pharmacy, School of Medical and Allied Science, Galgotias University, India*

Thota Ramathulasi, *Research Scholar, School of Computer Science and Engineering, Vellore Institute of Technology (VIT University), India*

Vandana Tyagi, *Research Scholar, Department of Pharmacy, Amity University, India*

Vetriselvan Subramaniyan, *Faculty of Medicine Bioscience and Nursing, MAHSA University, Malaysia*

Vikram Prajapat, *National Institute of Pharmaceutical Education and Research, India*

Yogita Kumari, *Assistant Professor, Department of Pharmacy, Arka, Jain University, India*

List of Figures

List of Tables

List of Abbreviations

ANDA	Abbreviated new drug application
ADME	Absorption, distribution, disposition, metabolism, and excretion
ADMET	Absorptions, distribution, metabolism, excretion, and toxicity
Ă	Accuracy
APIs	Active pharmaceutical ingredients
Adam	Adaptive moment estimation
AIA	AI combined with analytics
AD	Alzheimer's disease
AWS	Amazon Web Services
AAL	Ambient assisted living
ACR-DSI	American College of Radiology Data Science Institute
ACoS	American College of Surgeons
ATCC	American Type Culture Collection
ALS	Amyotrophic lateral sclerosis
ADNI	Anonymized medical image database
ADCC	Antibody-dependent cell-mediated cytotoxicity
ADCP	Antibody-dependent cellular phagocytosis
A4C	Apical four-chamber
API	Applications programming interface
AUC	Area under the curve
AGI	Artificial general intelligence
AI	Artificial intelligence
ANI	Artificial narrow intelligence
ANN	Artificial neural network
ASI	Artificial superintelligence
ASCO	American society of clinical oncology
ADHD	Attention deficit hyperactivity disorder
AD3D-MIL	Attention-based deep 3D multiple instance learning

AR	Augmented reality
BAFCOM	Bat Algorithm with Fuzzy C-Ordered Means
BN	Bayesian network
BCT	Behavior change technique
BD	Big data
BDA	Big data analytics
BKD	Binary kernel discrimination
BLA	Biologics license application
BMI	Body mass index
BAN	Body area network
BRCA	Breast cancer gene
BSD	Berkeley source distribution
CCLE	Cancer cell line encyclopedia
CRDC	Cancer research data commons
CMRI	Cardiac MRI
CUSI	Carotid ultrasound image
CU-net	Cascaded U-net
CFD	Computational fluid dynamics
CCT	Chest computed tomography
CXR	Chest X-ray
CAR	Chimeric antigen receptor
CML	Chronic myeloid leukemia
COPD	Chronic obstructive pulmonary disease
CID	Criminal Investigation Department
CiDD	Cancer in silico drug discovery
ctDNA	Circulation tumor DNA
CDRN	Clinical data research network
CDSS	Clinical decision support system
CDSS	Clinical decision support systems
CDA	Clinical Document Architecture
CC	Cloud computing
CSPs	Cloud service providers
CPSC	College of Physicians and Surgeons of Canada
CoC	Commission on Cancer
CDC	Complement-dependent cytotoxicity
CAGR	Compound annual growth rate
CS-MRI	Compressive sensing MRI

CMA	Computational modeling assistant
CT	Computed tomography
CAD	Computer-aided diagnostic
CADD	Computer-aided drug discovery
CAT	Computerized axial tomography
CIM	Computing information model
CCC	Concordance correlation coefficients
CGM	Continuous glucose monitoring
CEUS	Contrast-enhanced ultrasound
CAE	Convolutional autoencoder
CNN	Convolutional neural network
CSEN	Convolutional support estimator network
CNA	Copy number aberration
CPT	Current procedural terminology
CRISPR	Clustered regularly interspaced short palindromic repeats
CTC	CT colonography
cGMP	Current good manufacturing practices
CVNN	Complex-valued neural network
CYP	Cytochrome
DSRPAI	Dartmouth Summer Research Project on Artificial Intelligence
DWAS	Data wide association studies
DT	Decision trees
DNN	Deep neural network
DeepDTA	Deep drug–target binding affinity prediction
DEM	Discrete-element method
DenseNet	Densely connected convolutional network
DBSCAN	Density-based spatial clustering of applications with noise
DICOM RT	Digital imaging and communications in medicine-radiation therapy
DICOM	Digital imaging and communications in medicine
dRMSD	Distance-based root mean square deviation
DTI	Drug–target interaction
DHNA	Dutch Head and Neck Audit
DICA	Dutch Institute for Clinical Auditing

ECFP	Extended-connectivity fingerprints
ECCG	Echocardiography
EC2	Elastic Compute Cloud
ECG	Electrocardiogram
EEG	Electroencephalography
EHRs	Electronic health records
EMR	Electronic medical record
EDXL	Emergency Data Exchange Language
EMS	Emergency Medical System
EPHR	Emergency personal healthcare record
ED	End-diastolic
ES	End-systolic
ECN	Enhanced capsule network
ENISA	European Network and Information Security Agency
EPA	Environmental Protection Agency
EGFR	Epidermal growth factor receptor
EPI	Estimation program interface
EMBL-EBI	European Bioinformatics Institute
ES	Expert systems
XAI	Explainable artificial intelligence
xDNN	Explainable deep neural network
ECE	extranodal extension
FCONet	Fast-track COVID-19 classification network
FcRs	Fc receptors
FAERS	FDA Adverse Event Reporting System
5G	Fifth generation
FAIR	Findable, accessible, interoperable, reusable
FDG	Fluorodeoxyglucose
FNH	Focal nodular hyperplasia
FDA	Food and Drug Administration
Fc	Fragment crystallizable
Fab	Fragment-antigen binding
FC	Fully connected
GIST	Gastrointestinal stromal tumor
GRU	Gated recurrent unit
GM	Gaussian mixture
GEO	Gene expression omnibus

GAN	Generative adversarial network
GA	Genetic algorithm
GWAS	Genome-wide association studies
GDC	Genomic data commons
GDSC	Genomics of drug sensitivity
GPS	Global positioning system
GCP	Good clinical practices
GCP	Google cloud platform
GPRS	General packet radio services
gMCI	Gradual MCI
GCN	Graph convolutional network
GSM	Global system for mobile communication
GUI	Graphical user interface
HNSCC	Head and Neck Squamous Cell Carcinoma
HCX	Health Cloud eXchange
HIM	Health information management
HIT	Health information technology
HIPAA	Health Insurance Portability and Accountability Act
HL7	Health Level 7
HS	Healthy subject
HbA1c	Hemoglobin A1c
HMM	Hidden Markov model
HAU	High acuity units
HP APIs	High potency active pharmaceutical ingredients
HTS	HIGH throughput screening
HPC	High-performance computing
H2L	Hit to lead
HNC	Head and neck cancer
HIS	Hospital information system
HER2	Human epidermal growth factor receptor 2
HGP	Human Genome Project
HLA	Human leukocyte antigen
HMI	Human–machine interface
HI	Hybrid intelligence
IQVIA	I (IMS Health), Q (Quintiles), and VIA (by way of)
IDRI	Image database resource initiative
IEC	Independent Ethics Committee

IDP	Indigenous duple pattern
ICT	Information and communication technologies
IA	Information assurance
IT	Information technology
IaaS	Infrastructure as a Service
IMO	Inherent modal operation
IRB	Institutional Review Board
SALSA	Intelligent agent for activity limitation and safety awareness screening
IARC	International Agency for Research on Cancer
IBM	International Business Machines
ICGC	International Cancer Genome Consortium
IoMT	Internet of Medical Things
IM	Intramuscularly
IORT	Intraoperative radiation treatment
IP	Intraperitoneally
IRM	Information resources management
JCS	Joint classification and segmentation
KNN	K-nearest neighbor
KronRLS	Kronecker-regularized least squares
LIS	Laboratory information system
LRP	Layer-wise relevance propagation
LDA	Linear discriminant analysis
LO	Lead optimization
LeNet	It is a type of convolutional neural network
LS	Ligand SMILES
LDA	Linear discriminant analysis
LC	Liver cancer
LI	Liver injuries
LMCS	Lean manufacturing competitiveness scheme
LR	Logistic regression
LSTM	Long short-term memory
LIDC	Lung image database collaboration
LVQ	Learning vector quantization
ML	Machine learning
MLNeCh	Machine learning neuroimaging challenge for automated diagnosis of mild cognitive impairment

MRI	Magnetic resonance imaging
MHC	Major histocompatibility complex
MCI-C	MCI-converters
MCI-NC	MCI-non-converters
MRT	Mean residence time
MSE	Mean square error
MAI	Medical artificial intelligence
MI	Medical imaging
MIA	Medical imaging analysis
MEDi	Medicine and engineering designing intelligence
MCI	Mild cognitive impairment
miRNA	microRNA
MES	Model expert system
MCI	Moderate cognitive impairment
MW	Molecular weight
mAbs	Monoclonal antibodies
MSigDB	Molecular signatures database
MLP	Multi-layer perceptron
MLR	Multiple linear regression
MCECCG	Myocardial contrast ECCG
MS	Myocardial segmentation
NBC	Naive Bayesian classification
NPC	Nasopharyngeal carcinoma
NCDB	National Cancer Database
NCI	National Cancer Institute
NIST	National Institute of Standards and Technology
NIH	National Institutes of Health
NPCR	National Preventive Cardiology Registry
NLP	Natural language processing
NN	Neural network
NCE	New chemical entity
NDA	New drug application
NGS	Next-generation genome sequencing
NROR	National radiation oncology registry
NWHHT	Nederlandse werkgroep hoofd-hals tumoren
OHDSI	Observational Health Data Sciences and Informatics
OEL	Occupational exposure limit

OCR	Optical character recognition
ODDT	Open Drug Discovery Toolkit
OASIS	Organization for the Advancement of Structured Information Standards
OA	Osteoarthritis
POEM	Pareto-optimal embedded modeling
PLS	Partial least-squares
PCORI	Patient-Centered Outcomes Research Institute
pMCI	Persistent MCI
PHR	Personal health record
PIPEDA	Personal Information Protection and Electronic Documents Act
PD	Pharmacodynamics
PK	Pharmacokinetic
PV	Pharmacovigilance
PACS	Picture archiving and communication system
PK	Pharmacokinetic
PaaS	Platform as a Service
PLS	Polynomial least squares
PET	Positron emission tomography
PM	Precision medicine
PREM	Patient reported effectiveness measures
PCA	Principal Component Analysis
PNN	Probabilistic neural network
PROM	Patient-reported outcome measures
PoC	Proof of concept
PoP	Proof of principle
PHI	Protected health information
PADME	Protein and drug molecule association prediction
PSC	Protein arrangement composition
PDC	Proteomics Data Commons
PATRIOT	Providing Appropriate Tools Required to Intercept and Obstruct Terrorism Act
PELT	Pruned exact linear time
PSA	Prostate-specific antigen
PKI	Public key infrastructure
QOPI	Quality Oncology Practice Initiative
QSPR	Quantitative structure property relationship

QI	Quality improvement
QoS	Quality of service
QSAR	Quantitative structure activity relationship
QSM	Quantitative susceptibility mapping
RIS	Radiology information system
RANC	Reconfigurable architecture for neuromorphic computing
RF	Random forest
RWD	Real-world data
ReLU	Rectified linear unit
RFR	Recursive feature refining
RoI	Region of interest
R-CNN	Region proposal CNN
RNNs	Repetitive neural networks
R&D	Research and development
ResNet	Residual neural network
rs-fMRI	Resting-state functional MRI
RBM	Restricted Boltzmann machine
RT-PCR	Reverse transcription–polymerase chain reaction
RNAi	RNA interference
RPA	Robotic process automation
RCA	Root cause analysis
ROR-t	Retineic-acid-receptor-related orphan nuclear receptor
RUR	Rossum's Universal Robots
SEA	Self extracting archive
SOMs	Self-assembling manuals
Ŝ	Sensitivity
SARS-CoV-2	Severe acute respiratory syndrome Coronavirus 2
S3	Simple storage service
SNP	Single nucleotide polymorphism
SPECT	Single-photon emission computed tomography
SNOWMED-CT	Systematized Nomenclature of Medicine – Clinical Terms
SaaS	Software as a service
Ş	Specificity
SUV	Standardized uptake value
SBRT	Stereotactic body radiotherapy
SWOT	Strengths, weaknesses, opportunities, and threats

sMRI	Structural MRI
SSIM	Structural similarity index measure
SC	Subcutaneously
SVM	Support vector machine
SEER	Surveillance, Epidemiology, and End Results
SIB	Swiss Institute of Bioinformatics
TOUCH-AI	Technology Oriented Use Cases in Health Care-AI
TFLite	TensorFlow Lite
TCGA	The Cancer Genome Atlas
TCIA	The Cancer Imaging Archive
NCBI	The National Centre for Biotechnology Information
TARGET	Therapeutically Applicable Research to Generate Effective Treatments
TPSA	Topological polar surface area
THA	Total hip replacement
TL	Transfer learning
TP	Treatment planning
TRRUST	Transciptional regulatory relationship unraveled by sentence based text mining
TME	Tumor microenvironment
US	Ultrasound
UIS	Ultrasound information system
US	United States
UCSF	University of California San Francisco
FCFPs	Useful class fingerprints
VRT	Variance to mean residence time
VGGNet	A type of Convolutional neural network
VR	Virtual reality
VS	Virtual screening
VGG	Visual geometry group
VOC	Voice of the customer
VoxResNet	Voxel-wide residual network
WideDTA	Deep-learning based prediction model
WiMAX	Worldwide interoperability for microwave access
WHO	World Health Organization
YOLO	You Only Look Once
ZFNet	Zeiler furgus net

1

Healthcare 4.0: A Systematic Review and Its Impact Over Conventional Healthcare System

Sonali Vyas[1], Deepshikha Bhargava[*2], and Samiya Khan[3]

[1]School of Computer Science, University of Petroleum and Energy Studies, India
[2]Computer Science Engineering, Amity University, Greater Noida, UP, India
[3]Faculty of Science & Engineering, University of Wolverhampton, United Kingdom
*Corresponding Author: Deepshikha Bhargava
Computer Science Engineering, Amity University, Greater Noida, UP, India
deepshikhabhargava@gmail.com

Abstract

The healthcare industry is evolving and digitizing healthcare practices, tools, and techniques as part of the Industrial Revolution 4.0. Healthcare 4.0 includes big data analytics for healthcare data, AI for smart monitoring, and IoT for keeping track of patients' daily healthcare activities. Healthcare systems are embracing these innovative technologies in a fast and efficient manner. Healthcare 4.0 permits the handling and management of a large number of real-time patient data as well as the ability to make accurate and improved treatment decisions based on such data. It also helps doctors and medical practitioners to carry out predictive analyses of diseases in a better and efficient way. Furthermore, telemedicine and precision medicine are being used to streamline medical operations to ensure remote care. This chapter discusses the concept of Healthcare 4.0 and its benefits over conventional healthcare systems. It further describes the implementation of innovative technologies and the various data management approaches such as big data and virtualization. Subsequently, this chapter also highlights various tools, smart devices, and gadgets used for tracking and monitoring the healthcare vitals of a patient.

Moreover, various applications and a systematic review of different challenges and gaps related to the adoption of the Healthcare 4.0 have been discussed.

1.1 Introduction

The healthcare system has experienced different rollers of technology advancements, taking first from Healthcare 1.0, in which surgeons maintained hand-written records of patients. Then followed Healthcare 2.0, in which paper manual records were replaced by electronic records, and then came Healthcare 3.0, in which wearable devices were introduced to measure a person's real-time health status utilizing a large amount of computer power. An important technical feature that distinguishes Healthcare 4.0 is that numerous types of devices interconnect with each other [1], monitoring patient health [2] and performing other health activities based on IoT, cyber–physical systems, and the Internet of Services. According to various studies, Healthcare 4.0 can be defined as the combination of healthcare applications based on IoT, AI, and intelligent sensors. Its goal is to digitalize health services and organizations. As a result, the focus of this chapter is on demonstrating the effects of this technology on performance and healthcare service delivery. The Industrial Revolution 4.0 defines the transformation and advancement of industrial production through the integration of new digital technologies. If we look at the previous revolutions of industry in terms of technology, there were no automated systems like mechanical tools in the first revolution, electrification systems in the second revolution, and cyber–physical systems using IoT, big data, AI, blockchain, and cloud computing in the fourth revolution, to support humans in the healthcare sectors [1].

The main principles for the Healthcare Industry 4.0 are [2] as follows:

- Interoperability: Simplifies contextual information to continuously exchange the data.

- Virtualization: Monitoring and individualizing the different therapies to treat non-communicable and chronic diseases.

- Decentralization: Main aspect of intelligence and distribution into networks with dispersed fragments.

- Real-time capability: Also known as theragnostic [3], where real-time processes diagnostic, control, relief, and remediation approach.

- Service orientation: The services provided by the smart devices should not harm the patients.

- Modularity: The modular system can adapt and change settings and necessities with changing or increasing sub-systems. Therefore,

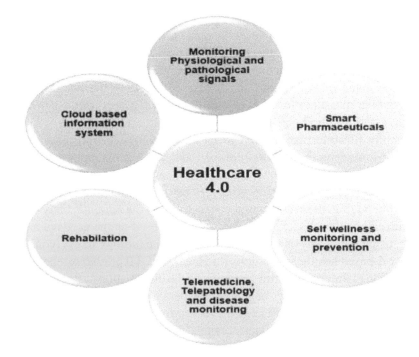

Figure 1.1 Applications of healthcare 4.0 [5].

modular systems can be easily configured to manage regular instabilities or changes in product/system features [4].

Healthcare 4.0 is a strategic deployment and a well-managed model in the health industry. Their principles permit the best value for the healthcare system. The virtualization allowance in healthcare satisfies the demand in the healthcare industry (for all patients, staff, doctors, formal, and informal caretakers). This system combines all the computational power from cyber–physical systems, cloud computing, and IoT. Various applications of Healthcare 4.0 are shown in Figure 1.1.

1.1.1 Application scenarios of healthcare 4.0

Healthcare is critical to survival, as evidenced by current marketing trends and classic science fiction. Therefore, healthcare and a variety of settings, such as long-term care and real-time response systems, can be implemented. The automatic procedures ensure a dramatic reduction in the risk of error submission in comparison to the methods that require manual involvement [6].

A 5G communication infrastructure serves as the backbone for Healthcare 4.0, which provides improved features to healthcare systems.

Implementing cloud and fog computing can manage the devices which can be implanted into human to enhance the data collection of every activity and movement happening inside the body [5]. To some extent, Healthcare 4.0 supports self-management, which includes wellness monitoring and illness prevention, management of chronic disease by tracking the health status, and preventive measures for obesity and diabetes. These systems can alert people to establish good nutritional food habits and build a knowledge base for a healthy lifestyle plan as well as assist them in achieving any fitness goals [7]. Monitoring the condition of the medication consumption by these systems adds value to disease management. The advancement of technology led to the development of the new advanced sensors (ingestible or wearable) that provide intelligent solutions and connectivity to the patient's body parts. Personalized healthcare seeks to make decisions based on the needs of a single user, using data analysis to better understand the biology of everyone.

Cloud-based healthcare systems for Healthcare 4.0 provide strength and simplification in the architectural designing of the information systems for collecting, analyzing, processing, or sharing the data to make the use of the data for the patient's treatment or to keep the record for monitoring any long-term consequences. It enhances the data storage efficiency; thus, the management of data becomes much easier for the officials. According to future projections, all surgeries will be made more transparent with the help of specific smart and IoT devices, transforming this approach into a feasible spectating by experts with the most virtualization. Teleconsultation, similar to telesurgery, can also be used, which does not require the consultant's physical presence [8].

1.1.2 The architecture of healthcare 4.0

Healthcare 4.0 comprises smart industry and smart engineering as the basic blocks to building smart healthcare systems. As per Figure 1.2, the usability of the smart devices precisely in the healthcare system broadens the scope of treatment available for the patients. The cyber–physical system monitors the real-world processes and produces the corresponding output which can assist in treating the diseases.

Cloud computing enhances the storage capacity of the healthcare system and analyzes the data at ease. The key element in the Healthcare 4.0 system design is the use of IoT devices, which is modern technology, including smart objects, sensors, and devices to track, monitor spectate, and supervise patients. In Healthcare 4.0, only medical-related IoT systems are evaluated to continuously improve healthcare. The data produced by the sensors in the devices are treated to form the results on which the main algorithm for

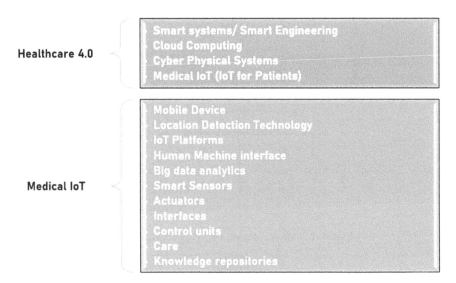

Figure 1.2 Framework for healthcare 4.0.

the treatment will work [9]. In Healthcare 4.0, there are two categories of sensors: clinical sensors and non-clinical sensors, with IoT steadily creating more and more space for automation.

Some of the best products given by IoT to healthcare for self-assessment in real time are respiratory sensors, pulse-oximeter sensors, pulse sensors, and BP sensors. Many prevailing healthcare resolutions utilize cloud systems to make decisions [9]. The problem of latency in the cloud-based environment arises in the time-based application systems. The network topology for accessing the data from the server should be capable of retrieving the amount of data transferred from source to destination. The security of data transfer is critical since any incorrect or inappropriate data can cause the device to malfunction, putting patients in danger. Big data techniques meet the need for extracting value from previously unmanageable or uncontrollable data. Healthcare 4.0 operators can test their procedures by looking for new ways for huge data collection, knowing that big data methods can fetch meaningful and valuable information from data.

1.1.3 Requirements and characteristics of healthcare 4.0

The functional requirements (shown in Figure 1.3) are very specific and vary from system to system, whereas non-functional requirements are not very specific and can be used to establish the system's quality [10]. Necessities

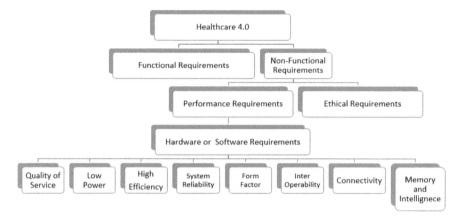

Figure 1.3 Requirements for the healthcare 4.0.

Figure 1.4 Characteristics of smart healthcare 4.0 [10].

of a smart healthcare system include less power, consistency, service quality, enhanced user experience, high effectiveness, interoperability across diverse platforms, easy deployment, intelligent health system, system robustness to advance to new technologies, and adequate communication.

The most significant features of a smart healthcare system are represented in Figure 1.4. The features of intelligent healthcare can be divided into three categories:

- App-oriented structures require reliable authentication and communication between sensors and applications on smartphones to create a personal connection between sensors and the device and protect data.

- Thing-oriented structures require application-based flexibility, instantaneous monitoring, timely distribution, high compassion, maintaining high effectiveness in low power degeneracy, and initiating intelligent deployment.

- Semantic-oriented organizations require to be capable of changing behavioral outlines depending on earlier attained knowledge, developing natural language execution strategies to augment user knowledge, and becoming omnipresent computer competencies [11, 12].

1.2 Evolution of Healthcare

Healthcare 1.0 signifies the first industrial revolution, from the early 18th century to the middle of the late 19th century. Over time, with the emergent technology of the first engines and railways, the transformation of machines has facilitated the social movement from agriculture to industry.

In the late 19th and early 20th centuries, a second industrial revolution, namely Industrial 2.0, demonstrated a standard shift from moderate to extensive production, due to high energy consumption. Henry Ford's integration lines revolutionized manufacturing processes by allowing for the efficient distribution and connecting of multiple operations and equipment [13].

Throughout the 20th century, fast technological advancements have led to a larger use of ever-changing, digital-reflecting systems, as well as network communication production and commercial procedures. Integrated and flexible production systems, also known as "enterprise resource planning," have gained widespread acceptance, bringing in the third industrial revolution, termed Industrial 3.0, and paving the way for the fourth, named Industry 4.0.

Entering the 21st century, Industry 4.0 signifies the fourth industrial revolution and aimed to fit in cyber–physical systems with data, procedures, devices, and operational technologies. IoT and related amenities are widely disseminated and utilized in combination with big data and methods of AI.

Through internet-based and cloud-oriented analysis, performance, controller, and storage of resources, goods, and business procedures, one can attain sustainable, intelligent, and affordable connections (i.e., amalgamation) of product and service. The industrial evolution of healthcare is shown in Table 1.1.

Table 1.1 Progression of healthcare 4.0.

Versions	Year	Characteristics
Healthcare 1.0	1975–1995	• In the field of health, "MRI" and "computerized axial tomography (CAT)" have been performed. • Modular computer programs have emerged in the health sector.
Healthcare 2.0	Subsequent period and half of 1.0	• IT healthcare connection systems continued to link and establish EHRs, which began to link to imaging and provide doctors with better knowledge.
Healthcare 3.0	From 2004 ahead	• Growth of genomic data, wearable, and implant development. • Healthcare 3.0 was established during such data integrated with interacted EHR programs.
Healthcare 4.0	Current period	• All these innovative methods, associated with real-time data processing, were included. Also, enhanced AI usage and an easy user interface were included. • United with the real-time data group, improved AI usage, the overlap of imperceptible interfaces, and integration of numerous technologies.

1.3 Need of Healthcare 5.0

Data is a powerful resource that has the potential to revolutionize the way healthcare is established, developed, and disseminated in the future. To find a model for fast change, Asian healthcare is adopting a digital healthcare system that will flourish as a precursor to Industry 4.0. Key to this challenge is the acceptance of a fast data communication network with smart devices, robots, distant monitoring, and other automatic tools in healthcare systems, benchmarking for Healthcare 5.0 and enabling both spectrum conclusions. Healthcare 5.0 will do authentication after the digital installation is complete. This revolution recognizes the key role of clients and the flexibility of working models in the industry to adapt customer procedures. It is a change in thinking from customer relationship management to a customer-managed relationship. In the healthcare industry, the fifth stage extends beyond patient-centered healthcare to include customer-focused health services.

In 5.0, health service benefactors inquire where they can get into the lives of their clients, rather than through other means (where clients can get into provider procedures). In this system, the main emphasis is on client models: truly knowing clients and not only patients in the healthcare industry. The planned difference presented in Healthcare 5.0 is lifetime association with clients, which is reflected in the focus on the well-being and quality of life of both patients as well as people who are often considered clients in the healthcare industry. Digital health will emerge as part of Healthcare 5.0. Industry stakeholders will move from assisting patients to strengthening lifelong relationships with people and manufacturing unique treatments. Automation and AI are two of the most popular concepts in Healthcare 5.0, designed to transform any type of work environment. The use of AI in Healthcare 5.0 includes the concepts of exact and automatic disease diagnosis and forecasting, remote-care patient monitoring and healing, robotic medical system, AI treatment that includes online courses for people with social anxiety disorder, and dealing with and verifying outcomes. Besides these, machine algorithms can measure the end consequences lacking human interference. Subsequently, the smart health system emerged as a result of the rapid growth and development of technology in healthcare [14]. The combination of AI methods for precise diagnosis and prediction makes healthcare smarter. Consequently, smart healthcare manages various modules of the healthcare organization intelligently. Through the integration of AI into healthcare units, large data groups produced by health components can be handled by various machine learning algorithms to make healthcare predictions and analyses.

1.4 Advances in the Healthcare Industry

Modern technology and inventions are making implications for the era in the medical and healthcare systems. The impact of flexible connected and wearable healthcare devices, as well as other innovations, has been so significant and influential for so long that any highly developed medical facility, when compared to previous years' healthcare services, is fortified with the latest gadgets and healthcare procedures. Advances in healthcare and digital technology enable us to live safer, healthier, and more fruitful lives. The increasing use of digital health applications has been mainly helpful in monitoring the five major diseases such as diabetes, asthma, heart disorders, and lung conditions that often have the greatest impact on human health. Therefore, their treatment and monitoring have benefited from such healthcare innovations. As per the IQVIA report, digital healthcare applications are assessed to save about $8 billion per year in US healthcare expenses consumed on treating and controlling these critical diseases [15].

1.4.1 M-Healthcare

Mobile applications have arisen as efficient ways to confirm the patient appointment. As a result of the health applications communicating with doctors and healthcare facilities per hour, the necessity is easier than before. In many countries having effective mobile applications for worldwide healthcare systems, it contributed to preventing, treating, and controlling severe diseases. For example, mobile applications play an important role in averting and monitoring type-2 diabetes. Certainly, the evolution of mobile applications for healthcare and their utilization transformed recent healthcare.

1.4.2 Healthcare data of patients

Nowadays, patients are well-educated and skilled in various diseases, diagnosis constraints, and healthcare procedures. Having easy-to-use healthcare gadgets that are convenient and linked to mobile applications for monitoring healthcare data is important for general analytic determinations such as blood sugar testing, measuring blood pressure, and using a heart rate calculator. Self-diagnostic technologies have resulted in a massive surge of patient medical data [16].

1.4.3 IoT and healthcare

Associated healthcare devices possess many new features to the healthcare system. From real-time diagnostic information to therapeutic approaches, IoT healthcare devices can convert health amenities in a way that has never been seen before. Such IoT devices combined with "geotagging" and "location tracking" help healthcare services accomplish better innovations, as mentioned below:

- Prompt diagnosis of patients in emergency.
- Promptly process important health data.
- Remotely monitoring patient's health.
- Easy access to sensitive healthcare hardware.
- Improved trailing of records, healthcare staff, and patients.
- Improved monitoring of medicines.

1.4.4 Blockchain technology and healthcare

The major concerns of health facilities around the world are to maintain and protect patient data and to prevent all types of threats and risks regarding

data security. There is no better and more sophisticated technology to date than blockchain when it comes to data security. Blockchain is a limited spot that permits open access to data for all participants inside the organization while monitoring protected encoding and preventing any information theft or disrupting the "write once, never delete, never change" policy. While blockchain enables healthcare organizations' data, patient data is stored in a highly secured cell. Database of blockchain may permit unrestricted access to customized patient records and information while averting all kinds of disruptive attempts [15].

1.4.5 Big data analytics and healthcare

Big data analysis of digital health applications has developed as important new technology to promote the advancements in carrying out treatment and monitoring. Big data analytics can handle a wide range of medical data, including patient information, medical history, records, medical equipment information and inventory, healthcare app data derived from user contacts, payment, insurance information, healthcare maintenance data, and many other types of data related to the healthcare industry. Additionally, many related details for dealing with numerous problems can be measured.

1.5 Telemedicine Services

According to research on mobile health applications in all forums, telemedicine apps are quite prevalent in every medical health application. Telemedicine applications have reformed the way of communication between doctors and patients. As a result of telemedicine applications, the physical connection between the surgeon and the patient is minimized and provides accessibility to appropriate attention and cure. Many healthcare experts believe that such applications may quickly treat chronic illnesses and a variety of daily health indicators, as well as recommend accurate medicines and therapy [14]. Telemedicine applications permit the patient to connect with physicians and simultaneous medication instructions with the help of a mobile device. Many telemedicine users think such applications are more efficient in getting standard treatment and care instead of visiting physical clinics.

1.5.1 Big data and IoT for healthcare 4.0

With recent technological advancements, many interconnected datasets cannot be analyzed, managed, or processed by the basic computational power carried by any of the computer systems for techniques planned for extracting value from large volumes of different data economically, by allowing fast capturing, discovering, and analyzing [16]. Healthcare 4.0 has the most concise classification associated with the big data, consisting of five Vs [17]:

• Volume: Scale on which data is increasing.

• Velocity: Time bounded collection and analysis.

• Variety: Structure of data, the creation of data into various types.

• Veracity: Data having various gradations as per the monitoring and processing.

• Value: Providing maximum value extraction by the big data architecture.

The existence of awareness about the great diversity of accessible databases is of the highest importance for comprehending the real competence of big data when applied to Healthcare 4.0.

1.5.2 Blockchain and healthcare 4.0

Contemporary healthcare systems are considered for being extremely complex and expensive. However, this can be focused on enhanced healthcare records organization, the use of insurance interferences, and blockchain technology. Blockchain was first presented to offer dispersed archives of financial transactions that did not depend on centralized financial institutions [18]. The introduction of blockchain technology has led to enhanced connectivity that includes medical records, insurance payments, and smart contracts, allowing for data security, in addition to offering a wide range of transaction databases. Another important benefit of utilizing blockchain technology in the healthcare system is that it can transform healthcare data interactions and provide additional access to medical records, tracking of the device, medical database, and hospital properties, as well as a complete device life cycle in blockchain structure. Furthermore, accessing the medical history of a patient is vital for recommending medicine effectively, and blockchain can significantly improve the framework of such healthcare services [18].

1.5.3 AI and healthcare 4.0

Initially, technology was used to perform routine and tedious tasks and to reduce paper usage by digitizing medical records while facilitating the flow of information between insurance companies, hospitals, and patients [19].

Although these activities endure being implemented, AI has extended its operations from prevention to back-office production and is now being used as a tool to expand healthcare outcomes. AI is being utilized to find associations between genetic codes, the use of surgical robots, or even to enhance hospital competence. Overall, AI has been recognized to be beneficial to the healthcare industry.

The major applications of AI in Healthcare 4.0 [19] are as follows:

- Support in clinical decisions: By the use of natural language processing (NLP).

- Improve primary care and prioritize by chatbots.

- Robotic operations.

- Virtual nursing subordinates.

- Assisting in the precise diagnosis.

- Minimizing the burden of electronic health records.

Threats of AI are unavoidable as there are many cons to using AI in healthcare. Some of them are as follows [19]:

- Errors and injuries

- Data availability

- Privacy and security concerns

- Bias and inequalities

- Unemployment

1.5.4 Cyber–physical system and healthcare 4.0

The cyber–physical system integrates with computing and physical components and processes. Computing components connect and interconnect with sensors, monitor online and physical signals, actuators, cyber-converters, and the physical environment. The cyber–physical system utilizes sensors to join all the distributed intelligence in the system to gain in-depth knowledge, resulting in more precise actions and functions [20]. Cyber–physical systems can be considered as one of the most important components of a medical device network. These programs are gradually being used in clinics to provide quality healthcare. Many types of data are obtained from real life using sensors. Furthermore, smart devices, together with smart meters, with great hearing ability and networks, are emerging. The evolution of the internet and the computer have unbolted a huge set of new regulatory powers that can influence human health through mobility, new health behaviors, new energy organization, and new facilities [21].

1.5.5 Smart medical devices

As per the market observation, Amazon provides an integrated healthcare display where users can fetch healthcare information, accessibility to the newest products, health insurance, and other services. As a result, smart wearable devices, like smartwatches, are modernizing the market. Furthermore, distinguished products like Fitbit, Proteus, Pebble Time, etc., are now available. Noteworthy to healthcare, smart devices are growing more and more common. The proposed annual sale is expected to reach around 80 million units at an annual growth rate of 20% by 2022. Apple is anticipated to have a significant market share; however, Android-enabled smart wearable devices are constantly evolving. Interestingly, Apple iWatch provides a built-in GPS package as well as heart rate sensors with a dual-core processor [10].

1.6 Opportunities and Challenges Involved in Healthcare

It is a fact that the application and dissemination of such technologies in the healthcare brings substantial profits to entire healthcare users. Although the healthcare industry is gradually becoming more concerned with the proper use of IoT and big data technologies, there are a few difficulties that must be addressed before digital healthcare becomes a widespread reality [22].

As data expands in quantity, interoperability, which is defined as the capacity for devices used by healthcare consumers to be networked rather than compatible, becomes a major barrier in smart healthcare [23]. The security of the smart devices used in Healthcare 4.0 opens the loops for many vulnerabilities. Therefore, putting a check and control upon the security of smart healthcare systems can nullify the major threats and challenges for healthcare.

1.7 Future Scope and Trends

Several efforts are being made to increase fog computing service quality, such as sensor network integration, to provide more precise disease forecasts and automatic prescriptions. Appropriate blockchain algorithms must be modified for devices at the network's edge to work in distributed areas with little computing [24]. Also, there is a need to understand and consider various deep and supervised learning approaches to detect malfunctions and failures and perform a quality check on the servers' services. Subsequently, there is a need for smart data analytics to analyze the data and formulate results based on it, which can help the community serve the commons more effectively.

1.8 Conclusion

This chapter discusses the concept of Healthcare 4.0 and its benefits over conventional healthcare systems in terms of energy, scalability, security, infrastructure, high-performance results and analysis, availability, and data privacy. Healthcare 4.0 benefits include smart devices that are more efficient than traditional healthcare systems. Healthcare 4.0 offers many opportunities and challenges in the engineering of healthcare systems. Additionally, human–machine interactions in the socio-technical system have proven to be significant, confirming active communication and safe usage of intelligent technology. The users of intelligent healthcare, which include patients, caregivers, and health professionals, are placed at the center. It is significant to reflect their features, requirements, skills, and issues when scheming and using intelligent and integrated healthcare systems. Moreover, it is important to discuss the problem of dissimilarity and to confirm that Healthcare 4.0 is intended to reduce and alleviate such inequalities. Altogether, Healthcare 4.0 permits all users to access high-standard healthcare services.

1.9 Acknowledgment

The authors would like to acknowledge the co-authors and editors for their kind support and, more specifically, the reviewers for their valued recommendations.

1.10 Funding

This research acknowledged no definite grant from any funding agency in the public, profitable, or non-profitable sectors.

1.11 Conflict of Interest

The authors have no conflicts of interest to declare. All co-authors have perceived and approved the insides of the manuscript and there is no financial interest to report.

References

[1] Paul, S., Riffat, M., Yasir, A., Mahim, M. N., Sharnali, B. Y., Naheen, I. T., ... & Kulkarni, A. (2021). Industry 4.0 applications for medical/healthcare services. Journal of Sensor and Actuator Networks, 10(3), 43.

[2] Hermann, M., Pentek, T., & Otto, B. (2016). Design Principles for Industrie 4.0 Scenarios. 2016 49th Hawaii International Conference on System Sciences (HICSS), 3928–3937.

[3] Jeelani S, Jagat Reddy RC, Maheswaran T, Asokan GS, Dany A, Anand B (2014) Theragnostics: a treasured tailor for tomorrow. J Pharm Bioallied Sci 6(Suppl 1): S6–S8. doi:10.4103/0975-7406.137249

[4] Ceselli A, Premoli M, Secci S. Mobile Edge Cloud Network Design Optimization. IEEE/ACM Transactions on Networking, 2017 DOI:10.1109/TNET.2017.2652850

[5] Aceto, G., Persico, V., & Pescapé, A. (2020). Industry 4.0 and health: Internet of things, big data, and cloud computing for healthcare 4.0. Journal of Industrial Information Integration, 18, 100129.

[6] Darshan, K. R., & Anandakumar, K. R. (2015, December). A comprehensive review on usage of Internet of Things (IoT) in healthcare system. In 2015 International Conference on Emerging Research in Electronics, Computer Science and Technology (ICERECT) (pp. 132–136). IEEE.

[7] Islam, S. R., Kwak, D., Kabir, M. H., Hossain, M., & Kwak, K. S. (2015). The internet of things for health care: a comprehensive survey. IEEE access, 3, 678–708.

[8] Thuemmler, C., & Bai, C. (Eds.). (2017). Health 4.0: How virtualization and big data are revolutionizing healthcare (pp. 2168–2194). Cham: Springer International Publishing.

[9] Naresh, V. S., Pericherla, S. S., Murty, P. S. R., & Sivaranjani, R. (2020). Internet of Things in Healthcare: Architecture, Applications, Challenges, and Solutions. Comput. Syst. Sci. Eng., 35(6), 411–421.

[10] Sundaravadivel, P., Kougianos, E., Mohanty, S. P., & Ganapathiraju, M. K. (2017). Everything you wanted to know about smart health care: Evaluating the different technologies and components of the internet of things for better health. IEEE Consumer Electronics Magazine, 7(1), 18–28.

[11] Bader, A., Ghazzai, H., Kadri, A., & Alouini, M. S. (2016). Front-end intelligence for large-scale application-oriented internet-of-things. IEEE Access, 4, 3257–3272.

[12] Banerjee, A., & Gupta, S. K. (2014). Analysis of smart mobile applications for healthcare under dynamic context changes. IEEE Transactions on Mobile Computing, 14(5), 904–919.

[13] Li, J., & Carayon, P. (2021). Health Care 4.0: A vision for smart and connected health care. IISE transactions on healthcare systems engineering, 11(3), 171–180.

[14] Mohanta, B., Das, P., & Patnaik, S. (2019, May). Healthcare 5.0: A paradigm shift in digital healthcare system using Artificial Intelligence, IOT and 5G Communication. In 2019 International Conference on Applied Machine Learning (ICAML) (pp. 191–196). IEEE.

[15] https://getreferralmd.com/2019/06/7-healthcare-digital-technology-trends-to-watch/

[16] Gantz, J., & Reinsel, D. (2011). Extracting value from chaos. IDC iview, 1142(2011), 1–12.

[17] Aceto, G., Persico, V., & Pescapé, A. (2020). Industry 4.0 and health: Internet of things, big data, and cloud computing for healthcare 4.0. Journal of Industrial Information Integration, 18, 100129.

[18] Tanwar, S., Parekh, K., & Evans, R. (2020). Blockchain-based electronic healthcare record system for healthcare 4.0 applications. Journal of Information Security and Applications, 50, 102407.

[19] https://www.analyticssteps.com/blogs/artificial-intelligence-healthcare-applications-and-threats

[20] Boulila, N. (2019). Cyber-physical systems and Industry 4.0: Properties, structure, communication, and behavior. Tech. report, Siemens Corp. Technol., (April).

[21] Dey, N., Ashour, A. S., Shi, F., Fong, S. J., & Tavares, J. M. R. (2018). Medical cyber-physical systems: A survey. Journal of medical systems, 42(4), 1–13.

[22] Zeadally, S., Siddiqui, F., Baig, Z., & Ibrahim, A. (2019). Smart healthcare: Challenges and potential solutions using internet of things (IoT) and big data analytics. PSU research review.

[23] Rehman, M. U., Andargoli, A. E., & Pousti, H. (2019). Healthcare 4.0: Trends, Challenges and Benefits.

[24] Tuli, S., Tuli, S., Wander, G., Wander, P., Gill, S. S., Dustdar, S., ... & Rana, O. (2020). Next generation technologies for smart healthcare: Challenges, vision, model, trends and future directions. Internet technology letters, 3(2), e145.

2

Data Imaging, Clinical Studies, and Disease Diagnosis using Artificial Intelligence in Healthcare

Vandana Tyagi[1*], Neelam Dhankher[2], Bhavna Tyagi[3], Iidiko Csoka[4], and Amrish Chandra[1]

[1]Amity Institute of Pharmacy, Amity University, India
[2]School of Pharmaceutical Sciences, Starex University, India
[3]Department of Oral and Maxillofacial Pathology and Microbiology, Sharda University, India
[4]Faculty of Pharmacy, Institute of Pharmaceutical Technology & Regulatory Affairs, University of Szeged, Hungary
*Corresponding Author
Vandana Tyagi
Research Scholar, Amity Institute of Pharmacy, Amity University, India
Email: tyagivandana01@gmail.com; ORCID ID: 0000-0001-5197-8082

Abstract

In recent years, artificial intelligence (AI) has been developing quickly, and it has become a breakthrough technology that uses computerized algorithms to analyze complex data. Diagnostic imaging is one of the most powerful clinical applications of AI and efforts are increasing toward improving performance to help detection and quantification of various disease conditions. AI tools are now being utilized in medical imaging to assess X-rays, CT scans, MRIs, and other images for lesions or other abnormalities that a human radiologist may overlook. The number of ways that AI can help physicians, researchers, and the patients they serve is continuously growing. AI is used to assist medical professionals with a wide range of activities, including administrative tasks, clinical documentation, and patient outreach, along with specialized support in areas such as image analysis, medical device automation, and patient monitoring. In addition to the various applications in healthcare settings, there is little dispute that AI will play a significant role in the digital healthcare systems that design and enable modern medicine. In addition, very few studies have described the specific benefits of AI in clinical settings, including data imaging and disease diagnosis. In line with the existing gap in the prevailing literature, this chapter attempts to discuss the role of AI in healthcare and clinical settings. Moreover, this chapter also presents the most important applications of AI in clinical trials, drug development, and biomedical science.

2.1 Introduction

The intelligence shown by the types of machinery systems is called artificial intelligence (AI) [1]. In recent years, artificial intelligence has been growing vastly concerning software algorithms, implementation of hardware, and application in numerous domains; for instance, healthcare, clinical studies, data imaging, and disease diagnosis. Hence, comprehensive learning algorithms can deal with up-surging data amounts rendered by mobile monitoring sensors, smartphones, and wearables [2]. Recently, researchers have claimed that AI can demonstrate an adequate and superior performance in the healthcare sector as compared to humans. Therefore, in this chapter, we discuss the benefits and drawbacks of using artificial intelligence (AI) to automate the medical field.

2.1.1 Classifications of artificial intelligence

AI is no single technology; however, it is a collection of various technologies whole together. Although most of the technologies have a significant and relevant role in the healthcare sector, the precise tasks and processes they support vary in nature. Therefore, we are choosing the specific AI technologies that are of great importance to the healthcare domain [3].

2.1.1.1 Machine learning: Deep learning and neural network

One type of AI is machine learning. Additionally, it is a statistical tactic to learn through exercising the data model [3], and the connection between AI, deep learning, and learning related to the machine is used to describe how computers learn [4]. The author stated that they (AI, deep learning, neural networks, and machine learning) are similar to Russian nesting dolls as every component is essential to the preceding team (demonstrated in Figure 2.1).

It is evident from Figure 2.1 that the learning of machines is a subspecialty of artificial intelligence. Similarly, deep knowledge is the subspecialty

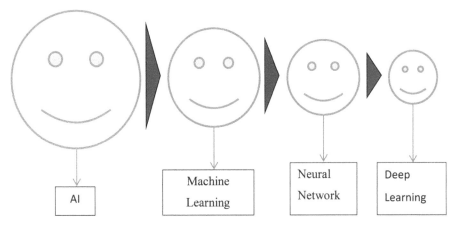

Figure 2.1 Russian nesting dolls model demonstrating the AI, machine learning, neural network, and deep learning relationship [3, 4].

of machine learning. Additionally, the networks of neurons are the spine of deep knowledge processes. Moreover, the number and depth of the node neural network layer differentiate the single neural network and the deep learning process [3, 4].

In the healthcare domain, precision medicine is the most collective presentation of outmoded machine learning. This consists of predicting the treatment protocols which are likely to succeed for the ill person grounded on the various disease attributes and the context of an intervention [5]. However, for precise machine learning, one needs to take the training for calculating datasets and variable outcomes, and the process of learning is designated as supervised learning. A neural network is a more advanced type of learning related to machines, although this tactic is available since the 1960s. Since then, this technology has been recognized in the healthcare domain and researched by various investigators to categorize its application including predicting the disease incidence rate in a healthy person. It evaluates the inputs, outputs, and variables used to solve problems that linked both input and output to present the results [6]. The most difficult type of learning related to the machine is deep knowledge which includes various variables to predict results. Moreover, there are thousands of veiled features and variables in this model that can be uncovered through faster processing of cloud architectures and graphing processing divisions. The common use of deeper learning in the healthcare domain is the identification of potential cancerous conditions and lesions identified by deploying radiological imagining [7]. More specifically, deep learning and radionics are frequently utilized together throughout the investigation of oncology-related images. Furthermore, deep learning is also gaining importance in speech recognition through natural language processing. Since the 1950s, the goal of the researchers is to make logic of the human being's language. However, this includes various applications, analysis of text, recognition of speech, translation, and goals associated with human language. NLP is the computer science branch, and, specifically, it is a part of AI which is concerned to render computers the ability to understand spoken words and texts; similarly to humans, it included two major approaches: semantic NLP and statistical NLP [8, 9]. Machine learning is the grounded principle of statistical NLP. Expanding more, NLP associates computational linguistics, deep learning models, machine learning, and rule grounded human language models with statics. Together with the above-mentioned dimensions, the computer system can access text and voice data containing human language format to fully comprehend the significance. Human language is complex due to ambiguities, which make it difficult to read and understand by computer software. Therefore, the NLP task divides the human voice and texts in a manner that aids the computer system to make sense (summarized in Table 2.1). In the healthcare sector, NLP creates, understands, and classifies the published research and clinical documents. Furthermore, the NLP system

Table 2.1   NLP tasks for processing the human language [10].

S. No	Task	Comments
1.	Speech recognition	• Also known as speech-to-text • Needed for understanding the voice commands and spoken questions
2.	Speech tagging	• Also known as grammatical tagging • This aids in identifying the part of speech
3.	Word sense	• Semantic analysis • It is recognizing the meaning of the word, which has multiple meanings
4.	Recognition of named entity	• This task helps in recognizing the phrases and words as useful units • For instance, NEM recognizes "Sam" as a name
5.	Resolution of co-reference	• This aids in accessing when and if two words need to refer to the same unit • Also, it assists to recognize idioms
6.	Analysis of sentiment	• This task is helpful in the extraction of subjective qualities, including confusion, sarcasm, attitudes, and emotions from the text
7.	Generation of natural language	• It is opposite to the first task, which is speech recognition • This task is specific to putting the structure of information and data into the human language [4]

can access unstructured patients' clinical notes, formulate reports, transcribe interactions of patients, and conduct AI.

2.1.1.2 Rule-based expert systems

It is the simplest type of AI and deploys prescribed knowledge-grounded rules to answer or explain the problem [11]. The goal of the expert system is to fetch the knowledge from the human proficiency and convert the same into the hardcoded rules to relate to the input data. The rules are the "if–then" type and which dominate AI technology since the 1980s. This expert system is used for clinical decision support; for instance, electronic health records are one of the expert systems [12]. Similar findings from the existing literature demonstrated that expert systems needed knowledge engineers and experts to form a rules series in a specific knowledge domain. The advantage of this system includes working well and being understandable. In contrast, when there are a lot of rules, then the rules begin to break down due to conflict between the numerous rules. Additionally, if the knowledge domain changes, then there is a requirement to change the rules set, which is a time-consuming process.

2.1.1.3 Physical robots and software robotics

In the recent scenario, around the world, more than 200,000 physical robots have been placed. Industrial robots are carrying out several functions including dislocating, assembling objects, welding, and lifting objects in place, including warehouses and factories. Furthermore, such robots are engaged in supplying items in hospitals. Recently, industrial robots have developed connections with humans and can work with humans in collaboration. One of the advantages is that these robots are easy to train by moving them by a required task. With time, robots become more intelligent because AI aptitudes are inserted into the robot's operating system (also known as the brain). In 2000, the USA government initially approved surgical robots. This initiative gives the power and vision to the surgeon as it helps in improving the ability to access wounds, create precise invasive incisions, and so on [13]. However, the important decisions are still taken by humans only. Some of the common surgical procedures deploying robotic surgery are prostrate surgery, head and neck surgery, and gynecologic surgery.

Robotic process automation (RPA) is the business procedure automation technology grounded on AI, metaphorical software bots (robots), and digital workers. It is also known as software robotics, although it is not associated with robot software. Traditionally, the workflow automation tools' software developer formulates the actions list to mechanize a task and deploy scripting language or internal applications programming interfaces (APIs) for outcomes. However, the RPA system formulates the list of actions by observing the task performed by the user and performing the same task by repeating it in the graphical user interface [14]. Such a technique aids in decreasing the barrier to deploying automation process in products, which is commonly not API's characteristic.

Additionally, the RPA tool demonstrated similarity with the GUI and automated the communication with GUI. In contrast, the RPA tool is not similar to such a system because it allows data handling in and between various applications, for example, getting mail including invoices, data extraction, and adding the same information in the book-keeping process system. The actual use of this system includes lending and mortgage process, OCR application, data extraction, fixed automation, customer care, and banking process automation. This technology will bring a new wave of efficiency and productivity to the market, specifically the labor market. Although we have discussed the above-mentioned techniques as separate topics, these techniques have been amalgamated, for instance, robots with AI grounded brains and image recognition combined with RPA [14].

2.2 Machine Learning for Typical Biomedical Data Types

2.2.1 Data from multiple omics

The biological process of merging and analyzing various "-omics" data, such as genomics, proteomics, transcriptomics, epigenomics, and microbiomics, is referred to as multi-omics data [15]. Traditional single-omics methodologies do not provide a comprehensive understanding of biological processes; however, multi-omics does. Typically, several omics datasets can describe the same or similar biological processes. Every omic is handled as a distinct viewpoint in ML [16], which is known as a multi-view configuration. Combining various sources necessitates that integration be data-driven or model-driven.

2.2.2 Integration based on data

A single model can be created by concatenating with or without transformation, data from all the perspectives. Data from single-nucleotide polymorphisms have been effectively combined using this integrative approach into a single matrix (SNPs) and gene expression of mRNA as well as the use of a Bayesian integrative model to investigate the link to estimate a quantitative outcome using SNPs and mRNA (e.g., cytotoxicity of drugs) [17].

2.2.3 Incorporating models

This method is used every perspective of data, which is subsequently combined in the model output, ATHENA [18, 19], for example, performed genomic studies by incorporating data from several omics sources, for example, changes in the number of copies, expression of genes, methylation, and miRNA to uncover connections, for example, ovarian cancer survival is a clinical outcome. During the stage of the investigation, based on each type of omic data, base models and neural networks are developed, and then comes the building of an integrative model. Wang *et al.* [20] presented a network convergence strategy to classify cancer that starts with the creation of resemblance matrices for patients. These are the resemblance matrices for patients built using data from the expression of mRNA, methylation of DNA, and expression of miRNA. Following the construction of these matrices, an iterative nonlinear approach is used to make the three work together to find patient groupings, and researchers combined equal matrices into a single matrix. Dr. Ghici and Dr. Potter [21] developed a multidisciplinary strategy to assist in the prediction of HIV protease mutations that are resistant to treatment. This strategy creates a foundation for the prediction model by using structural data from an HIV protease-drug inhibitor combination as well as differences in the sequence of DNA and then uses the underlying models' predictions to perform majority voting.

2.2.4 Data on behavior

Aside from multi-omics and clinical data, behavioral information is concerning one's health. When it comes to the utilization of behavior, information in applications related to health presents a few unique issues as a result of how information is acquired and stored; few study groups look into the link between information on behavior and related to health.

Sinnenberg *et al.* [22], for example, discovered links between Twitter tweets as well as the possibility of heart disease. This research discovered that individuals having coronary artery disease, maybe have tweets' tone, style, and perspective as well as some basic demographics, aid in identifying them, based on a sample of 4.9 million tweets through deploying AI system. Researchers [23] studied analytics on social media and mental health as well as indicators discovered that are related to social media usage that was associated with worsening symptoms of psychosis [24], hypomania [25], suicidal ideation [26], and sadness [27].

2.2.5 Data from video and conversations

Many people have become interested in the video and verbal data used, both within and without disciplines considering the field of medicine. The Chinese internet behemoth, Tencent, claims to be the inventor of a system for visual perception which is capable of detecting Parkinson's illness in about 3 minutes. Patients and professional medical workers in an investigational study conduct long discussions. and the use of language clues as a strategy for detecting cognitive impairment of mild severity (MCI) identification has recently shown potential [28, 29]. Tang *et al.* [30] used reinforcement learning techniques based on the transcripts from these clinical tests, and conversational AI has been developed [31]. This agent was taught to maximize MCI diagnostic accuracy with the fewest number of conversational events possible, and it outperformed supervised learning methods significantly.

2.2.6 Mobile sensor data

In recent years, many research studies have attempted to revolutionize healthcare by utilizing data from mobile sensors [32]. Mobile data insights could be tremendously valuable in chronic illnesses, for example, mental health difficulties along with persistent discomfort and mobility impairments. For example, Saeb *et al.* [33] looked into the link with global positioning system (GPS) location, data based on the phone utility, as well as the gravity of the situation of depression symptoms. Selter *et al.* [34] developed a self-management of chronic lower back pain with a mobile health app. Zhan *et al.* [35] used a machine learning approach to develop an app based on data from mobile sensors to predict the Parkinson's disease severity. Turakhia and Kaiser [36] imagined what impact mobile health could have to change atrioventricular fibrillation treatment. The National Institutes of Health [37] selected one of 11 Big Data Centers of Excellence as the Mobile Sensor Data-to-Knowledge (MD2K) Center, highlighting the significance of mobile data analysis in the field of health.

2.2.7 Data on the environment

Factors of the environment play a part in the development of several disorders, including coronary heart disease [38], chronic obstructive pulmonary disease (COPD) [39], paralysis agitans [40], psychological illnesses [41], and tumor [42]. Artificial intelligence technology has progressed to investigate data on the environment to gain a deeper understanding related to illness causes and raise the standard of care. Song *et al.* [43], for example, used time-series analyses to investigate the influence of the environment on hand, foot, and mouth disease. By using ML models, Stingone *et al.* [44] investigated air pollution and its relationship to exposure in the cognitive abilities of children of the United State. As per the results data driven machine learning can be utilized to assess the air pollutants impact on the helath outcomes. Similarly, Park *et al.* created environmental risk scores using sophisticated ML models and combine them with mental combinations, cardiovascular disease, and oxidative stress. Hahn *et al.* [46] created gene–gene and gene–environment interaction detection software; which was also boon in the healthcare system.

2.2.8 Pharmaceutical research and development data

Medicines have critical functions in healthcare. Data collected during the development of a drug frequently reveals fresh knowledge about disease mechanisms and prevention. To get information out of the data, artificially intelligent methods were employed.

2.2.8.1 Chemical compounds

PubChem [47] is a website that contains data about tiny compounds as well as their biological properties. Several researchers utilize the molecular structures in PubChem as a lexicon, followed by an examination of specific compounds using a zero-one footprint representation. Zhang *et al.* [48, 49], for example, employed a representation that is based on the footprint to determine medicinal parallels and related them to patient or illness resemblances to generate individualized therapy commendations. Graph convolutional networks (GCNs) [50] have recently been employed in the construction of molecular structures and analysis, which considers every molecule to be a graph, with atoms acting as nodes in a network. Using this approach, Duvenaud *et al.* [51] created a GCN structure for extracting features from molecules (also known as neural fingerprints), having a high level of predictability, comprehensibility, as well as parsimony. Molecular graph convolutions, according to Kearnes *et al.* [52], are "a novel approach to virtual screening using ligands."

2.2.8.2 Clinical trials

Clinical trials are a crucial part of the medication development process. Clinical trial participants are typically chosen based on precise criteria for acceptance and rejection. Data from clinical trials offer pharmaceutical companies a plethora of information. AI techniques have recently been employed in data mining and the design of clinical trials. Chekroud *et al.* [53], for example, in cross-trial depression treatment outcome prediction, can be utilized for gradient boosting and feed-forward feature selection. Kohannim *et al.* [54] studied the application of a machine that supports vectors to improve clinical trial strength and minimize its sample size.

2.2.9 Unintentional reports

The FDA Adverse Event Reporting System (FAERS) [55] gathers data on adverse events linked to specific drugs. Using FAERS data, Sakaeda *et al.* [56] assessed the efficacy of four specific methods related to data mining for forecasting the incidence of side effects from specific drugs. Tatonetti *et al.* [57] used FAERS to construct a method for detecting medication–drug interactions based on new signal detection. Zhang *et al.* [58] created a medication–drug interaction prediction approach depending on FAERS side-effect profiles and medication similarity graphs. Banda *et al.* [59] linked medicine names and results to RxNorm and SNOMED-CT standard vocabularies to improve FAERS utilization.

2.2.10 Literature in biomedicine data

Published findings in the literature on biomedicine are yet another major data source for artificial intelligence in healthcare. Technologies such as artificial intelligence (AI) and natural language processing (NLP) are extractable to inform health research, and useful information from the literature is needed. In the mining literature of biomedicine, numerous advanced AI methods have recently been developed and attained [60] cutting-edge performance due to which modern machine learning approaches have revolutionized as a result of this transformation, for example, deep understanding, particularly in natural language processing. Entity recognition and normalization are two major issues that are used to describe the mining of literature; recognizing and normalizing identified items of interest is a difficult task (e.g., illnesses, variations in genetics, etc.) in the text (e.g., if two separate textual descriptions are similar). Leaman *et al.* [61] created DNorm as a pairwise learning-to-rank machine learning technique for the normalization of disease names.

2.3 Application of AI

2.3.1 Biomedical information processing

In the sphere of medicine, the amazing advancement has been regarded as related to natural language interpretation and processing. Its means Question Answering (QA) is a benchmark Natural Language Processing (NLP) task in which models use linked documents, pictures, knowledge bases, and question-answer pairings to anticipate the answer to a given question. Hence, the NLP seems to be the appropriate technique for searching the answers. The whole process of searching for the answer is systematic and specific [62]. This process starts with classifying the asked question into a specific category to extract the specific information. More specifically, machine learning (ML) divides the question into four types with nearly 90% accuracy [63]. It is followed by the retrieval of the best answer to the question by an intelligent biomedical document/data recovery system. Sarrouti also documented that the "yes" and "no" answer formulator based on the sentiment analysis worked sufficiently while extracting binary answers [64]. This technique can be deployed for merging clinical information, conflict resolution, and comparison. Earlier, these were the time-consuming and laborious work done by a human. Recently, the AI is demonstrating the capability to perform the above-mentioned task effectively and accurately as the human professional evaluator can perform. Expanding more, NLP supports the medical narrative information which is required to the unrestricted human race from performing the challenging work including temporal events track while concurrently maintaining reasons and structures. Lastly, ML can be deployed in performing complex clinical information, for instance, biomedical data and tests, putting logical rationale in the datasets, and using knowledge for various purposes [65].

2.3.2 AI for living support

The author stated that AI can be used in assisting the disabled and elderly people as using smart robotic systems aids in up-surging life quality of the patients [66]. Moreover, the author added that, recently, various literature works have been published related to home tools and functions for the disabled person, and intelligent solution systems grounded on the AI, data mining, and wireless sensors for supporting the patients who are dependent on family members to sustain normal day-to-day work. These systems can learn through the transformation of images to access expression on the face of the human as per the instructions. Moreover, human–machine interfaces (HMIs) are grounded based on the investigation of a disabled person's facial expression which, in turn, guide robot-supported vehicles and wheelchairs without sensors or joysticks fixed to the body [67]. In a 2017 study, Hudec and Smutny, through their research work, elaborated that RUDO (ambient intelligent system) can support blind people to live and work together with the sighted individual in the specialized domains, including electronics and informatics. Furthermore, the author claimed that blind people can utilize intelligent assistance in various ways by a single operator interface [68]. The author Oprescu describes the role of AI in pregnant women. The author argued that women who reside in remote and rural areas are unable to access the proper consultations and medical facilities. Hence, considering this gap in the medical field, the author had planned to render support to the pregnant women through a mobile phone application. The key findings of this study suggested that AI can support pregnant women with crucial advice including diet, exercise, and medication during paramount maternity stages. This can be supported by the amalgamation of own intelligence of AI and cloud-dependent communication media between the entire individual concerned [69]. Oprescu claimed that the health and well-being enhancement of pregnant females can be attained through AI, including effective computer-based tools and devices, though further research is still required for strong suggestions [70].

The radar Doppler time-frequency system that the detecting sensor was anchored on was the fall's mechanism. The detecting sensor, which was grounded on a radar Doppler time-frequency system, was the fall's mechanism. The mechanism of the fall – the detection sensor which was grounded on a radar Doppler time-frequency system. The author claimed that radar is a paramount sensing modality for the people who have a higher risk of falls as this system worked on human motion monitoring. The general principle is to transfer an electromagnetic wave of some frequencies range and access the radar reoccurrence signals [71]. Hence, this system provides a clutter suppressed, noise-tolerant, and non-intrusive sensing system for evaluating moving motion [72]. Furthermore, the low-cost radar system with narrowband property can be deployed to access the moving objects' instantaneous velocity by calculating the backscattered waves' frequency shift, and this effect is called the Doppler effect [73]. A smart communication system has been developed for people with autonomy loss, based on the AI information processing system that collects the data from various communication technologies and channels. Hence, this system aids in determining the event that occurs within the network ambience and assists the elder people to lead a quality life. "Intelligent agent for activity limitation and safety awareness (SALSA) screening" helps the elderly patients by supporting daily medication processes and activities [74].

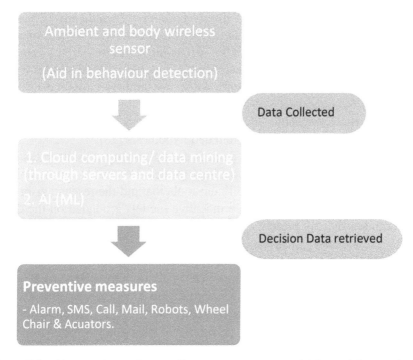

Figure 2.2   Alarm system and preventive measure process model for assisting aged people [75, 76].

Gait and motion study established that ML in the assessment of motion can raise an alarm system during the hazards act and support with activation of preventative measures [75]. In such cases, data (related to human behavior and ambient surrounding) is collected through the sensor and the same can be accessed through edge intelligence or cloud computing. It is followed by the activation of the alarm and rendering the preventative measures [76]. The same is illustrated in Figure 2.2. Furthermore, an AI-based expert system combined with personal digital assistance and mobile can aid the individual with memory damage by improving their memory competencies to achieve desired living standards.

2.3.3 Biomedical research

AI has been explored as a beneficial tool and application in biomedical studies. This is attributed to the fact that AI can increase indexing and screening of the pre-existing literature in the biomedical domain. This includes the research on the tumor-suppressor procedure, human genome, protein extraction, and genetic associated illness findings in healthcare [77]. The researcher can compile the literature as per the pre-decided subject or area of interest through the AI approach known as the semantic graph grounded approach. Expanded more, the researcher can rank the selected articles when the paper cannot be readable. Additionally, this process helps the investigators to test the hypothesis, which is considered the paramount step in the research process. For instance, the investigators can access and rank the literature through AI for testing the formulated hypothesis [78].

Intelligent medical tools and devices are up-surging consciousness and the same can be discovered in biomedical studies. Computational modeling assistant (CMA) is one of the intelligent tools that aid the researcher in the biomedical field to form executable incentive models grounded on the conceptual models that are present as a thought in their mind. The CMA renders the various databases, methods, and knowledge. The investigator's hypothesis and ideas are expressed as the biological model that is provided to the CMA as input. The CMA has a significant role to convert the hypothesis, ideas, and knowledge into the actual stimulation models. Then the investigator chooses the suitable and best model, which will be followed by the generation of stimulation code by CMA [79]. Throughout the whole process, the CMA helps the researchers by reducing time consumption and rendering accurate code without much hassle. Hence, CMA enhanced productivity and quickens the research procedure. The CMA procedure is demonstrated in Figure 2.3. Lastly, some of the intrusive tools can be beneficial in the field of oral surgery, plastic surgery, and biomedical imaging [80].

2.3.4 Medicine

This section is aiming to discuss AI's current application in medicine, which includes various fields like cardiology, gastroenterology, nephrology, neurology, pulmonary medicine, endocrinology, and so on. In the medicine field, early identification of arterial fibrillation was the first AI application. In 2014, Alive Cor got the approval from FDA to launch the mobile application which monitors ECG and aid in the early identification of atrial fibrillation [81]. The utility of the smartphone-driven Kardia application in arterial fibrillation identification. This REHEARSE-AF research demonstrated that ECG monitoring through Kardia application in ambulatory patients is beneficial in atrial fibrillation identification as compared to routine care [82]. Furthermore, Apple also gets approval from FDA for their latest watch which permits easy ECG acquirement and identification of arterial fibrillation. The same data can be shared with the healthcare worker with the help of a smartphone.

Figure 2.3   Demonstrating the interaction between CMA and researcher mind [79, 80].

There are various critiques of portable and wearable ECG technology, which includes uses limitation, adoption barriers in elderly patients, and false-positive results. AI can be applied through the electronic patient records to calculate the cardiovascular disease risk, for example, heart failure, and compared to standard measures for acute artery diseases [83]. The outcomes depend on the sample size included in the research [84]. Topalovic *et al.* [85] served to elaborate on the role of AI in the pulmonary medicine domain. The author stated that pulmonary function tests are an encouraging field for AI application. AI grounded software renders the more specific outcomes and serves as the decision tool for interpreting outcomes of the pulmonary function test [85]. However, Delclaux presented the critique for the study conducted by Topalovic *et al.* [85] as the results of this study demonstrated that accurate diagnosis rate in pulmonary functioning was significantly inferior to the country average [86].

Patients with diabetes mellitus should check their serum glucose levels to have a better check on their illness and to avoid the associated complications. FDA provided approval to "Medtronic" for measuring and monitoring serum glucose levels [87]. Medtronic tools can be linked with smartphones [88]. The company gets in conjunction with Watson to incorporate the AI property for the measurement of serum glucose levels to aid the clients to prevent or reduce the risk of hyperglycemic episodes grounded on the repeated examination values. It is evident from various studies [89] that regular assessment of the serum glucose level helps the patient to have better control over their condition. Moreover, it also helps patients to attain a positive attitude toward their illness as better glycemic control reduces the stigma and stress. More specifically, the personal failure in maintaining serum glucose levels can cause anxiety and the development of a negative attitude toward the disease. Decision assistance powered by artificial intelligence (AI) has the potential to lessen these issues and enhance clinical treatment and nephrology research. The author claimed that AI has a significant role in various settings of nephrology. AI aids in the estimation of the glomerular filtration rate in an individual suffering from polycystic kidneys disease. Expanding more, AI also helps in stating the risk of progressive IgA nephropathy. However, there is still a requirement for further research to critically evaluate the role of AI in clinical nephrology. AI technology is one of the vital techniques in gastroenterology. Gastroenterologists use the neural network amongst other types of endoscopy and ultrasound pictures are processed using deep learning algorithms. The images can assess the clinical conditions, like colonic polyps. Furthermore, a neural network is deployed for the diagnosis of atrophic gastritis and gastroesophageal reflux. AI can be deployed in the diagnosis of esophageal cancer, collateral cancer, metastasis, and inflammatory bowel disease [90]. Urinary function and storage of the bladder can be failed due to any neurological disorder or spinal cord injury. Moreover, aging can also be the potential cause of urinary function failure. The implantable neural trigger can be an effective treatment in treating bladder functioning failure in patients with the drug-refractory condition. Conditional neuro-stimulation can be an effective technique to enhance the safety and efficiency of neuroprostheses. In this process, the bladder sensor detects the filled bladder through a feedback mechanism [91]. The digital signal processor for accessing and sensing the urine fullness and pressure through the mechanoreceptors (bladder neural roots) shows the fluctuations during fillings. Such neuroprostheses have two basic units including an internal unit (inside the patient) and an external unit (wearable device). The internal unit accomplishes various functions including recording of neural signals, neuro-stimulation, on-chip processing as per the applications, controlling the external unit, and connecting to the external unit through sending the recorded signal [92]. The role of AI in neurology includes epilepsy and tremor assessment. The author claimed that intelligent seizure assessment devices may be an effective method for recognizing epilepsy early on as well as aiding in improving the management of the same. In 2018, FDA rendered the approval to "Empatica" for their wearable tool which connected with the ectodermal captors to detect the episode of epilepsy and send the report through the mobile application. This mobile application sends an alert to relatives and physicians when there is an epileptic attack experienced by the patients [93]. Furthermore, the wearable sensor can be an effective technique to quantitatively assess tremor, posture, and gait in Huntington's disease, Parkinsonism, and multiple sclerosis patients [94].

2.3.5 Cancer and miscellaneous

Several studies have been conducted to assess the applications of AI in healthcare along with algorithms for detection of cancer through mammograms, interpretation of radiographs for chest, analysis of various scans for computer tomography, identification of magnetic resonance images for brain tumors, and prediction for the development of Alzheimer's disease with the help of positron emission tomography [95]. Other applications of AI are in pathology, identification of cancerous skin lesions, explanation of retinal imagining, and detection of arrhythmias and hyperkalemia through electrocardiograms [96]. The other uses of AI in diagnosis are identification of polyp from colonoscopy, upgrading interpretation of genomics, identifications of genetic conditions with the help of the appearance of the face, and evaluation of the quality of embryo for maximizing the improvement of *in vitro* fertilization [97].

2.4 Assessment of AI Applications in Healthcare

Up-surge scrutiny of AI application in medicine and healthcare domain demands for evaluation of AI in the real-life situations for assessing unintended consequences and effectiveness. Healthcare is becoming increasingly complex, and AI applications are becoming more context- and user-dependent; so the traditional methods of evaluating AI must be changed. The author claimed that artificial intelligence (AI) is becoming more prevalently deployed in the care of the patient to diagnose the problem and choose the most appropriate course of action for the patient. However, there is a proper regulatory methodology for assessing the relevance of AI in healthcare. A comprehensive analysis of the role of AI was presented by the author. The phases are as follows.

2.4.1 Phase 0

This phase consists of two equivalent efforts, which include evaluation of the user's requirements and formulation of AI algorithms. This phase aims to identify users, ensure interpretability, and identify workflow and proto-typing of the initial design. To assess the user's requirements and explain-ability, the researcher can deploy algorithm grounded question-bank method for user-addressed AI design [98]. Evaluation of the data quality is the prime component of AI. For instance, the investigator must assess the data for com-pleteness (missing data extent and pattern); timeliness (data represent the recent practice); biases (data representativeness); and validity (erroneous input). Some of the open-basis toolkits including AI Fairness [99, 100] can be an effective tool in evaluating metrics of fairness and bias in AI algorithms. Moreover, statistical performance metrics should continue to be evaluated. More specifically, human performance measurement is paramount to finding the baseline from which the AI solutions' accuracy can be compared in line with the human task.

2.4.2 Phase 1

A researcher can evaluate the pros and cons of the intervention during this phase. For medicines, this phase evaluates the drug-associated toxicities and optimal dose. However, this stage improves model performance for AI systems, for instance, identifying the difference between recall and preci-sion. This phase requires the ability to evaluate the long-term implications of wrong negatives and positives. After the model is formulated based on the previous data, this phase can evaluate the real-world data. For instance, the AI model can render incorrect information based on algorithm biases grounded on improper information [101]. Users can also face the conse-quences of misunderstanding or misperception from the invalid model out-puts and designs of AI solutions. By the proper development of an AI model, users should have the necessary information on how to distribute the same and how to measure the sureness of the models. Applied social science meth-odologies ethnography and phase 0 can be deployed to understand interac-tions between consumers and AI solutions. Previous evaluation parameters are very helpful for the proper understanding, discovery, and use of system features by the users. Simulation studies are the usability testing that men-tions the hypothetical clinical scenarios and performance of certain tasks by the participants which can be involved in the detection and insurance of cognitive overload or overtrust issues. However, there is an identification of potential risks, including disruption of the flow of work, concerns as to patient safety or clinicians' insights can be contradicted by the model outputs are equally important as well. Errors such as poor model fit, instability in terms of numerical, malfunction of software and hardware, or human error should also be evaluated by the mechanisms for catching and correcting errors [102].

2.4.3 Phase 2

This phase intends to access the drug's efficacy and associated side effects. However, the device's effects and action mechanism are assessed through phase 2 trials. As AI solutions impact users and interests, results might be different from expectations or reality. Hence, these pages help in understanding both unintended benefits and concerns. It is essential to probe and note the participants' thought processes and activities to understand whether the intended efficacy of the AI solution is achieved or not. During the insight generation, the AI algorithms must be dynamic and usually include randomness. If the client does not develop faith in the AI system, then there is the possibility that the solution is under-evaluated. Kushniruk [102] stated that A/B testing can be an effective study design that aids in evaluating the relative efficiency of the AI solutions and discovering the inadvertent consequences. What is calculated in "phase 2" is equivalent to the intermediate results for the anticipated outcomes. For instance, decisions more attached to the pre-decided treatment guidelines would aid in improving the expected clients' outcomes and decreased time consumed in administrative procedures which indirectly reduces costs. To conclude, confirming intermediate measures' improvement is the vital step to rationalizing the subsequent phase and estimating the sample extents [102].

2.4.4 Phase 3

Clinical research works help in determining the AI solution's value in medicine by assessing whether AI can be useful in improving the health results in actual life. This phase's goal is to determine the safety and efficacy in comparison to the standard care by conducting large-scale and well-designed research. In most healthcare settings, AI tools are deployed to improve the client's performance but not to substitute them. Hence, it is vital to compare and contrast the decision-makers' or users' performance with or without using AI tools. Clinical research works of devices and medicines are extremely resource-demanding and needs several locations, where the investigators must collect the data and track the reliability of subjects. Usually, a trial at a large scale has been run by contacting the research organizations, but their operations are distinct from the healthcare domain. However, such approaches cannot be favorable for estimating the AI solution's efficacy because the information required for creating clinical practice will be based on actionable knowledge and the delivery system of AI should be a component of the clinical process. Hence, it is paramount to timely evaluate the study infrastructure to effectively process and store the gathered data, for instance, EHR.

2.4.5 Phase 4

AI technologies have unique characteristics for self-learning and self-improvement throughout their existence. Because, over time, the data and software components that make up the foundation might alter and evolve, procedures are needed to evaluate that the AI software's validity and quality are not jeopardized and that these adjustments have no detrimental implications. AI's efficacy and safety must be re-evaluated over time, much as the performance of antibiotics might be affected by growing insurgency [104].

To draw valid causal conclusions from observational data, any biases must be carefully adjusted. Confounding can be accounted for using a variety of epidemiological and machine learning approaches [105]. Many EHR systems can automate data gathering, submitting passively accumulated sets of data that reflect results and application. Such data can be used to assess the effectiveness of details transmission and inclusion in medical workflows. Furthermore, data collection could be programmed into AI software on how specific characteristics are utilized in practice, allowing for a more accurate assessment of their impact on results. It is a lot of fun to collect data on such a wide scale and in real time [106].

2.5 Artificial Intelligence's Challenges in the Use of Pharmaceutical R&D Data

Artificial intelligence's challenges, potential, and practical consequences in the use of pharmaceutical R&D data is crucial to investigate. Despite good studies, there are still obstacles to interpreting R&D statistics from the pharmaceutical industry, as described above. The following are a few of them.

1. Although graph convolution techniques are effective in de novo drug synthesis, interpretability remains an issue. Understanding and linking

the action mechanism of drug is crucial; besides being more effective, new therapeutic compounds should be identified. We should include a model-building method with biology and chemistry domain knowledge to attain this goal.

2. 2. Clinical trials have highly strict criteria for patient enrolment which include and exclude, which is one of its drawbacks. The idea is to get rid of the possibility of several complicating variables having an influence. As a result of the stringent requirements for recruiting, the recruited patients will be "perfect" and unlike patients in reality. FAERS data is made up of a collection of adverse drug reaction reports with insufficient information. It is critical to combine clinical trial and FAERS data using actual patient data from EHRs or insurance claims to make the insights gained more practical and usable. The FDA has announced a novel approach to promote and create the usage of evidence related to real-life drug and biological development [107]. This will open up a lot of possibilities for developing AI approaches for R&D in the pharmaceutical industry and applications in real life and facts.

2.6 Future Directions for AI in Healthcare

2.6.1 Analytical integration

There has been a surge in health-related AI research and activity in recent years, including the integration of many components of clinical data [108], building connections between pharmaceutical R&D and clinical evidence [109, 110], and connecting biorepository data to clinical data [111]. More exigently, the tonality is to design the health-related AI algorithms having to merge knowledge and data. Data related to patients is generally restricted and subject to significant fluctuations, unlike in other computer fields, such as vision and speech analysis, where large amounts of data may be collected.

2.6.2 Transparency in models

Systems that are based on rules, for example, are comprehensible traditional AI technology. Deep learning models, for example, are recent AI technologies that might yield excellent perceptible results; yet, they are usually ignored. Recently, there have been a lot of discussions about whether model interpretability is a necessary concern. In a recent interview [112], Geoff Hinton, a pioneer of deep learning, has suggested lawmakers should not impose their principles on people without understanding artificial intelligence algorithms because "Humans are unable to express how they function for the vast majority of what they do," says the author. Poursabzi-Sangdeh *et al.* [113] examined a randomized conducted study to find out the importance of model interpretability to users. Customer's faith in black-box and transparent models did not differ significantly. Furthermore, "The public's ability to discern critical errors in a model was compromised by greater transparency," according to the study. Holm [114] justified black-box models by comparing them to human decision-making, which is primarily subjective ("outcomes of their deep learning"). As a result, "neuroscience faces the equal level of interdependence problem as in the field of computer science today." Model interpretability may not be as critical in some cases, particularly in circumstances where algorithms related to artificial intelligence have already proven their worth to deliver accurate results on time. As a result, "Neuroscience, like computer science, is confronted with the same interpretability quandary" today. In this study, model interpretability may not be as critical in some cases, particularly in domains where AI algorithms have already proven their ability to be precise, repeatable, and generalizable results. However, with the recent level of analytics in healthcare using computational technology, this is not healthcare. Deep learning models, for example, have been found to perform similarly to EHRs [115] or claims [116] data to use logistic regression to solve hospital readmission problems. Models based on deep knowledge have even surpassed cutting-edge performance in medical picture analysis, and the model's generalization capacity is difficult to defend. If the model succeeds in a collection of medical picture data from one institution for radiology, it is difficult to argue that it will do as well in another. Furthermore, human clinicians will still make the final decisions in most healthcare settings, with AI algorithms supporting them. To make the doctor feel more at ease, it is critical to present explicit rationales for the AI algorithms' assertions. Furthermore, algorithms based on artificial intelligence should be included in everyday clinical procedures to improve their clinical value [117].

On the contrary, the current performance of algorithms of artificial intelligence in many health applications is far from ideal and cutting-edge. We ought to continue to push the use of models which is a black box to explore if they can improve rendition. *Post-hoc* explanation approaches [118] would be useful in this scenario to comprehend how the model operates. Knowledge distillation [119] is an example of such a technique, which used a teacher–student collaboration to acquire a simpler/easier-to-understand paradigm with measurable results that may mimic that the dark information is "distilled" from an intricate black-box model. Transparency of the model in another aspect to consider it a proprietorship. According to the perspective of

Shah *et al.* [120], there is a concerning tendency toward opaque proprietary algorithms, with developers "reluctant to publish" model specifics. When these models are used in clinical practice, this may increase the risk of injury [121]. As Parikh *et al.* [122] stated in their debate in the field of medical predictive analytics, "Professional and regulatory entities should work together and guarantee the powerful algorithms collide with clinical drugs and prognostic biomarkers, there are clinical benefit standards."

2.6.3 Model security

We frequently talk about how important it is to maintain the safety and privacy of personal health information, particularly information on individual patients. As in the field of health, there are a plethora of artificial intelligence models. We must be aware of the surveillance risks that these models pose as the healthcare field expands. In an adversarial assault, for example, data is gathered that can be a source of consternation for the machine's algorithms, leading to poor or even incorrect choices. Sitawarin *et al.* [123], for example, illustrated how self-driving cars are readily fooled by road pollution indicators. Sun *et al.* [124] showed that small changes in a patient's electronic health record and lab values can change the death rate estimate of a well-versed forecaster. Finlayson *et al.* [125] go over the potential problems in the healthcare industry; there are incentives for increasingly complex hostile attacks in further depth. According to the authors of this paper, (i) medical personnel must be informed of this potential danger; (ii) in the face of medical hostile attacks, AI researchers are working to design effective defense measures; and (iii) when building new regulatory frameworks, policymakers should think about the danger of model security.

2.6.4 Learning that is federated

Data on health is broadly disseminated within and among healthcare organizations, and each organization can be linked to a distinguished gathering of interested parties. These are private information that should not be shared. More data from a variety of sources to inform model training is desired in the context of model training. "Federated education is a machine learning environment that develops centralized workforce models based on massive client training data," according to Konecny *et al.* [126]. These clients' network connections are frequently unstable and slow. It is impossible to overestimate the importance and difficulty of developing federated health AI technology. Lee *et al.* [127] designed and tested leveraging MIMIC III data to develop a federated patient database that protects patient privacy method of learning based on similarity [128]. They were able to confirm that in a federated setting, adequate patient data is encrypted using a homomorphic algorithm that may maintain the level of care provided.

2.6.5 Data errors

Training data samples are required for all AI models. In most cases, the patient's size is usually a factor as a sample of training is insufficient to obtain the variances across the board of patients as well as the intricacies of their medical issues. Often, the prototype developed patients' perspectives that patients from a single institution are not eligible for services from another. Data bias is a term used to describe this issue, and one of the most popular significant obstacles to AI in health. As Khullar [129] points out, such bias can exacerbate health inequities. The collection of big and diverse patient datasets is one strategy to eliminate bias. Its means that it is a nonprofit research organisation that funds studies designed to help patients and health care consumers make better informed healthcare choices. These operations create the framework for assembling large-scale, diverse datasets, which are necessary for developing artificial intelligence models that are strong and generalizable. Further, researchers can eliminate prejudice from the entire process of creating models, which is biased [131]; for example, the Gaussian process in the counterfactual was made for hazard analysis and also for personalized treatment planning, i.e., to use "what-if" logic.

2.7 Conclusion

In the future, artificial intelligence (AI) will play a big role in healthcare. It is the primary capability propelling the advancement of precision medicine, which is widely accepted as a much-needed improvement in healthcare. Despite initial difficulties in providing diagnosis and therapy suggestions, we believe AI will eventually be able to master that field as well. With the advancement of artificial intelligence for image analysis, many radiology and pathology photos will likely be analyzed by a machine in the future. Speech and text recognitions are already utilized for things like patient communication and capturing healthcare notes, and their popularity will continue to rise in the future. In many healthcare disciplines, the biggest challenge for AI is not whether the technologies will be capable enough to be useful, but rather ensuring their adoption in everyday clinical practice.

For widespread acceptance to take place, regulators must provide their approval to AI systems, EHR systems are integrated, standardized to the point where equivalent items behave in the same way, physicians have been instructed, financed by public or private payer organizations, and continually updated in the field. These issues will be resolved in the end, but it will take considerably longer than it will take for the technologies themselves to mature. Furthermore, it seems that AI systems will not be able to completely replace human clinicians; rather than replacing their efforts to care for patients, they will augment them. Human therapists may someday move their focus to activities and job designs that demand distinctly human qualities like empathy, persuasion, and big-picture integration. Those healthcare providers who refuse to work with AI may be the only ones who lose their jobs in the long run. The work being done to address AI issues in healthcare is promising, and this progress will help AI play a bigger role in health in the future, both individually and collectively. In many circumstances, the main advantage of AI is that it is faster, unbiased, fairly accurate, and dependable than traditional approaches. AI has previously been used in medical science and will be used more frequently in the future. There are various risks, but additional studies and advancements are likely to alleviate them. Artificial intelligence's application in clinical practice is a promising study subject that is rapidly expanding alongside other modern sciences such as precision medicine, genomics, and teleconsultation. While scientific research into novel ways to improve modern healthcare should continue to be rigorous and transparent, the ethical and economical problems associated with this cornerstone of medical evolution should now be the focus of health policies. Humans are the most advanced machines that have ever been made. The human brain is working hard to create something that can accomplish any task far more efficiently than a human can. Watson for cancer, tug robots, and robotic pharmacy are just a few examples of AI technologies that have drastically transformed the area. The healthcare industry becomes increasingly sophisticated and technologically advanced as time goes on; the more infrastructure it will necessitate, the more it will cost. The creation and implementation of algorithms for data processing are known as artificial intelligence, learning, and interpretation. Healthcare AI implementation is a high-risk, high-reward venture. Similar to pharmaceuticals and medical equipment, clinical trials must be conducted in a sequential, long-term, and rigorous manner. To produce scientifically meaningful findings that can be replicated over time and across populations, AI research requires a sequential strategy and a long-term and rigorous investigation. In contrast to medications and medical devices, the effectiveness of AI technologies in medicine is determined by how well they are understood by their user and the trust and the actions that follow. Integration with the existing clinical setting as well as a data-gathering platform are also required for AI evaluation, storage, processing, and transmission of outputs to users on time. The analogy to the evaluation of drugs and medical equipment may aid in the clinical audience's understanding of AI evaluation; however, the framework has limits, particularly for adaptive AI systems. Learning-induced changes in the underlying data and model performance, for example, may need a reassessment of several phases at the same time. A thorough review of AI technologies throughout all stages of research will necessitate multidisciplinary teams with experience in computer science, healthcare, and the social sciences. Developers should ideally not review their technologies to prevent potential biases, especially in the later stages of evaluation. Collaborations between academic, public, and corporate institutions, as well as dedicated evaluation teams free of responsibility for solution development or sales, may be required. While some AI technologies in healthcare may be regulated, others are not; informatics professionals' ethical role should be to commit to comprehensive examination. By investing the time, skill, and money needed to conduct AI research, patients and healthcare systems may be able to reap the promised benefits. As a result of these achievements, we will experience along "AI Summer."

2.8 Acknowledgment

Nil.

2.9 Funding

Nil.

2.10 Conflicts of Interest

There are no conflicts of interest.

References

[1] Agrawal, A., Gans, J., & Goldfarb, A. (2018). *Prediction machines: the simple economics of artificial intelligence*. Harvard Business Press.

[2] Acemoglu, D., & Restrepo, P. (2020). The wrong kind of AI? Artificial intelligence and the future of labour demand. *Cambridge Journal of Regions, Economy and Society*, *13*(1), 25–35.

[3] Davenport, T., & Kalakota, R. (2019). The potential for artificial intelligence in healthcare. *Future healthcare journal*, *6*(2), 94.

[4] Kavlakoglu, E. (2020). Ai vs. machine learning vs. deep learning vs. neural networks: What's the difference?. *IBM*.

[5] Lee, S. I., Celik, S., Logsdon, B. A., Lundberg, S. M., Martins, T. J., Oehler, V. G., ... & Becker, P. S. (2018). A machine learning approach to integrate big data for precision medicine in acute myeloid leukaemia. *Nature communications*, *9*(1), 1–13.

[6] Sordo, M. (2002). Introduction to neural networks in healthcare. *Open Clinical: Knowledge Management for Medical Care*.

[7] Fakoor, R., Ladhak, F., Nazi, A., & Huber, M. (2013, June). Using deep learning to enhance cancer diagnosis and classification. In *Proceedings of the international conference on machine learning* (Vol. 28, pp. 3937–3949). ACM, New York, USA.

[8] Georgescu, T. M. (2020). Natural language processing model for automatic analysis of cybersecurity-related documents. *Symmetry*, *12*(3), 354.

[9] Furlan, R., Gatti, M., Menè, R., Shiffer, D., Marchiori, C., Levra, A. G., ... & Dipaola, F. (2021). A Natural language processing–based virtual patient simulator and intelligent tutoring system for the clinical diagnostic process: simulator development and case study. *JMIR medical informatics*, *9*(4), e24073.

[10] Savova, G. K., Danciu, I., Alamudun, F., Miller, T., Lin, C., Bitterman, D. S., ... & Warner, J. L. (2019). Use of natural language processing to extract clinical cancer phenotypes from electronic medical records. *Cancer Research*, *79*(21), 5463–5470.

[11] Ratnayake, R. C. (2016). Knowledge-based engineering approach for subsea pipeline systems' FFR assessment: A fuzzy expert system. *The TQM Journal*.

[12] Vial, A., Stirling, D., Field, M., Ros, M., Ritz, C., Carolan, M., ... & Miller, A. A. (2018). The role of deep learning and radiomic feature extraction in cancer-specific predictive modelling: a review. *Transl Cancer Res*, *7*(3), 803–816.

[13] Davenport, T. H., & Glaser, J. (2002). Just-in-time delivery comes to knowledge management. *Harvard business review*, *80*(7), 107–11.

[14] Frey, C. B., & Osborne, M. A. (2017). The future of employment: How susceptible are jobs to computerisation?. *Technological forecasting and social change*, *114*, 254–280.

[15] Hasin, Y., Seldin, M., & Lusis, A. (2017). Multi-omics approaches to disease. *Genome Biology*, *18*(1), 1–15.

[16] Sun, S. (2013). A survey of multi-view machine learning. *Neural computing and applications*, *23*(7), 2031–2038.

[17] Fridley, B. L., Lund, S., Jenkins, G. D., & Wang, L. (2012). AB ayesian Integrative Genomic Model for Pathway Analysis of Complex Traits. *Genetic epidemiology*, *36*(4), 352–359.

[18] Holzinger, E. R., Dudek, S. M., Frase, A. T., Pendergrass, S. A., & Ritchie, M. D. (2014). ATHENA: the analysis tool for heritable and environmental network associations. *Bioinformatics*, *30*(5), 698–705.

[19] Kim, S., Yeganova, L., Comeau, D. C., Wilbur, W. J., & Lu, Z. (2018). PubMed Phrases, an open set of coherent phrases for searching biomedical literature. *Scientific data*, *5*(1), 1–11.

[20] Wang, B., Mezlini, A. M., Demir, F., Fiume, M., Tu, Z., Brudno, M., ... & Goldenberg, A. (2014). Similarity network fusion for aggregating data types on a genomic scale. *Nature methods*, *11*(3), 333–337.

[21] Drăghici, S., & Potter, R. B. (2003). Predicting HIV drug resistance with neural networks. *Bioinformatics*, *19*(1), 98–107.

[22] Sinnenberg, L., DiSilvestro, C. L., Mancheno, C., Dailey, K., Tufts, C., Buttenheim, A. M., ... & Merchant, R. M. (2016). Twitter is a potential data source for cardiovascular disease research. *JAMA cardiology*, *1*(9), 1032–1036.

[23] Ra, C. K., Cho, J., Stone, M. D., De La Cerda, J., Goldenson, N. I., Moroney, E., ... & Leventhal, A. M. (2018). Association of digital media uses with subsequent symptoms of attention-deficit/hyperactivity disorder among adolescents. *Jama*, *320*(3), 255–263.

[24] Birnbaum, M., Rizvi, A., De Choudhury, M., Ernala, S., Cecchi, G., & Kane, J. (2018). O9. 2. identifying psychotic symptoms and predicting relapse through social media. *Schizophrenia Bulletin*, *44*(Suppl 1), S100.

[25] Birnbaum, M. L., Ernala, S. K., Rizvi, A. F., De Choudhury, M., & Kane, J. M. (2017). A collaborative approach to identifying social media markers of schizophrenia by employing machine learning and clinical appraisals. *Journal of medical Internet research*, *19*(8), e7956.

[26] De Choudhury, M., & Kiciman, E. (2017, May). The language of social support in social media and its effect on suicidal ideation risk. In *Proceedings of the International AAAI Conference on Web and Social Media* (Vol. 11, No. 1, pp. 32–41).

[27] De Choudhury, M., Gamon, M., Counts, S., & Horvitz, E. (2013, June). Predicting depression via social media. In *the Seventh International AAAI Conference on weblogs and social media*.

[28] Dodge, H. H., Zhu, J., Mattek, N. C., Bowman, M., Ybarra, O., Wild, K. V., ... & Kaye, J. A. (2015). Web-enabled conversational interactions as a method to improve cognitive functions: Results of a 6-week

randomized controlled trial. *Alzheimer's & Dementia: Translational Research & Clinical Interventions, 1*(1), 1–12.

[29] Asgari, M., Kaye, J., & Dodge, H. (2017). Predicting mild cognitive impairment from spontaneous spoken utterances. *Alzheimer's & Dementia: Translational Research & Clinical Interventions, 3*(2), 219–228.

[30] Tang, F., Lin, K., Uchendu, I., Dodge, H. H., & Zhou, J. (2018). Improving mild cognitive impairment prediction via reinforcement learning and dialogue simulation. *arXiv preprint arXiv:1802.06428.*

[31] Sutton, R. S., & Barto, A. G. (2018). *Reinforcement learning: An introduction.* MIT press.

[32] Kumar, S., Nilsen, W. J., Abernethy, A., Atienza, A., Patrick, K., Pavel, M., ... & Swendeman, D. (2013). Mobile health technology evaluation: the mHealth evidence workshop. *American journal of preventive medicine, 45*(2), 228–236.

[33] Saeb, S., Zhang, M., Karr, C. J., Schueller, S. M., Corden, M. E., Kording, K. P., & Mohr, D. C. (2015). Mobile phone sensor correlates of depressive symptom severity in daily-life behaviour: an exploratory study. *Journal of medical Internet research, 17*(7), e4273.

[34] Selter, A., Tsangouri, C., Ali, S. B., Freed, D., Vatchinsky, A., Kizer, J., ... & Estrin, D. (2018). A mHealth app for self-management of chronic lower back pain (Limbr): a pilot study. *JMIR mHealth and uHealth, 6*(9), e8256.

[35] Zhan, A., Mohan, S., Tarolli, C., Schneider, R. B., Adams, J. L., Sharma, S., ... & Saria, S. (2018). Using smartphones and machine learning to quantify Parkinson's disease severity: the mobile Parkinson's disease score. *JAMA neurology, 75*(7), 876–880.

[36] Turakhia, M. P., & Kaiser, D. W. (2016). Transforming the care of atrial fibrillation with mobile health. *Journal of Interventional Cardiac Electrophysiology, 47*(1), 45–50.

[37] Kumar, S., Abowd, G. D., Abraham, W. T., al'Absi, M., Gayle Beck, J., Chau, D. H., ... & Wetter, D. W. (2015). Center of excellence for mobile sensor data-to-knowledge (MD2K). *Journal of the American Medical Informatics Association, 22*(6), 1137–1142.

[38] Cosselman, K. E., Navas-Acien, A., & Kaufman, J. D. (2015). Environmental factors in cardiovascular disease. *Nature Reviews Cardiology, 12*(11), 627–642.

[39] MacNee, W., & Donaldson, K. (2000). Exacerbations of COPD: environmental mechanisms. *Chest, 117*(5), 390S-397S.

[40] Koller, W., Vetere-Overfield, B., Gray, C., Alexander, C., Chin, T., Dolezal, J., ... & Tanner, C. (1990). Environmental risk factors in Parkinson's disease. *Neurology, 40*(8), 1218–1218.

[41] Guloksuz, S., van Os, J., & Rutten, B. P. (2018). The exposome paradigm and the complexities of environmental research in psychiatry. *JAMA Psychiatry*, *75*(10), 985–986.

[42] Boffetta, P., & Nyberg, F. (2003). Contribution of environmental factors to cancer risk. *British medical bulletin*, *68*(1), 71–94.

[43] Song, Y., Wang, F., Wang, B., Tao, S., Zhang, H., Liu, S., ... & Zeng, Q. (2015). Time-series analyses of hand, foot and mouth disease integrating weather variables. *PloS one*, *10*(3), e0117296.

[44] Stingone, J. A., Pandey, O. P., Claudio, L., & Pandey, G. (2017). Using machine learning to identify air pollution exposure profiles associated with early cognitive skills among us children. *Environmental Pollution*, *230*, 730–740.

[45] Park, S. K., Zhao, Z., & Mukherjee, B. (2017). Construction of environmental risk score beyond standard linear models using machine learning methods: application to metal mixtures, oxidative stress and cardiovascular disease in NHANES. *Environmental Health*, *16*(1), 1–17.

[46] Hahn, L. W., Ritchie, M. D., & Moore, J. H. (2003). Multifactor dimensionality reduction software for detecting gene-gene and gene-environment interactions. *Bioinformatics*, *19*(3), 376–382.

[47] Wang, Y., Xiao, J., Suzek, T. O., Zhang, J., Wang, J., & Bryant, S. H. (2009). PubChem: a public information system for analyzing bioactivities of small molecules. *Nucleic acids research*, *37*(suppl_2), W623–W633.

[48] Zhang, P., Wang, F., Hu, J., & Sorrentino, R. (2014). Towards personalized medicine: leveraging patient similarity and drug similarity analytics. *AMIA Summits on Translational Science Proceedings*, *2014*, 132.

[49] Zhang, P., Wang, F., & Hu, J. (2014). Towards drug repositioning: a unified computational framework for integrating multiple aspects of drug similarity and disease similarity. In *AMIA annual symposium proceedings* (Vol. 2014, p. 1258). American Medical Informatics Association.

[50] Bronstein, M. M., Bruna, J., LeCun, Y., Szlam, A., & Vandergheynst, P. (2017). Geometric deep learning: going beyond euclidean data. *IEEE Signal Processing Magazine*, *34*(4), 18–42.

[51] Duvenaud, D. K., Maclaurin, D., Iparraguirre, J., Bombarell, R., Hirzel, T., Aspuru-Guzik, A., & Adams, R. P. (2015). Convolutional networks on graphs for learning molecular fingerprints. *Advances in neural information processing systems*, *28*.

[52] Kearnes, S., McCloskey, K., Berndl, M., Pande, V., & Riley, P. (2016). Molecular graph convolutions: moving beyond fingerprints. *Journal of computer-aided molecular design*, *30*(8), 595–608.

[53] Chekroud, A. M., Zotti, R. J., Shehzad, Z., Gueorguieva, R., Johnson, M. K., Trivedi, M. H., ... & Corlett, P. R. (2016). Cross-trial prediction of treatment outcome in depression: a machine learning approach. *The Lancet Psychiatry*, *3*(3), 243–250.

[54] Kohannim, O., Hua, X., Hibar, D. P., Lee, S., Chou, Y. Y., Toga, A. W., ... & Alzheimer's Disease Neuroimaging Initiative. (2010). Boosting power for clinical trials using classifiers based on multiple biomarkers. *Neurobiology of ageing*, *31*(8), 1429–1442.

[55] Wang, F., & Preininger, A. (2019). AI in health: state of the art, challenges, and future directions. *Yearbook of medical informatics*, *28*(01), 016–026.

[56] Sakaeda, T., Tamon, A., Kadoyama, K., & Okuno, Y. (2013). Data mining of the public version of the FDA Adverse Event Reporting System. *International journal of medical sciences*, *10*(7), 796.

[57] Tatonetti, N. P., Fernald, G. H., & Altman, R. B. (2012). A novel signal detection algorithm for identifying hidden drug-drug interactions in adverse event reports. *Journal of the American Medical Informatics Association*, *19*(1), 79–85.

[58] Zhang, P., Wang, F., Hu, J., & Sorrentino, R. (2015). Label propagation prediction of drug-drug interactions based on clinical side effects. *Scientific reports*, *5*(1), 1–10.

[59] Banda, J. M., Evans, L., Vanguri, R. S., Tatonetti, N. P., Ryan, P. B., & Shah, N. H. (2016). A curated and standardized adverse drug event resource to accelerate drug safety research. *Scientific data*, *3*(1), 1–11.

[60] Cohen, A. M., & Hersh, W. R. (2005). A survey of current work in biomedical text mining. *Briefings in bioinformatics*, *6*(1), 57–71.

[61] Leaman, R., Islamaj Doğan, R., & Lu, Z. (2013). DNorm: disease name normalization with pairwise learning to rank. *Bioinformatics*, *29*(22), 2909–2917.

[62] Abacha, A. B., & Zweigenbaum, P. (2015). MEANS: A medical question-answering system combining NLP techniques and semantic Web technologies. *Information processing & management*, *51*(5), 570–594.

[63] Sarrouti, M., & El Alaoui, S. O. (2017). A machine learning-based method for question type classification in biomedical question answering. *Methods of Information in Medicine*, *56*(03), 209–216.

[64] Sarrouti, M., & El Alaoui, S. O. (2017). A yes/no answer generator based on sentiment-word scores in biomedical question answering. *International Journal of Healthcare Information Systems and Informatics (IJHISI)*, *12*(3), 62–74.

[65] Rong, G., Mendez, A., Assi, E. B., Zhao, B., & Sawan, M. (2020). Artificial intelligence in healthcare: review and prediction case studies. *Engineering*, 6(3), 291–301.

[66] Rabhi, Y., Mrabet, M., & Fnaiech, F. (2018). A facial expression controlled wheelchair for people with disabilities. *Computer methods and programs in biomedicine*, 165, 89–105.

[67] Hudec, M., & Smutny, Z. (2017). RUDO: A home ambient intelligence system for blind people. *Sensors*, 17(8), 1926.

[68] Tumpa, S. N., Islam, A. B., & Ankon, M. T. M. (2017, September). Smart care: An intelligent assistant for pregnant mothers. In *2017 4th International Conference on Advances in Electrical Engineering (ICAEE)* (pp. 754–759). IEEE.

[69] Oprescu, A. M., Miro-Amarante, G., García-Díaz, L., Beltrán, L. M., Rey, V. E., & Romero-Ternero, M. (2020). Artificial intelligence in pregnancy: a scoping review. *IEEE Access*, 8, 181450–181484.

[70] Oprescu, A. M., Miro-Amarante, G., García-Díaz, L., Beltrán, L. M., Rey, V. E., & Romero-Ternero, M. (2020). Artificial intelligence in pregnancy: a scoping review. *IEEE Access*, 8, 181450–181484.

[71] Amin, M., Yoon, Y. S., & Kassam, S. (2017). Fast Acquisition and Compressive Sensing Techniques for Through-the-Wall Radar Imaging. In *Through-the-Wall Radar Imaging* (pp. 471–499). CRC Press.

[72] Zhang, Y., Amin, M. G., & Ahmad, F. (2008, April). Narrowband frequency-hopping radars for the range estimation of moving and vibrating targets. In *Radar Sensor Technology XII* (Vol. 6947, p. 694709). International Society for Optics and Photonics.

[73] García-Vázquez, J. P., Rodríguez, M. D., Tentori, M., Saldaña-Jimenez, D., Andrade, Á. G., & Espinoza, A. N. (2010). An Agent-based Architecture for Developing Activity-Aware Systems for Assisting Elderly. *J. Univers. Comput. Sci.*, 16(12), 1500–1520.

[74] Chin, L. C., Basah, S. N., Yaacob, S., Juan, Y. E., & Kadir, A. K. A. (2015, May). Camera systems in human motion analysis for biomedical applications. In *AIP Conference Proceedings* (Vol. 1660, No. 1, p. 090006). AIP Publishing LLC.

[75] Dahmani, K., Tahiri, A., Habert, O., & Elmeftouhi, Y. (2016, April). An intelligent model of home support for people with loss of autonomy: A novel approach. In *2016 International Conference on Control, Decision and Information Technologies (CoDIT)* (pp. 182–185). IEEE.

[76] Iezzi, R., Goldberg, S. N., Merlino, B., Posa, A., Valentini, V., & Manfredi, R. (2019). Artificial intelligence in interventional radiology: a literature review and future perspectives. *Journal of oncology*, 2019.

[77] Choi, B. K., Dayaram, T., Parikh, N., Wilkins, A. D., Nagarajan, M., Novikov, I. B., ... & Lichtarge, O. (2018). Literature-based automated discovery of tumour suppressor p53 phosphorylation and inhibition by NEK2. *Proceedings of the National Academy of Sciences*, *115*(42), 10666–10671.

[78] Ruffini, G. (2017). An algorithmic information theory of consciousness. *Neuroscience of Consciousness*, *3*(1).

[79] Kanevsky, J., Corban, J., Gaster, R., Kanevsky, A., Lin, S., & Gilardino, M. (2016). Big data and machine learning in plastic surgery: a new frontier in surgical innovation. *Plastic and reconstructive surgery*, *137*(5), 890e–897e.

[80] Briganti, G., & Le Moine, O. (2020). Artificial intelligence in medicine: today and tomorrow. *Frontiers in medicine*, *7*, 27.

[81] Benjamin, E. J., Blaha, M. J., Chiuve, S. E., Cushman, M., Das, S. R., Deo, R., ... & Muntner, P. (2017). Heart disease and stroke statistics—2017 update: a report from the American Heart Association. *circulation*, *135*(10), e146–e603.

[82] Raja, J. M., Elsakr, C., Roman, S., Cave, B., Pour-Ghaz, I., Nanda, A., ... & Khouzam, R. N. (2019). Apple Watch, wearables, and heart rhythm: where do we stand?. *Annals of translational medicine*, *7*(17).

[83] Bouadjenek, M. R., Verspoor, K., & Zobel, J. (2017). Automated detection of records in biological sequence databases that are inconsistent with the literature. *Journal of biomedical informatics*, *71*, 229–240.

[84] Dorado-Díaz, P. I., Sampedro-Gómez, J., Vicente-Palacios, V., & Sánchez, P. L. (2019). Applications of artificial intelligence in cardiology. The future is already here. *Revista Española de Cardiología (English Edition)*, *72*(12), 1065–1075.

[85] Topalovic, M., Das, N., Burgel, P. R., Daenen, M., Derom, E., Haenebalcke, C., ... & Janssens, W. (2019). Artificial intelligence outperforms pulmonologists in the interpretation of pulmonary function tests. *European Respiratory Journal*, *53*(4).

[86] Delclaux, C. (2019). No need for pulmonologists to interpret pulmonary function tests. *European Respiratory Journal*, *54*(1).

[87] Lawton, J., Blackburn, M., Allen, J., Campbell, F., Elleri, D., Leelarathna, L., ... & Hovorka, R. (2018). Patients' and caregivers' experiences of using continuous glucose monitoring to support diabetes self-management: a qualitative study. *BMC endocrine disorders*, *18*(1), 1–10.

[88] Christiansen, M. P., Garg, S. K., Bragg, R., Bode, B. W., Bailey, T. S., Slover, R. H., ... & Kaufman, F. R. (2017). Accuracy of a fourth-generation subcutaneous continuous glucose sensor. *Diabetes technology & therapeutics*, *19*(8), 446–456.

[89] Saad, A. M., Younes, Z. M., Ahmed, H., Brown, J. A., Al Owesie, R. M., & Hassoun, A. A. (2018). Self-efficacy, self-care and glycemic control in Saudi Arabian patients with type 2 diabetes mellitus: A cross-sectional survey. *Diabetes research and clinical practice, 137*, 28–36.

[90] Niel, O., Boussard, C., & Bastard, P. (2018). Artificial Intelligence Can Predict GFR Decline During the Course of ADPKD. *American journal of kidney diseases: the official journal of the National Kidney Foundation, 71*(6), 911–912.

[91] Martens, F. M. J., Van Kuppevelt, H. J. M., Beekman, J. A. C., Rijkhoff, N. J. M., & Heesakkers, J. P. F. A. (2010). Limited value of bladder sensation as a trigger for conditional neurostimulation in spinal cord injury patients. *Neurourology and Urodynamics: Official Journal of the International Continence Society, 29*(3), 395–400.

[92] Mendez, A., Belghith, A., & Sawan, M. (2013). A DSP for sensing the bladder volume through afferent neural pathways. *IEEE transactions on biomedical circuits and systems, 8*(4), 552–564.

[93] Regalia, G., Onorati, F., Lai, M., Caborni, C., & Picard, R. W. (2019). Multimodal wrist-worn devices for seizure detection and advancing research: focus on the Empatica wristbands. *Epilepsy research, 153*, 79–82.

[94] Dorsey, E. R., Glidden, A. M., Holloway, M. R., Birbeck, G. L., & Schwamm, L. H. (2018). Teleneurology and mobile technologies: the future of neurological care. *Nature Reviews Neurology, 14*(5), 285–297.

[95] Campanella, G., Hanna, M. G., Geneslaw, L., Miraflor, A., Werneck Krauss Silva, V., Busam, K. J., ... & Fuchs, T. J. (2019). Clinical-grade computational pathology using weakly supervised deep learning on whole slide images. *Nature medicine, 25*(8), 1301–1309.

[96] Chang, H. Y., Jung, C. K., Woo, J. I., Lee, S., Cho, J., Kim, S. W., & Kwak, T. Y. (2019). Artificial intelligence in pathology. *Journal of pathology and translational medicine, 53*(1), 1.

[97] Park, Y., Jackson, G. P., Foreman, M. A., Gruen, D., Hu, J., & Das, A. K. (2020). Evaluating artificial intelligence in medicine: phases of clinical research. *JAMIA Open, 3*(3), 326–331.

[98] Liao, Q. V., Gruen, D., & Miller, S. (2020, April). Questioning the AI: informing design practices for explainable AI user experiences. In *Proceedings of the 2020 CHI Conference on Human Factors in Computing Systems* (pp. 1–15).

[99] Saleiro, P., Kuester, B., Hinkson, L., London, J., Stevens, A., Anisfeld, A., ... & Ghani, R. (2018). Aequitas: A bias and fairness audit toolkit. *arXiv preprint arXiv:1811.05577.*

[100] Bellamy, R. K., Dey, K., Hind, M., Hoffman, S. C., Houde, S., Kannan, K., ... & Zhang, Y. (2018). AI Fairness 360: An extensible toolkit for detecting, understanding, and mitigating unwanted algorithmic bias. *arXiv preprint arXiv:1810.01943*.

[101] Osoba, O. A., & Welser IV, W. (2017). *An intelligence in our image: The risks of bias and errors in artificial intelligence*. Rand Corporation.

[102] Kushniruk, A. W., & Patel, V. L. (2004). Cognitive and usability engineering methods for the evaluation of clinical information systems. *Journal of biomedical informatics*, *37*(1), 56–76.

[103] Austin, P. C. (2011). An introduction to propensity score methods for reducing the effects of confounding in observational studies. *Multivariate behavioural research*, *46*(3), 399–424.

[104] Stuart, E. A. (2010). Matching methods for causal inference: A review and a look forward. *Statistical science: a review journal of the Institute of Mathematical Statistics*, *25*(1), 1.

[105] Hernán, M. A., & Robins, J. M. (2006). Estimating causal effects from epidemiological data. *Journal of Epidemiology & Community Health*, *60*(7), 578–586.

[106] Robins, J. M., Hernan, M. A., & Brumback, B. (2000). Marginal structural models and causal inference in epidemiology. *Epidemiology*, *11*(5), 550–560.

[107] US Food and Drug Administration. (2018). Framework for FDA's real-world evidence program. *Silver Spring, MD: US Food and Drug Administration*.

[108] Zhang, X., Chou, J., & Wang, F. (2018, November). Integrative analysis of patient health records and neuroimages via memory-based graph convolutional network. In *2018 IEEE International Conference on Data Mining (ICDM)* (pp. 767–776). IEEE.

[109] McCarty, C. A., Chisholm, R. L., Chute, C. G., Kullo, I. J., Jarvik, G. P., Larson, E. B., ... & Wolf, W. A. (2011). The eMERGE Network: a consortium of biorepositories linked to electronic medical records data for conducting genomic studies. *BMC medical genomics*, *4*(1), 1–11.

[110] Li, L., Cheng, W. Y., Glicksberg, B. S., Gottesman, O., Tamler, R., Chen, R., ... & Dudley, J. T. (2015). Identification of type 2 diabetes subgroups through topological analysis of patient similarity. *Science translational medicine*, *7*(311), 311ra174–311ra174.

[111] Zhang, P., Wang, F., Hu, J., & Sorrentino, R. (2014). Towards personalized medicine: leveraging patient similarity and drug similarity analytics. *AMIA Summits on Translational Science Proceedings*, *2014*, 132.

[112] Simonite, T. (2018). Google's AI guru wants computers to think more like brains. *Wired Magazine*.

[113] Poursabzi-Sangdeh, F., Goldstein, D. G., Hofman, J. M., Wortman Vaughan, J. W., & Wallach, H. (2021, May). Manipulating and measuring model interpretability. In *Proceedings of the 2021 CHI conference on human factors in computing systems* (pp. 1–52).

[114] Holm, E. A. (2019). In defence of the black box. *Science, 364*(6435), 26–27.

[115] Rajkomar, A., Oren, E., Chen, K., Dai, A. M., Hajaj, N., Hardt, M., ... & Dean, J. (2018). Scalable and accurate deep learning with electronic health records. *NPJ Digital Medicine, 1*(1), 1–10.

[116] Min, X., Yu, B., & Wang, F. (2019). Predictive modelling of the hospital readmission risk from patients' claims data using machine learning: a case study on COPD. *Scientific reports, 9*(1), 1–10.

[117] Wang, F., Casalino, L. P., & Khullar, D. (2019). Deep learning in medicine—promise, progress, and challenges. *JAMA internal medicine, 179*(3), 293–294.

[118] Wang, F., & Preininger, A. (2019). AI in health: state of the art, challenges, and future directions. *Yearbook of medical informatics, 28*(01), 016–026.

[119] Hinton, G., Vinyals, O., & Dean, J. (2015). Distilling the knowledge in a neural network. *arXiv preprint arXiv:1503.02531, 2*(7).

[120] Shah, N. D., Steyerberg, E. W., & Kent, D. M. (2018). Big data and predictive analytics: recalibrating expectations. *Jama, 320*(1), 27–28.

[121] Topol, E. J. (2019). High-performance medicine: the convergence of human and artificial intelligence. *Nature medicine, 25*(1), 44–56.

[122] Parikh, R. B., Obermeyer, Z., & Navathe, A. S. (2019). Regulation of predictive analytics in medicine. *Science, 363*(6429), 810–812.

[123] Sitawarin, C., Bhagoji, A. N., Mosenia, A., Chiang, M., & Mittal, P. (2018). Darts: Deceiving autonomous cars with toxic signs. *arXiv preprint arXiv:1802.06430*.

[124] Sun, M., Tang, F., Yi, J., Wang, F., & Zhou, J. (2018, July). Identify susceptible locations in medical records via adversarial attacks on deep predictive models. In *Proceedings of the 24th ACM SIGKDD international conference on knowledge discovery & data mining* (pp. 793–801).

[125] Finlayson, S. G., Bowers, J. D., Ito, J., Zittrain, J. L., Beam, A. L., & Kohane, I. S. (2019). Adversarial attacks on medical machine learning. *Science, 363*(6433), 1287–1289.

[126] Konečný, J., McMahan, H. B., Yu, F. X., Richtárik, P., Suresh, A. T., & Bacon, D. (2016). Federated learning: Strategies for improving communication efficiency. *arXiv preprint arXiv:1610.05492*.

[127] Lee, J., Sun, J., Wang, F., Wang, S., Jun, C. H., & Jiang, X. (2018). Privacy-preserving patient similarity learning in a federated environment: development and analysis. *JMIR medical informatics*, *6*(2), e7744.

[128] Johnson, A. E., Pollard, T. J., Shen, L., Lehman, L. W. H., Feng, M., Ghassemi, M., ... & Mark, R. G. (2016). MIMIC-III, a freely accessible critical care database. *Scientific data*, *3*(1), 1–9.

[129] Khullar, D. A. I. (2019). Could Worsen Health Disparities. *New York Times*, *2*(2), 2019.

[130] Hripcsak, G., Duke, J. D., Shah, N. H., Reich, C. G., Huser, V., Schuemie, M. J., ... & Ryan, P. B. (2015). Observational Health Data Sciences and Informatics (OHDSI): opportunities for observational researchers. *Studies in health technology and informatics*, *216*, 574.

[131] Schulam, P., & Saria, S. (2017). Reliable decision support using counterfactual models. *Advances in neural information processing systems*, *30*.

3

Leveraging Artificial Intelligence in Patient Care

**Yogita Kumari[1], Khushboo Raj[2*], Dilip Kumar Pal[2],
Ankita Moharana[1], and Vetriselvan Subramaniyan[3]**

[1]Department of Pharmacy, Arka Jain University, India
[2]Department of Pharmacy, Guru Ghasidas Vishwavidyalaya, India
[3]Faculty of Medicine, Bioscience and Nursing, MAHSA University,
Malaysia
*Corresponding Author
Khushboo Raj
Department of Pharmacy, Guru Ghasidas Vishwavidyalaya, India
Email: khushboo048raj@gmail.com

Abstract

Artificial intelligence (AI) is a prominent tool that enables people to rethink
how they consolidate information, analyze data, and use the observations
to improve decision making, and it is already revolutionizing every walk
of life. The objective of AI is to model human intellectual functions. It is
causing a fundamental change in healthcare, thanks to the growing availability of healthcare data and the rapid advancement of analytics techniques.
The healthcare market for AI is rapidly increasing at a rate of 40%, and by
the end of 2021, it is expected to reach $6.6 billion. Deep neural networks,
natural language processing, computer vision, and robotics have all made
significant advances in artificial intelligence (AI) in recent years. These techniques are already being used in healthcare, with AI anticipated to take over
many of the tasks currently performed by clinicians and administrators in
the future. Patient administration, clinical decision support, patient monitoring, and healthcare treatments are the four primary areas where AI will have
the largest impact. Many elements of patient care, as well as administrative
operations inside providers, payers, and pharmaceutical companies, could be

transformed by these technologies. The approach to medicine is progressing with the advancement of new (AI) methods of machine learning. Conjoined with rapid improvements in computer processing, these AI-based systems are already enhancing the accuracy and efficiency of diagnosis and treatment across various specializations. The developing focus of AI in radiology has led some experts to suggest that someday AI may even substitute radiologists. A number of studies have already shown that AI can perform as well as or better than humans at crucial healthcare activities like disease diagnosis. Algorithms are already surpassing radiologists in terms of detecting dangerous tumors and advising researchers on how to build cohorts for expensive clinical trials. However, we believe it will be several years before AI replaces humans in large medical process domains for a variety of reasons. Unquestionably, AI is the most considered issue today in medical imaging research, both in diagnostic and therapeutic areas. Scientists have enforced AI to automatically analyze complex patterns in imaging data and help in quantitative assessments of radiographic characteristics. In radiation oncology, AI has been applied to different image procedures that are used at different stages of the treatment, i.e., tumor declination and treatment assessment. For example, AI is essential for boosting power for processing a huge number of medical images and therefore brings to light disease characteristics that are not seen by the naked eye. The utilization of AI within the diagnostic process aiding medical specialists could be of great potential for the healthcare sector and the overall patient's well-being. The assimilation of AI into the current technical framework stimulates the identification of relevant medical data from multiple sources, which is tailored to the needs of the patient and the treatment process. Simultaneously, AI unchains silo thinking, such as sharing knowledge across departmental boundaries, as information from all involved areas is taken into account. Furthermore, AI develops results based on a larger community rather than on subjective experiences and achieves equal outcomes when using similar medical data and does not depend on situations, emotions, or time of day.

3.1 Introduction

A subject of computing science known as "artificial intelligence" (AI) is concerned with the creation of intelligent computers that function and behave like human beings. Some examples of AI-enhanced computer functions include voice assistants, learning and planning, and decision making. We will discuss the following topics in this section: deep learning, machine learning, computer programming, and the medical field. Deep learning enables a variety of practical applications of machine learning and, by extension, the whole domain of artificial intelligence [1]. Deep learning deconstructs work in such a manner that all sorts of machine assistance look conceivable, if not probable. Driverless automobiles, improved preventative healthcare, and even more accurate movie suggestions are all available now or in the near future. Artificial intelligence is both the present and the future. With the assistance of deep learning, AI may potentially achieve the science fiction state we have always envisaged. At its most fundamental level, machine learning is the process of parsing data, learning from it, and then making a decision or prediction about anything in the environment. Instead of writing computer programs that follow a set of rules to do a job, with the help of the superior data, the software can be trained to solve a difficult problem [2].

In computer science, an area known as artificial intelligence (AI) focuses on the creation of artificially intelligent devices. It has grown to be a vital part of the technology industry throughout time. The study of artificial intelligence is a highly specialized and technological endeavor. Computers may be programmed to demonstrate a wide range of human-like abilities, such as understanding of the world around them, reasoning and problem solving, as well as the capacity to learn and plan. An important part of AI research is studying knowledge engineering. For machines to act and respond like humans, they must have a vast amount of information about their surroundings [3]. It takes a long time and a lot of effort to give robots intelligence, reasoning, and problem-solving skills. Another important part of artificial intelligence is machine learning. Without sufficient supervision, learning requires the ability to recognize patterns in streams of inputs, but with good supervision, learning requires categorization and numerical regressions [4–6]. Regression identifies a set of numerical input or output samples, therefore discovering functions capable of producing acceptable outputs from related inputs, whereas classification determines the category to which an item belongs. Computational learning theory is the branch of computational science that deals with the learning and execution of machines through an algorithm. When it comes to computer vision and machine perception, there are a few sub-problems, such as face and object recognition and gesture recognition, which are addressed in both fields of study. Another important field of AI is robotics. Object handling and navigation, as well as the sub-problems of identification, trajectory planning, and cartography, need the use of intelligent robots. Intelligence is undetectable because it is so subtle [5]. It is made of the following components:

- Reasoning

- Education

- Resolving issues
- Observation

Knowledge representation, planning, reinforcement learning, natural language processing, self-awareness, and the ability to manipulate objects are all aims of AI research. The long-term goals of the general intelligence sector are well known. In many ways, today's reality resembles the fictional Wonderland described in Lewis Carroll's book [6]. In Lewis Carroll's famous works by British mathematician Charles Lutwidge Dodgson, artificial intelligence (AI) is described as "a system's capacity to successfully take information from various sources, learn from that data, and utilise that learning to fulfil specified objectives and tasks via successful modifications." It has made significant advancements throughout the years. This is all possible. More than half a century after its academic beginnings in the 1950s, artificial intelligence has remained a mostly unexplored area of research. Due to the expansion of big data and improvements in computer power, big data has become a part of the landscape in today's business and public debate. An AI system that is based on humans can be called analysis, human-inspired, or humanized. It can also be called artificial narrow or artificial general intelligence based on what kind of intelligence it has (intellectual, emotional, or societal intelligence).When AI is employed in a widespread manner, it is no longer referred to as "artificial intelligence." The "AI effect" is a phenomenon that occurs when onlookers dismiss an AI program's behavior by stating that it lacks actual intelligence. Author Arthur Clarke famously quipped, "Any sufficiently advanced technology is imperceptible from magic." Magic, on the other hand, is gone as soon as one masters the technology. Artificial general intelligence (AGI) has been predicted by academics since the 1950s as being just a matter of years away. AGI would be systems that behave exactly like humans in all aspects and exhibit intellectual, emotional, and social intelligence. If this is the case, we will have to wait and see. For a greater knowledge of what is practical, one may approach AI from two perspectives: the route we have already taken and the one we will take in the future. As an example, consider the history of artificial intelligence. Then we return to the present to learn more about the difficulties businesses have in accurately forecasting future occurrences. [8]

3.2 Advancement in Artificial Intelligence

3.2.1 AI spring: artificial intelligence's inception

Isaac Asimov released his short story "Runaround" in 1942, which many consider a prelude to today's self-driving autos. There are three laws of robotics that guide the story of "Runaround," a robot story created by Gregory Powell and Mike Donavan. The first law states that no robot may harm a human being, and the second states that no robot may allow harm to come to a human being through inaction. The third law states that robots must obey human commands, except when doing so conflicts with the first or second law [9]. He had a profound impact on generations of roboticists, artificial intelligence researchers, and computer scientists, notably Marvin Minsky of the United States (who later co-founded the MIT AI laboratory). The British government commissioned Alan Turing, an English mathematician, to embark on a somewhat less fantasy project: a code-breaking computer codenamed The Bombe, which he developed to crack the German army's Enigma code during World War II. For its size and weight, the Bombe is commonly considered to be the first operational electro-mechanical computer [10]. After seeing The Bombe break the Enigma, Turing was astonished. Even the best human mathematicians had hitherto been unable to do this. "Computers and Intelligence" was published by him in 1950, which outlined how to develop intelligent computers and the suitable ways to quantify their intellect. This Turing Test is still used today to figure out how smart an artificial system is: if a person interacts with a human and a machine and cannot tell them apart, the computer is thought to be intelligent.

When Stanford computer scientists John McCarthy and Marvin Minsky conducted the roughly eight-week-long Dartmouth Summer Research Project on Artificial Intelligence (DSRPAI) at Dartmouth College in New Hampshire six years later, in 1956, they formally invented the phrase "artificial intelligence." The Rockefeller Foundation funded this workshop, which kicked off the AI Spring and brought together individuals who would go on to be known as the founding fathers of AI. Claude Shannon, the founder of information theory, was among the participants. With DSRPAI, researchers from several fields were brought together under one roof to create robots that could mimic human intelligence [11].

3.2.2 AI summer and winter: Artificial intelligence's highs and lows

Before the Dartmouth Conference, there had been almost two decades of significant progress in artificial intelligence. Joseph Weizenbaum created the ELIZA computer program at MIT between 1964 and 1966 while he was a student. It was one of the first computers capable of passing the Turing Test using ELIZA, a natural language processing program that was able to simulate a human-to-human conversation. Herbert Simon, Cliff Shaw, and Allen Newell of the RAND Corporation created the General Problem Solver software, which was one of the first successes in artificial intelligence. The Towers of Hanoi and other easy problems might be solved automatically by it. In light of these positive success stories, substantial funding has been devoted to AI research, resulting in a rise in the number of programs. In a 1970 interview with Life Magazine, Marvin Minsky predicted that it would take three to eight years to build a computer with the general intelligence of an average human being. However, this was not the case [12]. Just three years later, in 1973, Congress started to voice substantial concerns about AI research funding. He also questioned the optimism of AI researchers that year by writing an article for the British Science Research Council that was commissioned by British mathematician James Lighthill. Lighthill believes that robots will never be able to play chess at the level of a "seasoned amateur" and that they will never be able to use common sense. After the British government stopped supporting AI research at all but three universities (Edinburgh, Sussex, and Essex), the United States government shortly followed. The AI Winter began at this time. Because of this, despite governmental funding for AI research being increased by both Japan and the United States in 1980, little progress was made in the following years [13].

3.2.3 AI's fall: The harvest

Early artificial intelligence systems such as ELIZA and the General Problem Solver aimed to replicate human cognition in a unique way, which may have contributed to a lack of progress and the fact that reality fell far short of expectations. Assuming that the human mind can be codified and reconstructed as a series of "if–then" statements using a top-down strategy, they were all expert systems. In fields that lend themselves to formalization, expert systems are capable of extraordinary performance. This is shown by the 1997 loss of Gary Kasparov to IBM's Deep Blue chess supercomputer, which proved a nearly 25-year-old claim by James Lighthill to be false. A technique known as "tree search" was used by Deep Blue to analyze 200 million possible movements per second and choose the best course of action [14]. Expert systems, on the other hand, struggle in fields where formalization is difficult or impossible. Such tasks like identifying people or distinguishing between muffins and Chihuahuas are beyond the capabilities of an expert system. Artificial intelligence (AI) is the hallmark of these types of activities since AI systems can effectively comprehend and learn from external data, as well as adjust those learnings to meet specific goals and tasks [15]. As a result, expert systems are not considered true artificial intelligence under this definition. Real artificial intelligence has been contested since Canadian psychologist Donald

Hebb first proposed Hebbian Learning in the 1940s, a theory of learning that attempts to replicate the neural processes seen in the human brain. A study on artificial neural networks was born as a consequence of this. When Marvin Minsky and Seymour Papert proved in 1969 that computers lacked the processing ability to perform the tasks required by artificial neural networks, this project came to an abrupt end. In 2015, the term "deep learning" was coined when Google's AlphaGo AI beat the world Go champion using artificial neural networks. It was long considered that computers would never be able to beat humans in the game of Go, which is significantly more intricate than chess (for example, there are 20 possible moves in chess, but 361 in Go). Deep learning, a kind of artificial neural network, is how AlphaGo was able to achieve such a high level of skill. Artificial neural networks and deep learning are at the core of most AI applications today. Facebook's photo recognition and speech recognition algorithms, speakers, and self-driving cars are built on top of them. The moment of AI fall that we are now experiencing is the harvest of previous statistical breakthroughs [16].

3.2.4 The future: The importance of regulation

It raises the question of whether anyhow ordinance is essential and, if so, what shape it will take as AI systems become further integrated into our everyday lives. A system driven by AI does not have to be completely objective and free of prejudice to be slanted. An AI system's inherent biases are preserved and, in some cases, amplified by the raw data it is trained on. Self-driving car sensors are better at distinguishing bright skin colors than gloomy ones because of the photos that are utilized to direct such algorithms, for example, or judges' capability of decision making may be racially biased because they are trained on photographs of bright skin colors rather than gloomy ones (since they are grounded on the scrutiny of past rulings) [17]. To avoid such errors, rather than attempting to regulate AI directly, the best approach is likely to be the adoption of universally agreed standards for training and testing AI algorithms, possibly in conjunction with some type of guarantee, similar to consumer and safety testing processes for physical goods. Even if the technological components of AI systems continued to advance, this would allow for constant control. If businesses should be held liable for algorithmic mistakes, or if AI programmers should pledge allegiance to a moral code of conduct, these are related issues worth exploring. There is no guarantee that such rules would prevent malicious hacking of AI systems, the misuse of such systems for microtargeting based on personality traits, or the production of fake news. This is further complicated by the fact that deep learning, a fundamental AI technique, is itself a black box. Many of these systems' output metrics (e.g., the percentage of correctly classified photos) are easily measured, but the method by which these outputs are generated remains largely unknown [18]. There are several reasons for this opacity, including a company's desire to hold an algorithm unrevealed, a lack of technical knowledge, or the size of the application (e.g., in cases where a multitude of programmers and methods are involved). In certain cases, this is OK, but not always. For example, few people care about how Facebook decides who to tag in a picture. On the other hand, AI systems employed to provide skin cancer diagnosis suggestions may provide an understanding of how these suggestions are made and are essential for automated image processing [19].

3.3 Artificial Intelligence's Health Benefits

Artificial intelligence (AI) was first proposed in 1956, but noteworthy evolution has occurred in the last 12 years. Many medical records should be examined to provide patients with more efficient and effective treatment. Using machine-like computer systems, artificial intelligence (AI) mimics human cognition and processes. With this technology, you will be able to learn rapidly, forecast and analyze your future, and generate your conclusions. Various medical difficulties, including planning, imaging, speech recognition, and the acquisition of a specific feature [20], were meant to be addressed by it. Based on data, AI systems can forecast and achieve a cut above outcomes with the ability to solve difficult problems accurately. Patients' medical records may be digitized and assembled into a digital database that can be utilized for diagnosis, treatment, and regular maintenance by artificial intelligence. Data collection and routine tasks must be developed by medical specialists in partnership with software and hardware professionals depending on the final requirements. Customizations are made to generic software to satisfy the needs of specific applications. Patients' specific requirements will be taken into account while developing modules for diagnosis, treatment, and follow-up care. However, the AI system's success depends on the analysis of the data it collects. The inventiveness of doctors and surgeons is enhanced by AI. For example, these smart machines understand medical and financial information in the same way that human beings do. As a result, these robots are capable of understanding human speech and making informed decisions [21]. Because of the correct information it provides, it is possible to perform precise surgery on the patient. Patient data gathered by this technology may be utilized in the future to predict and reduce the risk of joint replacement surgery, hospitalization, or recuperation. At this time, artificial intelligence (AI) seems to have the greatest promise for extending human life expectancy. In a difficult condition, it uses artificial intelligence to aid in robotic surgery. The patient is regularly contacted by this technology, which produces data via different virtual help. As a result of the lack of healthcare experts in rural areas, the quality of healthcare suffers.

This is a problem that technology can help solve. Satisfying any urgent demand in rural areas raises the quality of medical students in such areas. The main role of this technology is to upgrade the productivity of medical practitioners and also lower the cost of medical management while improving its quality [22]. Diagnostic accuracy is improved as a result of its use by medical professionals. Scanners such as X-rays, computed tomography (CT), magnetic resonance imaging (MRI), and three-dimensional scanners rely heavily on artificial intelligence. Using this information, you may build a more accurate opinion of the patient. For optimum health, artificial intelligence (AI) promotes a balanced diet and good eating habits. Appointment reminders are sent to patients through text message or email when they are due. With the adoption of this technology, the medical industry has become more efficient in dealing with a wide range of issues [23]. Artificial intelligence in the medical industry has several advantages. When it comes to executing complicated surgeries, AI has the potential to enhance both the excellence of the procedure and the excellence of the results. Various applications of AI are mentioned in Table 3.1. Most patients may now take pleasure in the quick and precise judgment that was made.

3.3.1 Advantages

AI has a variety of advantages in the medical profession, which include the following:

- to look for abnormalities and recommend medical treatment if necessary;
- foretell the onset of new illnesses;
- accurate and time-saving diagnostic procedures;
- complex and novel treatments benefit from the use of this tool;
- ensure that the patient's blood glucose levels are in check;
- give careful attention to the patient's condition;
- make both doctors and patients feel at ease;
- medical students must get enough training;

- enhance hospital security;

- enhance the experience of physicians and surgeons;

- improve the quality of healthcare;

- improved pathological outcome;

- cost savings in diagnostics: maintaining a clinical record is an important part of this;

- ensure that the patient receives the best possible care.

Self-operating quantification activities like measuring carina angle, the aortic valve, and pulmonary artery diameter are all made easier using artificial intelligence in the medical industry. In orthopedics, it is now utilized to assess a patient's degree of fracture and trauma [24].

Orthopedics, neurology, cardiology, and cancer are just a few of the fields where these technologies are used. As a result, the patient receives a better and more accurate service as a result of this. Physicians may now reduce their reliance on manual labor while improving the quality of their treatment plans, clinical judgments, and treatment methods. It is now possible to identify a patient's medical history, and the patient's relatives may be informed. With the help of backend processing and data storage, AI can easily manage routine queries. The patient will be notified if a lab test is running late.

The advancement of artificial intelligence: When it comes to medical technology, artificial intelligence is going to be a game-changer. Faster and more accurate results are the result of better data analysis and more digital automation. Digital consultations and medication management are made easier with this technology [25].

It assists physicians in achieving their goals, which are mentioned below:

1. **Medication:** The use of artificial intelligence (AI) has the power to refine diagnosis, treatment personalization, and the discovery of new medications. It completes a time-consuming pharmaceutical process. To acquire an accurate result, this technology is functional in clinical trials and effective monitoring. It can keep an eye on the patient and efficiently convey information.

2. **Surgical procedure:** Doctors and surgeons are efficiently implementing AI into surgery by capturing data at all stages of the procedure. It provides the highest possible patient attention in the near future. It

Table 3.1 Medical applications of several forms of artificial intelligence technology.

S. No.	Technologies	Description
1.	Machine learning (ML)	• It is possible for these systems to automatically look at medical results and figure out how accurate they are based on statistics.
		• Machine learning (ML) algorithms can use a variety of algorithms and methods to make decisions, such as guided, unguided, semi-guided, or bolstered learning, to make them more likely to make good decisions.
		• Medical professionals utilize this technology to predict the likelihood of sickness.
		• In addition, ML is useful for archiving data for patients to get better care.
2.	Artificial neural networks (ANN)	• ANN system is similar to the human brain and works on the principle of back-propagation and layers. ANN works like neurons and is connected similarly to each other.
		• It may be used to predict the occurrence of sickness and make decisions.
3.	Natural language processing	• It relates to voice recognition and language assessment using a variety of methodologies.
		• Hidden Markov model (HMM) based tagging is one of several NLP algorithms that may be used in the medical field. People can use this system to help with clinical trials as well as to support and analyze data that is not organized.
		• Automated coding and patient documentation are also performed using it.
4.	Support vector machines	• Support vector machines classify the data based on its primary requirements.
		• After training the SVM classifier, fresh and previously unknown data points may be utilized for future correlations in e-mail spam filters.
		• They have been extensively used for data collection.
5.	Heuristics analysis	• Input data is fed to an assisting vector machine, which identifies the groupings of data based on the provided input data.
		• When they are trained, SVM classifiers may utilize unknown data points for future correlations.
		• It is used to gather and analyze medical data.
		• Adequately manage the patient and assist in making an evidence-based decision.

allows the surgeon and the patient to make an informed treatment decision. It performs complex surgeries with ease.

3. **Diagnostic radiology:** With the help of artificial intelligence, surgical procedures may be performed more consistently and accurately. They may help the surgeon get better results during surgery and treatment. The patient's active recovery and enhanced surgical options are facilitated by these changes and advances. To help with conceptualization, this technology may also pre- or post-set data variables connected with the approach. AI has achieved substantial advances that have made it possible to represent and understand complex facts with greater accuracy. Managing the hospital and its patients' health records, this technology digitizes records in the healthcare industry to improve competence and reliability. For example, AI in hospital management systems improves the quality of medical records as well as the collection and storage of customer and patient data. Tracking patients' vital data is made easier with the use of this technology, which provides real-time information to doctors as well as the patient's family. As a result, assessing health systems and deciding who effectively controls the institution has been made easier thanks to this method of verification. It accurately identifies a person's underlying cause. Artificial intelligence (AI) helps doctors, surgeons, and other people in hospitals work more quickly.

3.4 Application

3.4.1 Cardiology

AI may also be used to reduce the threat of unforeseen heart-related problems in the field of cardiology. It includes information about cardiac problems which is supported by scientific data. To prevent a heart attack, this gadget warns the user when the heart valve is blocked. In addition, it provides reliable data on blood flow. When AI is used at every step of a patient's stay in a hospital, it makes things better for them, from admission to treatment and recovery.

Precision and speed are made possible by AI throughout even the most intricate surgical operations. Automated follow-up procedures are planned, verified, and created automatically by the software. Treatment efficiency is increased while the risk of misdiagnosis is reduced. Medical professionals and academics alike have reaped the benefits of these advancements. Patient test results may be reviewed by artificial intelligence, which can then alert or remind patients at the right time. Medical imaging, electroencephalography, respiratory monitoring, and anesthesia are all viable application areas for automatic electrocardiograms (ECGs). With the help of this emerging technology, it is possible to analyze blood tests, glucose levels, medical photographs, and a wide range of other activities in real time. When patient information is included in algorithms, AI can extract the data needed to treat a medical condition. With the use of artificial intelligence, computers can understand speech and writing and use that information to better manage and analyze patients. Doctors, surgeons, and other medical professionals may use it to learn and improve their abilities in real time. To improve the surgeon's performance and provide better outcomes, AI analyzes the surgeon's every step. Furthermore, it highlights the kinds of medical advancements that may be produced. Physician adherence and emergent concerns may be tracked and treated with the help of this system, which is constantly updated. Artificial intelligence (AI) can boost efficiency in medical settings while offering minimal hazards. It is very good at getting data with the help of neural networks, high-resolution images, and NLP [27].

3.4.2 Applications of artificial intelligence in the medical field

Innovating technologies that have a positive influence on human life are required in our daily lives. Artificial intelligence (AI) has several benefits for medical innovation. With the use of this technology, a doctor may evaluate a patient without having to go to a clinic or hospital. It is now possible to provide online patient service using this technology. Inquiries about a wide range of health issues may be answered instantly. In terms of treatment planning and attaining better results, it has several applications (Table 3.2) [28].

In the medical business, artificial intelligence has amazing potential for doing activities that need little human intervention. Clinical judgment, analysis, and training seem to be best served by artificial intelligence. It has been shown that an accurate and speedy diagnosis can be achieved by correctly using this technology. Artificial intelligence (AI) is capable of removing the threat of possible human error in treatment and surgery, which is the main goal. It is possible that the medical team will investigate the in-depth medical tests and the data collected. It is utilized to determine the genetic profile of a patient. In this system, medical concerns, case studies, and patient histories are all stored. When used properly, it may make a patient aware of the need to follow a healthy diet, get enough exercise, and take the correct medications [29].

Table 3.2   Use of AI in the medical industry.

S. No.	Technologies	Description
1.	Prevention of data record	• Data that is stored digitally can be used for the identification of disease causes and to aid research and development activities. • It gathers, maintains, and analyzes medical data to make it more readily available for use and better informed medical decisions. • To enhance the record of previous treatment to felicitate the diagnostic, in a patient's day-to-day history.
2.	Proper diagnosis and treatment	• When it comes to clinical diagnosis and treatment, artificial intelligence (AI) relies mostly on computer tools. • In the healthcare industry, all data and information is maintained digitally, which aids in treatment.
3.	Medications alert	• An app-based personal virtual assistant is available to remind patients to take their medicines. • Assisting patients with their unique medical needs is a job that needs a lot of attention and education. • In the future, artificial intelligence (AI) will have a significant impact on healthcare.

3.4.3 Image and disease diagnosis using artificial intelligence

After AIA's conception, the healthcare industry was considered one of its most promising application sectors. Utilizing cutting-edge algorithms from several fields of information technology, AI-assisted analytics is revolutionizing medical practice and healthcare delivery in groundbreaking ways (IT). Computer algorithms are used to understand complicated data in artificial intelligence (AI), a game-changing technology. An increasing amount of attention has been spent on establishing and fine-tuning diagnostic imaging's performance so that a wide range of clinical disorders may be identified and quantified more easily. Recent computer-assisted diagnostics studies suggest that they can better detect tiny radiographic abnormalities with greater precision, sensitivity, and specificity. But in AI imaging research, outcome measurement is usually based on finding lesions without taking into account what kind of lesions they are or how aggressive they are, which gives an inaccurate picture of AI's performance [30].

Computer scientist Alan Turing coined the term "artificial intelligence" to characterize the science and engineering involved in building robots that can think for themselves. AI systems are computer programs that mimic the cognitive capacities of humans. As a result of these influences, artificial intelligence may be traced back to the philosophy of possibilities as well as the demonstrations of dreams. Several parallel demands, opportunities, and interests led to the development and growth of the discipline of information assurance (IA). Several sectors, including healthcare, are turning to artificial intelligence (AI) combined with analytics (AIA). Medical analytics has been one of the most successful ones and is now a viable sector for artificial intelligence. Applications were developed and made accessible to clinicians to improve their practice when research first started in the middle of the twentieth century.

A few of the possible applications include those in the fields of medical systems and automated surgery, as well as in the management of healthcare [31].

Diagnostic medical imaging uses artificial intelligence (AI) and is now undergoing intensive testing. Imaging abnormalities may now be detected with exceptional precision and sensitivity thanks to AI, and this technology has the potential to revolutionize tissue-based detection and characterization. There are certain drawbacks to this, such as the following: 1) the identification of small changes that may or may not be substantial; 2) while artificial neural networks may not be as good as radiologists in spotting cancer in mammograms, one study found that they are more sensitive to aberrant results in general and smaller lesions in particular; 3) to ensure that AI-assisted diagnostic imaging has a successful and safe introduction into clinical practice, the medical community must anticipate the unknowns that will be associated with this technology. Defining the role of AI in clinical care will require a comprehensive study of its potential hazards in light of its unique capabilities, and straddling the line between greater detection and overdiagnosis will be difficult. External validation and well-defined cohorts are key components of this assessment, which requires ongoing external validation.

There are now several AI image processing studies that measure diagnostic performance using precision and recall computations, while others focus on clinically relevant results. New diagnoses of severe illness, sickness requiring treatment, or illnesses linked to poor long-term survival are key outcome determinants since AI typically detects subtle image alterations. Patients' quality of life is directly impacted by clinically important events, such as the presence of symptoms, the requirement for disease-modifying medicine, and death. Despite several studies showing that AI has a higher level of specificity and a lower level of recall than conventional reading, these studies frequently fail to take into consideration the kind and biological aggressiveness of a lesion. Endpoint selection that is not patient-centric may increase sensitivity, but at the cost of potentially increasing false positives and overdiagnosis by identifying early or indolent disease. AI may be able to find imaging pattern changes that humans cannot see, compared to the clear results that come from sophisticated radiography studies [32].

Machines can identify tissue changes that are suggestive of an early ischemic stroke within a very short time from the beginning of symptoms. Despite the potential of artificial intelligence, the link between relatively minor parenchymal brain anomalies detected by AI, whether in the natural history of microscopic growing infarcts or non-ischemic processes, and large neurological sequelae is still unknown based on early detection. If AI-defined cerebral changes suggestive of early ischemia are linked with a specific profile of neurologic damage or benefit following thrombolysis, this has to be investigated further. A treatment recommendation may be made without a clear abnormality being seen during regular imaging, which presents a unique set of challenges. This might lead to confusion and even scepticism among patients, which necessitates a public education campaign on the new notion of deep learning in image analysis. It also raises medical liability problems (such as missed diagnosis or maybe unneeded surgery) if AI becomes the norm of therapy. Early detection of Alzheimer's disease (AD) necessitates the development of noninvasive and quantitative assessment methods. It is essential to use positron emission tomography (PET) imaging to identify, categorize, and quantify tumors. Various methods for segmenting medical image data have been created via quantitative examination. When using quantitative methods to analyze medical images, a large amount of computer time is required to accurately analyze large amounts of data. Using artificial-intelligence-based algorithms, diagnostic accuracy and efficiency can both be improved [33].

3.5 Recent Advancements in the Field of Artificial Intelligence

An automated database system capable of analyzing medical images and configuring enormous volumes of data using computers has begun to emerge. Back-propagation deep-learning artificial intelligence (AI) systems are now being used in medical imaging, and it has been said that accurate diagnosis may be achieved with this technology. It is the "convolutional neural network" (CNN) that is the most effective image processing model. Several optimization approaches are used to build CNNs, including LeNet, ZFNet, GoogleLeNet, VGGNet, and ResNet (see Figures 3.1 and 3.2). When it comes to extracting features from images, CNN's deep layer is quite successful.

Figure 3.1 Medical image diagnosis method using complex neural network.

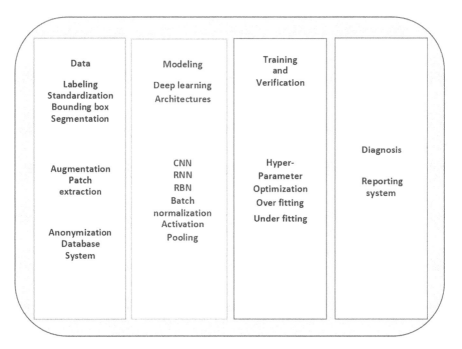

Figure 3.2 Diagram showing the process of developing a medical imaging AI diagnostic.

Convolutional networks, in particular, are rapidly gaining popularity in deep learning medical image analysis, using traction as a technique. With the advent of deep-learning architecture, medical imaging analysis of nerves, retina, lungs, and digital pathology, as well as the musculoskeletal system and the breast, has increased tremendously in recent years [34].

3.5.1 For medical imaging, the use of artificial intelligence is essential

The development of noninvasive and quantitative evaluation methods is critical for the effective treatment of Alzheimer's disease (AD). Because of its capacity to detect, categorize, and measure tumors, PET imaging is critical for early diagnosis and therapy planning. Utilizing quantitative analysis, segmenting medical image data has been made easier. The problem is that many quantitative techniques for analyzing medical images are inexact and require a large amount of computer time to analyze enormous amounts of data. There may be ways to speed up diagnosis and decrease costs based on artificial intelligence systems [35].

3.5.2 Artificial intelligence science and technology

The AI method, a deep-learning technology, improves diagnosis accuracy by automatically gathering and comparing features from highly detailed medical images. Images may now be analyzed using artificial intelligence (AI). The diversity of skin lesion types makes it difficult to automatically detect skin lesions using photos; yet, they have effectively identified skin cancer. Deep learning employing the Inception-v3 architecture was very sensitive and exact in the diagnosis of diabetic retinopathy when looking at diabetic retinal fundus images. The lung image database collaboration (LIDC) and the image database resource project were successful in detecting pulmonary nodules on a chest CT scan for the diagnosis of lung cancer early (IDRI). Deep learning taught on big mammography lesions beats a currently used computer-assisted diagnosis (CAD) method [36]. MR imaging has been effectively used to identify cartilage defects in the knee joint, resulting in the development of an automated detection method. Although PET images with inadequate spatial resolution overestimated the volume, the optimal volume was recovered using an artificial neural network method. AI algorithms are expected to be used in medical picture detection systems, even though other quantitative analysis methods have been developed. This is because they have a better diagnostic performance index than any other quantitative analysis method.

3.6 Artificial Intelligence and its Applications in Diagnostics

For medical imaging diagnosis, massive datasets need to be tagged and harmonized. Depending on the characteristics of the images captured in the database, there are several approaches to investigate artificial intelligence, which incorporate both machine learning and deep learning. Preprocessing and deep-learning architecture are used in the same order to arrive at the final diagnostic results. Deep-learning results are checked and fine-tuned as appropriate once the hyper-parameter has been changed. Techniques like rotation left and right flipping and generative adversarial networks (GANs) are employed to overcome the problem of overfitting caused by a lack of adequate labeled data [37].

3.6.1 Sets of data

There is a picture archiving and communication system (PACS) that stores medical images, but they are of little use to AI researchers. It is necessary to identify, standardize, boundary box, and segment medical images before they can be used for AI. Human effort is required to produce these high-quality images, which may be time-consuming and challenging depending on the skill level of the photographer. Building and testing AI algorithms that need huge volumes of data necessitates the use of standardized and anonymized databases with well-labeled data. A large-scale database system has been built, and diagnostic performance has been improved via competition using these difficult processes. As a result of these efforts, diagnostic accuracy seems to have significantly improved [38].

Researchers may now approach and present study results with more freedom because of the recent availability of an anonymized medical image database. ADNI participants' F-18 fluorodeoxyglucose (FDG) PET scans are being compiled into a database for the research of early detection of Alzheimer's disease (AD). With an accuracy of 88.24%, a specificity of 88.64%, a sensitivity of 87%, and a specificity of 87%, F-18 FDG PET images from the ADNI database were used to distinguish AD patients. For the most accurate diagnosis of knee injuries, magnetic resonance imaging (MRI) is the best option. However, MRI processing is time-consuming and the risk of mistakenly diagnosing an injury is high. A collection of 1370 knee MRI images collected at Stanford University Medical Center was analyzed using MRNet, a CNN, and logistic regression. Analyses were undertaken by nine Stanford University Medical Center clinical specialists, and the validation results were quite similar. As of today's (January 09, 2019) research on abnormal detection, anterior cruciate ligament injuries, and meniscal tears, the leaderboard indicates an average area under the receiver operating characteristic curve (AUC; 0.917). A radiologist looked at 13,292 anterior chest X-ray photos to make a database for testing for severe pneumothorax.

Digital imaging and communications in medicine were used to store potentially analyzed images in a clinical PACS (DICOM; National Electrical Manufacturers Association, Arlington, VA, USA). It is appropriate to use these well-defined datasets to learn about and test out different network configurations. It takes a long time and a lot of effort to build a database of this caliber. The chest X-ray 14 set has been made publicly available by the National Institutes of Health (NIH). To help students learn, the National Institutes of Health (NIH) provides students with well-organized and annotated datasets and a database that anybody may use to challenge and improve algorithms. To make it easier to spot lung cancer, the LIDC and IDRI databases were created and made freely accessible to the public. Skin cancer diagnostic competition participants may use the database. Anybody with an interest in the MR knee database is invited to participate in the AI analysis [39]. The Department of Nuclear Medicine at Dong–A University Medical Center collaborated with IRM to build SortDB. Researchers may use SortDB to download DICOM files in a certain format (nii, JPEG, or GIF, for example) by entering an Excel file list. A vast quantity of data is needed to employ an AI algorithm. For small- and medium-sized hospitals, it is an automatic database generation tool that stores and retains data in the needed format.

3.6.2 A medical image's preprocessing

Using an affine model with 12 parameters, an image was normalized for quantitative analysis using the SPM5 program. Images must be normalized as part of the preprocessing step for machine-learning-based quantitative evaluation. There is a wide range of opinions among researchers since the standard template or the process used to normalize an image affects its accuracy. An AI system's accuracy improves when it is applied to a normalized image. There is, however, a lack of evidence to support the claim that normalized images are more accurate than non-normalized images. The proper preprocessing method for picture analysis using AI requires both theoretical and empirical research [40]. The quality of the preprocessed image transformation is strongly influenced by the accuracy of the quantitative analysis. The quality of the medical images used to apply the algorithm has a substantial impact on the accuracy of AI analysis in particular. The use of a library's GetOuputData function results in a linear change in the DICOM file image. It was scaled linearly to the 0–255 range for the eighth bit and the standardized uptake values (SUV). A warped jpeg file of the SUV may be created using this method of linear manipulation. It is essential to explore the image conversion technology to guarantee that medical images are not distorted throughout the conversion process. It is feasible to guarantee the consistency of AI learning by utilizing image-aware CNNs by downsampling frequently used photographs [41].

3.6.3 Optimization of models and parameters based on improved data

It is tough to establish a database of massive data using PET images since they do not provide a lot of information. Randomly dispersing data for AI analysis is achieved by applying image inversions, enlargements and reductions, shearing, and rotations to reshape it. Over-tuning of hyper-parameters is a major roadblock to achieving the highest possible accuracy. The Future Gadget Laboratory framework may be used to do this. With respect to AUC, sensitivity, specificity, and positive predictive value with scikit-learn, the whole validation set can be computed. To plot the receiver operating characteristic curves, you may use Matplotlib. Currently, there are several ways to build CNN algorithms. The recent trend is to build a CNN model on top of TensorFlow, utilizing the Keras deep-learning package. Deep learning architectures like VGG16/19, Xception, Inception, and ResNet are the most effective for AI analysis because of their high performance. The best image classification system may be one that correctly depicts the unique qualities of each medical image. This feature's location is currently estimated using an activation map. As things stand right now, the only way to discover the optimal model is to carry out a slew of imaging tests in succession. Five types of chest X-rays could be classified using the AlexNet method, which won the ImageNet Large Scale Visual Recognition Challenge (20) in 2012. Using a softmax activation function in the final layer, AlexNet has three fully connected (FC) layers and five convolutional layers. The researcher (21) used a 256,256-pixel png file as the AlexNet input image after downsampling the DICOM image. The adaptive moment estimation (Adam) optimizer has a mini-batch size of 128 and 100 training iterations, an adaptive learning rate of 1103, and a momentum of 0.5. ReLU activation functions were used in the previous stage of the max-pooling layers, and L2 regularization was set to 104 [42].

3.6.4 The principal component analysis (PCA)

It is one of the best ways to use Eigen image analysis. The principal component analysis (PCA) method is one of them.

To improve the accuracy of diagnosis and analysis, computer-assisted classification (PCA) was used to analyze images of people with early dementia [36]. PCA was used to categorize SPECT pictures, utilizing single-photon emission computed tomography (SPECT) and three-dimensional (3D) scattering to show the results. A quantitative evaluation and classification is an effective and sensible method. When PCA-SVM was applied to SPECT and PET scans, the classification accuracy for AD was 96.7% and 89.52%, respectively. High and low information can be used to automatically choose the area of interest (ROI) in a 3D functional brain image.

Gaussian mixture models were used to determine activation zones. Using brain image analysis and pattern illness discrimination, we discovered that PCA/independent component analysis is an effective tool for early AD diagnosis. Classification and idea extraction using PCA/independent component analysis (CAD) may be achieved using eigenvector decomposition and SVM [11].

3.6.5 Analyzing medical images using artificial intelligence

Medical practitioners are enthralled by AI and machine learning because of their increasing accuracy and applicability in a broad variety of jobs. In the healthcare business, artificial intelligence (AI) is being used to uncover previously unseen clinical insights and connect them to the tools patients need to better manage their health. Imaging data from patients' medical records is a rich and complex source of information. It may be difficult even for the most seasoned healthcare professional to sift through high-resolution images from X-ray, CAT, MRI, and other examinations since they contain so much information. With the aid of artificial intelligence, radiologists and pathologists will be able to increase their efficiency and accuracy. Recent studies have demonstrated that AI technology can recognize features in photographs quicker and more precisely than human doctors, if not faster. According to society authorities, the American College of Radiology does not consider AI's entrance as a problem for physicians in diagnosis. As a means of promoting uniformity, security, and efficiency in clinical decision support and diagnostics, the American College of Radiology Data Science Institute (ACR-DSI) has announced many high-value use cases for AI in medical imaging [42]. This list of use cases is always being updated as new possibilities arise.

AI developers may use the ACR's DSI use cases to assist them in tackling healthcare challenges holistically that translate ideas for AI explanations into harmless and active tools to support radiologists well serve patients, according to DSI Chief Medical Officer, Bibb Allen Jr., MD, FACR, ACR. According to the ACR DSI Technology Oriented Use Cases in Health Care-AI (TOUCH-AI) Framework, AI-enhanced medical images may be utilized to improve patient attention across a wide range of disorders and organ groups.

3.6.6 Imaging the brain via artificial intelligence

Alzheimer's disease classification, anatomical brain segmentation, and AI tumor identification have all been the subjects of several investigations. Researchers were able to accurately classify patients as having Alzheimer's disease (AD), moderate cognitive impairment (MCI), or healthy controls by locating feature expressions in MRI and PET volume patches using the Gaussian restricted Boltzmann machine (RBM). Three-dimensional convolutional neural networks (3D CNN) beat other algorithms in classifying AD. To segment MR pictures of the human brain, it uses CNN technology. Deep CNN was used to segment the striatum, and the results were compared to those obtained with FreeSurfer to see which was better. For complex structures like those seen in the brain, automated segmentation is outmatched by the time and variability required for human segmentation. The voxel-wide residual network (VoxResNet) is a network of 25 deep layers that were successfully segregated automatically. A nonlinear mapping from MR images to CT images was shown using a 3D fully CNN in a real pelvic CT/MRI dataset. To improve performance, researchers used two-volume CNNs, which produced excellent results when analyzing MRI and PET images from the ADNI database [43]. As a result of the incurable nature of amyotrophic lateral sclerosis (ALS), victims may be surprised when they are diagnosed. Unfortunately, many neurological diseases, such as ALS, are now incurable, but an accurate diagnosis may help patients plan for their long-term care and last wishes. To distinguish ALS from PLS, imaging studies are required, according to the College of Physicians and Surgeons of Canada (CPSC). It is common for radiologists, who are tasked with determining whether or not lesions resemble disease-causing entities, to report false positives. Recent research has focused on developing new biomarkers to improve diagnostic speed and accuracy. According to ACR DSI, manual separation and quantitative susceptibility mapping (QSM) examinations of the motor cortex are now mandatory, complex, and time-consuming. Programming this process could lead to the development of imaging biomarkers. An algorithm may be able to speed up this procedure by highlighting suspicious pictures and providing risk ratios for ALS or PLS. To ease the burden on service providers, algorithmic reporting may also be possible [41].

3.6.7 Chest imaging with artificial intelligence

The various learning architectures are used to create a de-CNN that shows suspicious areas on a heat map. Radiologic datasets of publicly accessible chest X-rays and reports were used to identify and classify 17 distinct patterns. On a dataset of individuals with interstitial lung disease, it has been claimed that a segmentation-based label propagation technique can detect interstitial patterns and that CNN can identify lung texture patterns. Automated metadata annotations were shown to distinguish between frontal and side chest radiography. Using a 3D CNN, researchers have developed a new method for minimizing the number of false positives in the automated detection of lung nodules. More spatial information and more representative characteristics can be extracted from a 3D CNN thanks to a hierarchical architecture trained on 3D examples. The recommended technique obtained good competition metric scores in the LUNA16 Challenge and is capable of processing 3D PET scans [44].

3.6.8 In breast imaging, artificial intelligence is being used

Deep learning in natural photography may be used to analyze AI photos rapidly since most mammograms are two-dimensional and include a huge amount of data. It is possible to use CNN or RBM approaches to efficiently assess the identification and classification of tumor lesions as well as the detection and classification of microcalcifications and risk-scoring tasks. The estimation of breast density was made possible through the use of CNNs. The estimation of breast density was made possible through the use of CNNs, which is a common strategy which has been modified. We utilized a modified version of CNN's area proposal and R-CNN to pinpoint our location. An MRI dataset was segmented between breast and fibro-gluteal tissue using U-net, and the right breast density calculation results were seen. According to one study, using mammograms as classifiers for various layers of perception, a short-term risk assessment model with a projected accuracy of 71.4% was developed [45].

3.6.9 The use of AI in cardiac imaging

Cardiac artificial research is now focusing on the left ventricle, slice categorization, image quality evaluation, automated calcium scoring, and coronary centerline tracking. The classification relies on deep learning algorithms like the U-net segmentation method, whereas segmentation relies on 2D and 3D CNN techniques. With the use of an innovative image SR method, the reconstruction of the high-resolution 3D volume in the 2D picture stack has been accomplished. However, the SR-CNN approach is better at picture segmentation and motion tracking than the SR method because of the CNN model's efficiency (43). When the ROI is considered a candidate for coronary artery calcification and multi-stream CNN is used, deep learning is useful in recognizing low-dose chest CT with great accuracy (three views) (44). The 3D CNN and multi-stream 2D CNN were used to detect coronary calcium during gated computed tomography angiography. Measuring the individual's heart could reveal their risk of cardiovascular disease or identify problems that may necessitate surgery or medication. Systematizing the finding of abnormalities in imaging studies, such as chest X-rays, may decrease diagnostic errors. One of the most common imaging studies that physicians order when someone comes in complaining of shortness of breath is a chest X-ray. This simple test may detect cardiomegaly, an early sign of heart disease that other tests may miss. The "rapid visual examination" by a radiologist is not always accurate. Medical professionals may be better equipped to diagnose and treat their patients if they employ artificial intelligence in the interpretation of chest X-rays to look for left atrial enlargement. Analyses of the aortic valve, coronary angle, and pulmonary artery's diameter all benefit from the use of artificial intelligence approaches of a similar kind. Images of the heart and arteries may also be used to identify changes in the thickness of the muscle tissues, for example, the left ventricle wall, or monitor fluctuations in the flow of blood. ACR DSI made the statement that "Manual evaluations would be eliminated, which would free resources for the interpreter while also reducing the possibility of finding errors and providing designed quantitative data for use in future studies or risk categorization systems". Algorithms might save time and effort for doctors by filling out reports and spotting irregularities, which could save time and effort [45].

3.6.10 Artificial intelligence in bone imaging

Artificial neural networks (ANNs) may be used to detect abnormalities in musculoskeletal images. Segmenting vertebral bodies from three-dimensional magnetic resonance imaging (MRI) scans automatically yielded a "Dice" similarity coefficient of 93.4%. Since there are so many different types of spines and postures, it is difficult to do automatic spine detection without a large amount of photographic data to work with. A deep-learning architecture known as a modified deep convolution network was used to automatically correct the picture's spine posture before analysis [46–50]. Using a CNN regression to generate an intensity-based 2D/3D registration method has the disadvantage of needing a long computation time and a restricted capture range. Even if the capture range is much bigger, real-time 2D/3D registration with extreme precision is said to be possible.

Many deep-learning X-ray age estimation methods have been developed, and their performance has been confirmed by averaging a discrepancy of a little over one year. Fractures and musculoskeletal injuries may cause long-term pain if they are not treated quickly and efficiently. Because of the decreased mobility and subsequent hospitalization that hip fractures cause, they have been linked to worse overall outcomes in older patients. As a result of using artificial intelligence, surgeons and experts may have greater confidence in their treatment decisions [51–54]. According to the ACR DSI, fractures are considered less critical in the aftermath of a traumatic incident than internal bleeding or organ injury. Because of this, fractures can be overlooked during a trauma-related imaging assessment. Patients who have suffered head and neck trauma and have been brought to the emergency room could have an odontoid fracture diagnosed with the use of an AI radiology tool. For example, however, AI may be better able to identify subtle abnormalities in the picture that might indicate an unstable fracture requiring surgery than the obvious fracture itself, which would require human intervention. An independent algorithm can help ensure that the patient receives the proper treatment and has a positive outcome. Providers may find AI to be a useful safety net in the case of regular continuations of popular hip procedures, for example, replacing the hip joint of a patient. According to the ACR DSI [55], there are an estimated 400,000 total hip replacements (THAs) made each year. Around 100 follow-up exams are done each day by an arthroplasty radiologist who works with musculoskeletal radiologists. If a joint replacement device turns out to be loose or if the tissue around the device reacts badly to the implant, patients may need an expensive and surgical revision. As a result, finding potential problems on the land might be a challenge. For example, "results are not evident on the X-ray image and need comparison with several earlier tests to determine the development of abnormality over time." Therapy may be delayed for years if the diagnosis is delayed. The legal and medical dangers that radiologists face are lessened since AI that fits the criteria of this use case is capable of doing so. Patients with elevated amounts of cobalt in their blood may be sent for an MRI for additional investigation [56].

3.6.11 The use of Artificial Intelligence (AI) in stroke imaging

Stroke is a long-term disorder that manifests itself in a series of acute episodes. To treat a stroke, doctors must use their clinical judgment in several different ways. Research in clinical medicine has always been limited in scope to one or a few clinical issues, ignoring the continuing aspects of stroke treatment. Artificial intelligence (AI) is anticipated to assist researchers in the study of vastly more complex, but equally fascinating, topics in the future, including pertinent clinical issues, leading to better decision-making in stroke therapy. A recent study using this strategy has shown intriguing early findings [57].

3.6.12 Using AI to treat diseases of the lungs

Pneumonia and pneumothorax are two conditions that need immediate medical attention. It is possible that artificial intelligence algorithms will be interested in both. Pneumonia may be fatal if left untreated. Pneumonia may be life-threatening whether it is acquired in the community or during medical care. If you are sick, radiation scans may help doctors determine if you have bronchitis or pneumonia. Radiologists may have difficulty identifying pneumonia in patients with preexisting lung conditions, for example, cancer or cystic fibrosis, since they are not always available to analyze images. Subtle pneumonia, for example, those protruding below the dome of diaphragms on anterior chest radiographs, the ACR DSI states, may also help reduce unnecessary CT scans through the use of artificial intelligence (AI). To speed up treatment, X-rays and other images might be examined by an AI algorithm for indicators of opacities that signal pneumonia [58]. When a pneumothorax is suspected, artificial intelligence (AI) may be able to assist in identifying at-risk individuals. Trauma or invasive procedures can cause pneumothorax, a condition that occurs when air pockets form between the lung and the chest wall. A more serious condition might arise if an illness is left untreated or identified too late. According to the American College of Radiology (ACR) DSI, pneumothorax can be detected by non-radiologists. Treatments for pneumothoraces may be more urgent if artificial intelligence can help prioritize the different types and severity of pneumothoraces. In the long term, artificial intelligence (AI) may be useful to healthcare providers. It is proposed that this use case could be extended to monitor the size of previously discovered and treated pneumothoraces [59].

3.6.13 Artificial intelligence in the treatment of cancer

The use of imaging methods is becoming more common in cancer screenings, such as those for breast and colon cancers. In many circumstances, microcalcification in breast tissue could be difficult to differentiate as malignant or benign. When malignancies go undetected, unnecessary intrusive testing and treatment might be carried out, which can have negative consequences. When it comes to interpreting microcalcifications on diagnostic imaging, "radiologist interpretations vary," the ACR DSI adds. To reduce the number of unnecessary biopsies, artificial intelligence (AI) may be able to improve the accuracy of microcalcifications using quantitative imaging characteristics. Physicians and patients might make better decisions regarding testing and treatment if they were given risk ratings for areas of concern. People may have better connections with their physicians if polyps are found through routine colon cancer screenings. Polyps are the first signs of cancer. ACR DSIS claims that "CT colonography (CTC) enables a nominally invasive structural examination of the colon and rectum" to identify clinically relevant polyps. In contrast, less experienced radiographers may miss polyps or take too long to complete the test. It is possible that employing artificial intelligence (AI) to improve polyp detection accuracy and efficiency might lessen the risk of medical negligence for radiologists at CTC [60].

Patients with advanced cancer may benefit from the use of artificial intelligence (AI) in the detection of metastasized malignant tumors. After surgery, extranodal extension (ECE), a cancer with a poor prognosis, is often discovered. There may be a way to find ECE in diseases that do not often lead to surgery using a high-performance algorithm. According to ACR DIS, "treatment optimization for postoperative imaging-detected nodal disease might benefit from automated ECE categorization and identification." According to society, artificial intelligence may help treat cancers of the head and neck, colorectal, prostate, and cervical types. This algorithmic or moderate approach may improve cancer consequences and decrease illness, according to the ACR DSI, "even if not proven" [61, 62]. Medical imaging, according to the ACR DIS, is ready for artificial intelligence, even though further study is needed. There are a lot of illnesses, injuries, and ailments that are hard to see with the human eye alone. Artificial intelligence (AI) could help healthcare professionals and patients see things that are not visible to the human eye.

3.7 Conclusion

While artificial intelligence technologies continue to gain substantial attention in medical research, their practical application faces major hurdles. If they are to perform well, artificial intelligence (AI) systems must be taught regularly using data from clinical research. Future diagnostic systems will rely heavily on artificial intelligence (AI) for their analysis. Nevertheless, there are a few instances where AI diagnosis might need some refinement. A deep-learning architecture for AI learning necessitates a significant quantity of data. Although the bulk of medical images are technical and manual, developing large data systems is difficult. It also takes time to establish a collection of medical images that are standardized and tagged. With AI diagnosis, previously unanalyzed data may be detected, enhancing diagnostic accuracy and speed as well as the quality of medical treatment.

3.8 Acknowledgment

I would like to thank my co-authors for contributing their knowledge and time and giving their support in compiling the work.

3.9 Funding

The authors received no specific grants from any funding agencies.

3.10 Conflicts of Interest

The authors declare no conflict of interest.

References

[1] Shen, J., Zhang, C. J., Jiang, B., Chen, J., Song, J., Liu, Z., ... & Ming, W. K. (2019). Artificial intelligence versus clinicians in disease diagnosis: systematic review. *JMIR medical informatics*, *7*(3), e10010.

[2] Ao, C., Jin, S., Ding, H., Zou, Q., & Yu, L. (2020). Application and development of artificial intelligence and intelligent disease diagnosis. *Current Pharmaceutical Design*, 26(26),3069–3075.

[3] Hosny, A., Parmar, C., Quackenbush, J., Schwartz, L. H., & Aerts, H. J. (2018). Artificial intelligence in radiology. *Nature Reviews Cancer*, *18*(8), 500–510.

[4] Mintz, Y., & Brodie, R. (2019). Introduction to artificial intelligence in medicine. *Minimally Invasive Therapy & Allied Technologies*, *28*(2), 73–81.

[5] Jiang, F., Jiang, Y., Zhi, H., Dong, Y., Li, H., Ma, S., ... & Wang, Y. (2017). Artificial intelligence in healthcare: past, present and future. *Stroke and vascular neurology*, *2*(4).

[6] Kim, D. W., Jang, H. Y., Kim, K. W., Shin, Y., & Park, S. H. (2019). Design characteristics of studies reporting the performance of artificial

intelligence algorithms for diagnostic analysis of medical images: results from recently published papers. *Korean journal of radiology*, *20*(3), 405–410.

[7] Ramesh, A. N., Kambhampati, C., Monson, J. R., & Drew, P. J. (2004). Artificial intelligence in medicine. *Annals of the Royal College of Surgeons of England*, *86*(5), 334.

[8] England, J. R., & Cheng, P. M. (2019). Artificial intelligence for medical image analysis: a guide for authors and reviewers. *American journal of roentgenology*, *212*(3), 513–519.

[9] Raghavendra, U., Acharya, U. R., & Adeli, H. (2019). Artificial intelligence techniques for automated diagnosis of neurological disorders. *European neurology*, *82*(1–3), 41–64.

[10] Deepa, S. N., & Devi, B. A. (2011). A survey on artificial intelligence approaches for medical image classification. *Indian Journal of Science and Technology*, *4*(11), 1583–1595.

[11] Chang, H. Y., Jung, C. K., Woo, J. I., Lee, S., Cho, J., Kim, S. W., & Kwak, T. Y. (2019). Artificial intelligence in pathology. *Journal of pathology and translational medicine*, *53*(1), 1.

[12] Thrall, J. H., Fessell, D., & Pandharipande, P. V. (2021). Rethinking the approach to artificial intelligence for medical image analysis: the case for precision diagnosis. *Journal of the American College of Radiology*, *18*(1), 174–179.

[13] Owais, M., Arsalan, M., Choi, J., & Park, K. R. (2019). Effective diagnosis and treatment through content-based medical image retrieval (CBMIR) by using artificial intelligence. *Journal of clinical medicine*, *8*(4), 462.

[14] Park, S. H., & Han, K. (2018). Methodologic guide for evaluating clinical performance and effect of artificial intelligence technology for medical diagnosis and prediction. *Radiology*, *286*(3), 800–809.

[15] Jones, L. D., Golan, D., Hanna, S. A., & Ramachandran, M. (2018). Artificial intelligence, machine learning and the evolution of healthcare: A bright future or cause for concern?. *Bone & joint research*, *7*(3), 223–225.

[16] Stoitsis, J., Valavanis, I., Mougiakakou, S. G., Golemati, S., Nikita, A., & Nikita, K. S. (2006). Computer aided diagnosis based on medical image processing and artificial intelligence methods. *Nuclear Instruments and Methods in Physics Research Section A: Accelerators, Spectrometers, Detectors and Associated Equipment*, *569*(2), 591–595.

[17] Parekh, V., Shah, D., & Shah, M. (2020). Fatigue detection using artificial intelligence framework. *Augmented Human Research*, *5*(1), 1–17.

[18] Dilsizian, S. E., & Siegel, E. L. (2014). Artificial intelligence in medicine and cardiac imaging: harnessing big data and advanced computing to provide personalized medical diagnosis and treatment. *Current cardiology reports*, *16*(1), 1–8.

[19] Hasanzad, M., Aghaei Meybodi, H. R., Sarhangi, N., & Larijani, B. (2022). Artificial intelligence perspective in the future of endocrine diseases. *Journal of Diabetes & Metabolic Disorders*, 1–8.

[20] Yu, K. H., Beam, A. L., & Kohane, I. S. (2018). Artificial intelligence in healthcare. *Nature biomedical engineering*, *2*(10), 719–731.

[21] Palomar, M., & Martínez-Barco, P. (2001). Computational approach to anaphora resolution in Spanish dialogues. *Journal of Artificial Intelligence Research*, *15*, 263–287.

[22] Oke, S. A. (2008). A literature review on artificial intelligence. *International journal of information and management sciences*, *19*(4), 535–570.

[23] Wilkins, D. E., Lee, T. J., & Berry, P. (2003). Interactive execution monitoring of agent teams. *Journal of Artificial Intelligence Research*, *18*, 217–261.

[24] Wray, R. E., & Laird, J. E. (2003). An architectural approach to ensuring consistency in hierarchical execution. *Journal of Artificial Intelligence Research*, *19*, 355–398.

[25] Xu, X., He, H. G., & Hu, D. (2002). Efficient reinforcement learning using recursive least-squares methods. *Journal of Artificial Intelligence Research*, *16*, 259–292.

[26] Yu, K. H., Fitzpatrick, M. R., Pappas, L., Chan, W., Kung, J., & Snyder, M. (2018). Omics AnalySIs System for PRecision Oncology (OASISPRO): a web-based omics analysis tool for clinical phenotype prediction. *Bioinformatics*, *34*(2), 319–320.

[27] Zanuttini, B. (2003). New polynomial classes for logic-based abduction. *Journal of Artificial Intelligence Research*, *19*, 1–10.

[28] Zhang, N. L., & Zhang, W. (2001). Speeding up the convergence of value iteration in partially observable Markov decision processes. *Journal of Artificial Intelligence Research*, *14*, 29–51.

[29] Zucker, J. D. (2003). A grounded theory of abstraction in artificial intelligence. *Philosophical Transactions of the Royal Society of London. Series B: Biological Sciences*, *358*(1435), 1293–1309.

[30] Vashistha, R., Chhabra, D., & Shukla, P. (2018). Integrated artificial intelligence approaches for disease diagnostics. *Indian journal of microbiology*, *58*(2), 252–255.

[31] Shiraishi, J., Li, Q., Appelbaum, D., & Doi, K. (2011, November). Computer-aided diagnosis and artificial intelligence in clinical imaging. In *Seminars in nuclear medicine* (Vol. 41, No. 6, pp. 449–462). WB Saunders.

[32] Noorbakhsh-Sabet, N., Zand, R., Zhang, Y., & Abedi, V. (2019). Artificial intelligence transforms the future of health care. *The American journal of medicine, 132*(7), 795–801.

[33] Joloudari, J. H., Mojrian, S., Nodehi, I., Mashmool, A., Zadegan, Z. K., Shirkharkolaie, S. K., ... & Mosavi, A. (2021). A Survey of Applications of Artificial Intelligence for Myocardial Infarction Disease Diagnosis. *arXiv preprint arXiv:2107.06179.*

[34] Yadav, S. S., & Jadhav, S. M. (2019). Deep convolutional neural network based medical image classification for disease diagnosis. *Journal of Big Data, 6*(1), 1–18.

[35] Mongan, J., Moy, L., & Kahn Jr, C. E. (2020). Checklist for artificial intelligence in medical imaging (CLAIM): a guide for authors and reviewers. *Radiology: Artificial Intelligence, 2*(2), e200029.

[36] Krittanawong, C., Zhang, H., Wang, Z., Aydar, M., & Kitai, T. (2017). Artificial intelligence in precision cardiovascular medicine. *Journal of the American College of Cardiology, 69*(21),2657–2664.

[37] Wong, Z. S., Zhou, J., & Zhang, Q. (2019). Artificial intelligence for infectious disease big data analytics. *Infection, disease & health, 24*(1), 44–48.

[38] Ahn, J. C., Connell, A., Simonetto, D. A., Hughes, C., & Shah, V. H. (2021). Application of artificial intelligence for the diagnosis and treatment of liver diseases. *Hepatology, 73*(6), 2546–2563.

[39] Holzinger, A., Langs, G., Denk, H., Zatloukal, K., & Müller, H. (2019). Causability and explainability of artificial intelligence in medicine. *Wiley Interdisciplinary Reviews: Data Mining and Knowledge Discovery, 9*(4), e1312.

[40] Young, A. T., Fernandez, K., Pfau, J., Reddy, R., Cao, N. A., von Franque, M. Y., ... & Wei, M. L. (2021). Stress testing reveals gaps in clinic readiness of image-based diagnostic artificial intelligence models. *NPJ digital medicine, 4*(1), 1–8.

[41] Neri, E., Coppola, F., Miele, V., Bibbolino, C., & Grassi, R. (2020). Artificial intelligence: Who is responsible for the diagnosis?. *La radiologia medica, 125*(6), 517–521.

[42] Drummond, C. (2002). Accelerating reinforcement learning by composing solutions of automatically identified subtasks. *Journal of Artificial Intelligence Research, 16*, 59–104.

[43] Elomaa, T., & Kaariainen, M. (2001). An analysis of reduced error pruning. *Journal of Artificial Intelligence Research*, *15*, 163–187.

[44] Fox, M., & Long, D. (2003). PDDL2. 1: An extension to PDDL for expressing temporal planning domains. *Journal of artificial intelligence research*, *20*, 61–124.

[45] Frantz, R. (2003). Herbert Simon. Artificial intelligence as a framework for understanding intuition. *Journal of Economic Psychology*, *24*(2), 265–277.

[46] Kundu, M., Nasipuri, M., & Basu, D. K. (2000). Knowledge-based ECG interpretation: a critical review. *Pattern Recognition*, *33*(3), 351–373.

[47] Lang, J., Liberatore, P., & Marquis, P. (2003). Propositional independence-formula-variable independence and forgetting. *Journal of Artificial Intelligence Research*, *18*, 391–443.

[48] Kassens-Noor, E., Wilson, M., Kotval-Karamchandani, Z., Cai, M., & Decaminada, T. (2021). Living with autonomy: Public perceptions of an AI-mediated future. *Journal of Planning Education and Research*, 0739456X20984529.

[49] Malik, M., Tariq, M. I., Kamran, M., & Naqvi, M. R. (2021). Artificial intelligence in medicine. In *Advances in Smart Vehicular Technology, Transportation, Communication and Applications* (pp. 159–170). Springer, Singapore.

[50] Vearrier, L., Derse, A. R., Basford, J. B., Larkin, G. L., & Moskop, J. C. (2022). Artificial Intelligence in Emergency Medicine: Benefits, Risks, and Recommendations. *The Journal of Emergency Medicine*.

[51] Jotterand, F., & Bosco, C. (2022). Artificial Intelligence in Medicine: A Sword of Damocles?. *Journal of Medical Systems*, *46*(1), 1–5.

[52] Bazoukis, G., Hall, J., Loscalzo, J., Antman, E. M., Fuster, V., & Armoundas, A. A. (2022). The inclusion of augmented intelligence in medicine: A framework for successful implementation. *Cell Reports Medicine*, *3*(1), 100485.

[53] Pham, D. T., & Pham, P. T. N. (1999). Artificial intelligence in engineering. *International Journal of Machine Tools and Manufacture*, *39*(6), 937–949.

[54] Amisha, P. M., Pathania, M., & Rathaur, V. K. (2019). Overview of artificial intelligence in medicine. *Journal of family medicine and primary care*, *8*(7), 2328.

[55] Musib, M., Wang, F., Tarselli, M. A., Yoho, R., Yu, K. H., Andrés, R. M., ... & Sharafeldin, I. M. (2017). Artificial intelligence in research. *science*, *357*(6346), 28–30.

[56] Gutierrez, G. (2020). Artificial intelligence in the intensive care unit. *Critical Care*, *24*(1), 1–9.

[57] Harrer, S., Shah, P., Antony, B., & Hu, J. (2019). Artificial intelligence for clinical trial design. *Trends in pharmacological sciences*, *40*(8), 577–591.

[58] Shelmerdine, S. C., Arthurs, O. J., Denniston, A., & Sebire, N. J. (2021). Review of study reporting guidelines for clinical studies using artificial intelligence in healthcare. *BMJ Health & Care Informatics*, *28*(1).

[59] Li, J., & Qian, J. M. (2020). Artificial intelligence in inflammatory bowel disease: current status and opportunities. *Chinese Medical Journal*, *133*(07), 757–759.

[60] Mirbabaie, M., Stieglitz, S., & Frick, N. R. (2021). Artificial intelligence in disease diagnostics: a critical review and classification on the current state of research guiding future direction. *Health and Technology*, *11*(4), 693–731.

[61] Robertson, S., Azizpour, H., Smith, K., & Hartman, J. (2018). Digital image analysis in breast pathology—from image processing techniques to artificial intelligence. *Translational Research*, *194*, 19–35.

[62] Osoba, O. A., & Welser IV, W. (2017). *An intelligence in our image: The risks of bias and errors in artificial intelligence*. Rand Corporation.

[63] England, J. R., & Cheng, P. M. (2019). Artificial intelligence for medical image analysis: a guide for authors and reviewers. *American journal of roentgenology*, *212*(3), 513–519.

[64] Niazi, M. K. K., Parwani, A. V., & Gurcan, M. N. (2019). Digital pathology and artificial intelligence. *The lancet oncology*, *20*(5), e253–e261.

[65] Prevedello, L. M., Halabi, S. S., Shih, G., Wu, C. C., Kohli, M. D., Chokshi, F. H., ... & Flanders, A. E. (2019). Challenges related to artificial intelligence research in medical imaging and the importance of image analysis competitions. *Radiology: Artificial Intelligence*, *1*(1), e180031.

4

Patient Monitoring Through Artificial Intelligence

Thota Ramathulasi*, Rajasekhara Babu, and Mohamed Yousuff

School of Computer Science and Engineering, VIT University, India
*Corresponding Author
Thota Ramathulasi
School of Computer Science and Engineering, VIT University, India.
Email: ramathulasimca@gmail.com

Abstract

Chronic diseases are now the leading cause of death and illness worldwide, replacing infectious diseases. There is no immediate cure for chronic conditions, which is a significant challenge. To prevent severe symptoms, people with chronic illnesses need to constantly manage their illnesses. There is a growing interest and need for patient monitoring solutions to facilitate chronic disease management and, in particular, to improve self-management of chronic disease. With the implementation of information technology and telecommunication tools, remote patient monitoring has become an emerging healthcare sector. Different sensors or devices, such as smartphones with built-in sensors and wireless transmissions, make up the data collection system. The data processing system continues to receive and send information. Large datasets collected from remote monitoring solutions such as medical devices, wearables, and apps can be processed with machine learning (ML) tools, which improve assessment, classification, and decision support. In this context, significant advances are being made in artificial intelligence (AI), big data, and deep learning (DL) to meet customer demands. For patients with chronic conditions, the integration and use of AI and the Internet of Things (IoT) for sensors, mobile apps, social media, and location-tracking technologies can enable early detection, treatment, and better self-management. As a result, there is growing interest in AI solutions for future remote patient monitoring to facilitate chronic disease management and improve self-management.

4.1 Introduction

Healthcare is a constantly changing sector that offers numerous research opportunities. Such evolution is predicated on the adoption of IoT technologies. It integrates information and communication technologies (ICT), sensor technology, massive data generation, big data, ML techniques, and AI. The application of novel skills is primarily aimed at constantly monitoring affected roles with chronic sicknesses (Mukhopadhyay, 2014), which has improved over the years. Thus, the technology IoT enables the development of novel solutions for diabetes patients. Chronic diseases are those that last an extended period and necessitate long-term therapy. Patients with stable conditions frequently occupy vast amounts of time in the hospital to be monitored daily.

Heart disease, cancer, and diabetes are all examples of common chronic diseases. Currently, diabetes condition is quite severe, as it claims thousands of lives each year. As a result, the diabetes patient's blood sugar level must be maintained to live a regular everyday life. It is defined by persistent hyperglycemia caused by pancreas dysfunction when the organ either does not insulin properly or the organism does not utilize enough insulin adequately. Low or high blood glucose levels in the blood can impair the function and degradation of various organs, including the eyes, neurons, and blood vessels. Thus, continuous and everyday monitoring is essential to prevent the diabetic patient's healthiness from deteriorating. Over the former limited years, the increasing count of diabetes patients consumed necessitated the deployment of different technologies to monitor these individuals. Monitoring policies for diabetes patients are designed to measure blood glucose levels regularly.

Consequently, patients, families, and physicians may monitor interpretations at all moments and respond swiftly when an alarming reading occurs. Transferrable monitoring systems for people living with diabetes provide some advantages, including improving diabetic patients' quality of life by minimizing inpatient time. As a result, the adoption of wireless communication with extensive coverage which enables data transmission from patient to clinicians is quite intriguing. In this perceive, fifth generation (5G) technology, also referred to as the resulting generation of mobile, enables faster transmission, higher bandwidth capacity, and networks the calling. However, this technique is currently being evaluated to boost data transfer rates (Pantelopoulos & Bourbakis, 2009). It is possible to treat diabetes using deterministic mathematical models. However, there have been few investigations on scientific simulations of diabetes mellitus published so far. The stochastic numerical study continues to be an intriguing tool for studying diabetes mellitus epidemic illness propensity (Neuman, 2010). To collect data for categorization, an essential glucose monitoring device with efficiency, minimal, and low cost was employed. Each day, the cloud was updated with new patient data. Clinicians used the acquired facts to track blood glucose variability and administer appropriate medical care in an erroneous glucose level. The forecast was made using a combination of ML techniques. In order to obtain the best level of accuracy, many classification algorithms were analyzed, tested, and compared using a variety of parameters.

Asthma, tumor, cardiac, diabetic, and other mental health-related problems have already surpassed viral infections as the primary source of impermanence and illness worldwide (Wickramasinghe, John, George, & Vogel, 2019). It is estimated that nearly all participants will increase. An increase in suffering from various chronic diseases will rise as the population gets older and lives longer in poverty. Chronic illnesses present a significant problem because there is no rapid treatment. Individuals with chronic illnesses are forced to handle their sickness continually to avoid keen symptoms. It means that food, physical activity, and medical management must all be monitored regularly. As a result, patient monitoring systems are becoming more popular and necessary as a crucial doorway to facilitating chronic disease care, particularly improving self-management (Wickramasinghe, Essential considerations for successful consumer health informatics solutions, 2019). Remote patient monitoring is a new discipline of medicine that focuses on managing health and illness to treat or diagnose illness through information technology and telecommunications. It employs various tools to gather health data from users at home or during everyday activities and then retains or sends that data to healthcare specialists for review and recommendations (Ashok Vegesna, Melody Tran, Michele Angelaccio, & Steve Arcona, 2017). Patient monitoring, telehealth, mobile health, and telemedicine are wholly terms that encompass numerous aspects of patient monitoring beyond the hospital surroundings with the help of information technology.

Patient monitoring is the process of analyzing the indications and performance of individuals who have long-lasting illnesses and are in danger of acquiring acute symptoms of those illnesses. Increased patient care, assurances that unanticipated health crises will be monitored, and a reduction in hospital visits are all benefits of the program. Many advantages of remote monitoring can be realized, including genuine or rather fairly constant symptom analysis, early identification or prevalence of the disease, lower healthcare costs, improved knowledge of health issues, and, in the end, expanded support and emergency care opportunities (Malasinghe, Ramzan, & Daha, 2019). It is cost reduction that is the primary impetus for the deployment of remote health monitoring, which is being pushed by an aging population, increasing needs, limited facilities, and the possibility of healthy aging (Albahri, *et al.*, 2018). A monitoring system is typically composed of the following components: a heart rate monitor, data capture, and end-user administer, also a transmission module. The data gathering arrangement is made up of a variety of sensors or equipment, such as handsets with built-in devices and wireless gets turned.

The procedure involves receiving and transmitting data, and the terminus can be anything from a hospital computer to a gadget to a smartphone database to a tablet computer. Frequently, a communication network is utilized to connect this same user and data order fulfillment with a practitioner (Catherine Klersy, Annalisa De Silvestri, & Gabriella, 2009). Patients may collect and use data continuously and only submit it to a professional on an *ad hoc* basis in some situations, whereas in others, data gets delivered to the registered dietitian for review in all instances. All of this is dependent on the patient's condition; the information will be stored, as well as the complexity of the data. It may be recommended that the patient goes to the clinic, remove harm, take medication, or contact a healthcare professional. Technology, communication channels, and sensor integration are all significant differences between the current monitoring systems, as illustrated in Figure 4.1.

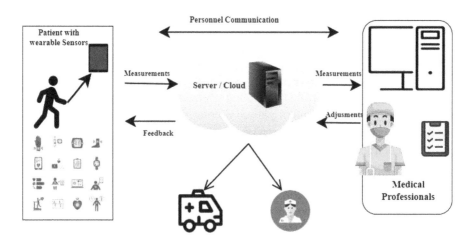

Figure 4.1 Patient monitoring architecture using AI.

A review of previously published research on generation wireless systems for the control of type-2 diabetic glucose levels is presented in this section. The section contains various previously published studies on large healthcare datasets that use classification to predict potential episodes of reactive insulin resistance or reduced blood sugar levels. These studies include the following: the ability to segment data in the maintenance of e-health is vital for tracking and treating the disease. Ahad *et al.* (Ahad, Tahir, & Yau, 2019) present a brief introduction of 5G technologies and intelligent smart systems brought about by Digital of Things, as well as a case study of a smart healthcare application. In addition, the authors analyze the problems, research areas, and future aspirations in the technology field in the context of 5G technology deployment. According to Lloret *et al.* (Lloret, Parra, Taha, & Tomás, 2017), an infrastructure and protocol based on 5G technology would be used to enable continuous and sophisticated online monitoring. The plan is meant to make use of the current scenario, which includes 5G mobile phones and wearables, to collect 16 critical indications from patients. Each piece of data collected is stored in a database and evaluated by employing business intelligence and ML technologies to deliver intelligent reactions that arouse an alarm if the system detects an unexpected event. Chen *et al.* (Chen, Yang, Zhou, Hao, & Zhang, 2018) are proposing a 5G mobile health system to continuously measure the effectiveness of diabetic patients. For starters, the authors explain the 5G-Smart Diabetes setup, which includes recent technology such as Wearable tech 2.0, algorithms, and predictive analysis to provide diabetics with comprehensive evaluation and control. Next, the authors present the data-sharing mechanism and analytic paradigm for 5G-Smart Diabetes, both of which are described in detail below. Finally, a testbed for the 5G diabetes has been developed.

According to the findings, the approach is capable of offering patients with tailored diagnoses and treatment recommendations. Another device that relies on the IoT is the indoor anti-collision warning system (Xiao, Miao, Xie, Sun, & Wang, 2018). Through the use of radio frequency identification, the system can recognize and track electronic labels by evaluating back-broadcast signals that have been received from labels.To assist blinded users in avoiding obstacles, scientists have developed wireless signal indicators and phase patterns that act as fingerprints. In terms of obstacle avoidance, experiments have shown that the systems function admirably, with 94% effectiveness.

Goyal *et al.* (Chatrati, et al., 2020) offer a smart home prototype system for the early diagnosis and prediction of diabetes and hypertension in people who live alone. This technology is intended to be used in the home to monitor patients' blood pressure and glucose levels. Aside from that, the system predicts the presence or absence of hypertension and diabetes by employing typical supervised learning classification methods. Najm *et al.* (Najm, Hamoud, Lloret, & Bosch, 2019) present a unique ML technique based on the decision tree technique that is proposed for estimating the best rise in network problems in 5G IoT wireless sensors. A unique approach to forecasting the blood glucose levels of diabetic patients has been developed by Ahmed *et al.* (Ahmed & Serener, 2016). The glucose application is used by the authors to analyze the patient data. As a result, there has been a reduction in the intensity of noise. In this system, continuous glucose measuring sensors plus an additional Kalman filter would have been used to measure glucose levels continuously. This method aids in the prevention of serious complications linked with hypoglycemia. Kannadasan *et al.* (Kannadasan, Edla, & Kuppili, 2019) aim to increase the success rate and other evaluation criteria for identifying the Pima Indians type-2 diabetes dataset to improve the overall accuracy of the dataset. They propose a method for diabetes data classification that is based on deep architectures and stacked autoencoders, as described in the paper. High accuracy, recall, responsiveness, and $F1$-score are used to evaluate the effectiveness of the investigations, which are conducted using these methodologies.

Wang *et al.* (Wang, Wang, Chen, Jin, & Che, 2020) propose a string section learning algorithm, extreme gradient boosting, to predict type-2 diabetes risks. They contrast it to artificial neural networks (ANNs), support vector machine (SVM), the random forest (RF), and the K-nearest neighbor (KNN) algorithm to support the computer model effect of existing models. When forecasting chronic renal illness using clinical data, Charleonnan *et al.* (Charleonnan, et al., 2016) recommend further use of KNN, SVM, regression models, and tree-based classifiers, among other methods of prediction. Several models are evaluated to establish the best effective technique for anticipating chronic renal illness, which the authors do in this paper. A model for customized heart problem classification is presented by Yoo *et al.* (Yoo, Han, & Chung, 2020), which is combined with a rapid and convenient preprocessing phase and a convolutional neural network to process the proper cumulative biosensor input data. Based on a handoff approach, González-Valenzuela *et al.* (González-Valenzuela, Chen, & Leung, 2011) present a constant monitoring program for ambulatory patients in their own homes. This system is built on two-tier internet services, with this layer of biosensors accessing vital signs and constructing an intersection connection between both the bodily sensor nodes coordinator devices and a port number, and another layer of sensing devices collecting patient's condition and trying to establish a level linkage between some physiologic sensor network coordinator devices and a destination address (AP). The sensor is carried on certain wrists and the service user moves at a velocity of 0.5 m/s, the amount of packet loss is decreased to 20% to 25% of the value achieved when using only the time in history, administrator link.

Finally, we look at several papers that are concerned with the use of statistical models. In (Izonin, et al., 2018), Izonin presents two ways for identifying medical implant materials that are consistent with the use of the Fourier polynomials and SVM. These approaches are consistent with the use of the Fourier polynomials and SVM. When comparing the proposed approach to existing available algorithms, the author makes several observations. To reduce the chance of erroneous alloy detection in medical devices, Temple *et al.* (Tepla, et al., 2018) developed a classification biomaterial technique based on classification, regression analysis to create materials wearable imaging. In their paper (Tkachenko, Doroshenko,, Izonin, Tsymbal, & Havrysh, 2018), Tkachenko *et al.* examine the outcomes of tackling data classification problems using the most extensively used different classifiers and propose a unique classification methodology based on the neural-like properties of the geometric transformation model.

4.2 Purpose of Patient Monitoring

Patient monitoring organizations collect biological information from affected roles over sensors and supplementary data causes that are required. Most commonly obtained by current systems are pulse, cardiovascular system, respiratory rate, air circulation, oxygen levels, volume, electrocardiograph, electroencephalography (EEG), electromyography, mass and strength heating rate, and blood levels, among other things. It is possible to include information like the steps taken and caloric burn, sleeping statistics, position, and the patient's weight. Most traditional monitoring systems collect information through sensors fixed in place, for example, electrodes attached to the skin, while few are better invasive – in particular, factors that can affect brain monitoring (Gopalsami, et al., 2007). With the continued advancement of technology and wireless devices, sensors and surveillance equipment are becoming thinner and more wireless. Devices, such as smartphones or tablets, are powering many new gadgets.

4.2.1 Patient monitoring involvement in today's healthcare

Patient monitoring is essential in today's healthcare environment. In addition, remote patient monitoring can reduce hospitalizations, decrease mortality, and improve treatment and service in a range of infections (e.g., diabetic, cardiovascular, and neurological illnesses). Diabetes comes to a widespread long-term illness that necessitates meticulous supervision to keep blood levels back to normal and reduce the likelihood of complications. Diabetes patients' insulin levels can vary significantly during the day; thus, it is critical to monitor them multiple times to ensure that they are within the proper range when taking insulin therapy. It is difficult and intrusive to use traditional glucose sensors, which use electromechanical methods to evaluate blood glucose levels from a bit of blood sample taken from the finger prick (Yoo & Lee, 2010). Continuous glucose monitoring (CGM) turned out to have emerged as a specific miracle for individuals over type-1 diabetes, and also type-2, allowing them to improve their lives and health outcomes significantly. These CGMs monitor blood glucose scales during the day and, provide analysis reports and hollow blood sugar detection.

The primary advantage of CGM is that they allow patients to spend less time in hyperglycemia as they are often constantly monitored (Rodbard, 2016). It has been shown that they are connected with lower hemoglobin A1c (HbA1c) levels, which is an important diabetic biomarker. CGMs, when used in conjunction with an insulin pump, are particularly effective for regulating blood sugar levels since family members may provide more frequent monitoring and changes in glucose levels. Like Medtronic and Dexcom (which includes Tandem), few manufacturers provide integrated systems, which are unsuitable for immense monitoring (Chen, et al., 2017). Furthermore, despite the tremendous benefits of CGM devices, clinics associated with the implementation have remained modest in recent years. Heat loss is a serious medical condition that affects approximately 26 million individuals worldwide and is related to higher rates of hospitalization and mortality (Savarese & Lund, 2017). The inability to follow patients with heart failure after treatment episodes is likely to increase the likelihood of re-hospitalization and morbidity in these patients. The cost-effectiveness of patient monitoring is demonstrated by the comparison of visits by healthcare experts. An electrocardiogram (ECG), heart rate, and mass measurements are routinely taken as part of the monitoring process. Phone-based tracking and consultations are also available in some cases. Many approaches are used, including telemonitoring, others of which include conferencing or cellphone surveillance and assistance (Bashi, Karunanithi, Fatehi, Ding,, & Walters, 2017).

Several clinical studies have been conducted to document the outcomes of such monitoring programmers, and the conclusions have varied in terms of their significant influence on patients. However, according to the literature, some home surveillance can minimize cardiac hospitalizations and death (Ong, et al., 2016) if appropriately used. Vascular dementia, seizures, neurodegenerative disorders, and dementia are just a few of the neurological conditions that can be treated using remote patient monitoring technology. In patients with these illnesses, two clinical issues must be addressed: identifying and measuring changes in everyday functioning over time, as well as evaluating the efficacy of medicine on symptoms and function. Because of the link between being physically inactive and a variety of morbidities, assessing physical activity is beneficial in treating many neurological conditions. In remote monitoring, it is possible to assess aerobic exercise evaluate determinants and outcomes, and assess determinants, results, and decision-support mechanisms. A technique called regular exercise monitoring, which uses accelerometer sensors to forecast fall risk and detect mistakes in Parkinson's patients, has proven to be beneficial (Block, et al., 2016). Various devices, like those of the Omron monitoring, which is widely viable, can simulate the human movement upon the leg and arm while a person is reaching for something while walking. They appear to be legitimate in terms of posture lifting time and rate of rotation. Users and physicians obtain feedback from these systems, enabling them to evaluate the situation accurately and provide better rehabilitation and therapeutic management alternatives (Lonini, et al., 2018) to the patients.

4.2.2 Improving healthcare outcomes by using patient monitoring

Improving healthcare results through patient monitoring based on clinical trials and pilot projects has established that patient monitoring applications can improve clinical outcomes in a wide range of circumstances. While this is going on, numerous studies show no improvements in clinical results, even when using the same sort of monitoring equipment or treating the same condition. Because of the variety of applications and diseases that can be monitored, it is difficult to make broad statements regarding patient monitoring systems. The fact that most market monitoring systems are based on a particular invention or ailment illustrates the fragmentation of the market. Historically, the majority of discussions about clinical outcomes have been on lower morbidity and mortality and physiological indicators like suHbA1c diabetes abdominal obesity.

Whereas, on the other side, some critical indications of therapeutic success, such as tolerance, are frequently disregarded. The term "adherence" applies to a patient's willingness to comply with medical recommendations and medications. It entails adhering to prescribed medicine and accessing additional forms of care that have been recommended. Adherence should be acknowledged as a crucial inpatient monitoring measure to maintain correct control and improve patient outcomes. Following up on medication compliance with remote monitoring devices, including health technologies, has risen in recent months and is currently one of the most often used applications for new software applications. Various methods of encouraging compliance are available, ranging from text message reminders to sensor-based surveillance systems that include built-in medical support. After compiling data on patient outcomes and medication compliance from 107 research trials in which both medication compliance and traditional clinical outcomes were recorded, Hamine *et al.* concluded that revisions required a significant favorable influence on adherence in 57% of cases.

However, the revisions required such a heavy negative impact proceeding health consequence (Hamine, Gerth-Guyette, Faulx, & Green, Journal of medical Internet research) that they were significant. Also demonstrated was the text message alerts are the furthermost extensively and successfully used tool for loyalty facilitation then that admittance improvement is very error-prone, requiring a high degree of active patient contact and information sharing. Increasing the level of active patient satisfaction is essential for keeping users and enhancing commitment and therapeutic outcomes. Because it requires additional effort and may not yield immediate results, patients are usually reluctant to improve their adherence to their medications. Successfully implementing habits is challenging, and this is made considerably more difficult when the behavior lacks any entertainment factor or an immediate benefit.

On the other hand, these adherent measures must go beyond clinical studies and case studies to significantly impact society. For various reasons, most systems have been reluctant to gain widespread adoption to date (Baig, GholamHosseini,, Moqeem, Mirza, & Lindén, 2017). Body-worn monitoring devices are commonly used to obtain reliable data. Because they are typically substantial in size, they must be precisely positioned on the body to give reliable results. Many detection systems also have connectivity issues caused by delays, losing data, with the addition of network communication, all of which contribute to a negative user experience for the user. Clinical decision-making precision and consistency assist could also exist a problem for physicians who utilize the tracking if the framework was established in addition to assessed primarily following the following organizational and simulation conditions.

Finally, involvement and communication with clients frequently lack quality. As aggressive specifics receivers with limited access with the intention of this information collected, users frequently perceive themselves as having a limited interpretation of the information they collect, which inhibits their acceptance and uptake (Deshmukh & Shilaskar, 2015). The acknowledgment must be increased, and monitoring materials essentially remain improved by stressing end-user engagement, even more guaranteeing that the service delivers worth to users. The receiving and approval of each invention across the healthcare business depend significantly on the user's awareness and appreciation, including the patient and the physician.

4.3 Wearable Patient Monitoring Sensors

The value of health monitoring has grown in recent years, as has most individuals who use it. It can be defined as a method of looking at the difference in an individual's well-being overview period. Furthermore, it has been defined as the recognition of changes in one's circumstances that impact one's health. Indeed, vital signs should be closely checked thoroughly that the hospital's life quality is not adversely impacted. It has been demonstrated that on-body sensors are particularly effective when considerations such as service user mobility and convenience of use are considered (Teichmann, De Matteis, Bartelt, Walter, & Leonhardt, 2015). The development of wearable monitoring devices has attracted significant interest from industry and academics over the last decade. Because of the rising expense of medical services and the changing demographics, there has been an increase in the desire for patients' health conditions to be monitored outside of the clinical setting. In this case, the goal is to offer essential info about a person's current health and well-being, to the elected official, a hospital, or exclusively to the medical team, with the capability of notifying when life-threatening conditions are detected (Pantelopoulos & Bourbakis, 2009). Due to their multiple advantages and the usage of devices for data gathering, wearable sensors are incredibly advantageous since they continuously monitor a patients' life while limiting human involvement and results in a low price whenever attached to the human body.

Ubiquitous healthcare systems are composed of a variety of tiny sensors, power supply, processing elements, and actuators that are worn or implanted to provide monitoring and treatment. Each of the multiple biosensors can send data to something like a Wi-Fi module, a personal digital assistant, and go to a health center in its entirety, depending on the data collection method. Important health indicators, including heartbeat, pulse rate, respiration rate, and blood pressure, can be monitored using biosensors. When treating patients with sleep problems, heart attack, vascular dementia, and other conditions, wearable sensors have made it possible to treat them at home rather than requiring them to be admitted to the hospital following the attack. To build intelligent sensing devices that can detect falls in the family environment of older people (Clifton, Clifton, Pimentel, Watkinson,

& Tarassenko, 2012), researchers are conducting research (Yan, Yoo, Kim, & Yoo, 2010) in this area. Falling can result in physical concerns that require immediate attention to avoid more complications from occurring. If immediate aid is not provided, symptoms such as coldness, dehydration, and acute discomfort may deteriorate and get worse.

4.3.1 Wireless health monitoring specifications

By implementing a health monitoring system, data obtained from biosensors can be transmitted as far as the structure's central node immediately to a distant pharmaceutical station or the physician's cell phone. Data transfer can be accomplished via a wired medium, which restricts the user's movement and comfort, or via a wireless medium. In the wireless medium (Pantelopoulos & Bourbakis, 2009), data transmission can be accomplished by establishing a catalog of sensor networks known as a body area network that is configured in a star network to facilitate data flow to the BAN's central node, which can be a smartphone, a microcontroller-based device, or a smart device. When data transmission to a remote medical station is required, a variety of technologies can be employed. Among them are Wi-Fi, Bluetooth, WiMAX, GSM, and GPRS (Mukhopadhyay, 2014).

The network's design is critical since it should be competitively priced, flexible in terms of configuration, and allow for additional nodes. As shown in Table 4.1, ZigBee (IEEE 802.15.4), Bluetooth is the most extensively used

Table 4.1 Comparison of different wireless specifications.

Wireless specifications	Area	Data	Bandwidth	Application
IrDA	1 m	115 Kbps to 1 Mbps	–	A wireless monitoring system is not a viable option.
ZigBee	10–100 m	250 Kbps	868 GHZ, 915 GHz, 2.4 GHz	Applications with lower power usage and data transfer rates
Bluetooth	100 m	1–3 Mbps	2.14 GHz	Control and monitoring on a short range
Wi-Fi	5 Km	1–450 Mbps	2.4, 2.5 GHz	Acquires data from a personal computer and gives internet service
WiMAX	15 Km	75 Mbps	2.3 GHz, 2.5 GHz, 3.5 GHz	Provides access to the internet for mobile devices

standard for body area networks (IEEE 802.15.1). ZigBee is typically preferred in applications that demand network security, long battery life, and a modest data rate (Ramathulasi & Babu, 2020). For authentication and encryption, ZigBee makes use of advanced encryption standard with a 128-bit key. If data needs to be transported over a long distance, it can be transmitted via a communication network of gateway devices. Bluetooth is frequently used for short-range data transmission by masking mobile devices' further use of personal area network. Bluetooth eliminates synchronization issues by connecting several devices. As Bluetooth transmission occurs in the presence of a master and slave device, a total of seven devices can communicate via the master device. Bluetooth comes in various versions, as stated in the chart, including data rates of 1, 3, 24, and 24 Mbps, respectively.

4.3.2 Different types of sensors

According to the manufacturer, physical signs should be monitored with precision and for an extended length of time. Because it is based on monitoring systems, sensors contribute to the achievement of the goals mentioned above. There is a plethora of biosensors available for multi-tracking activities. As dynamic behavior, advanced materials, and other analogous technologies evolve, the formation of acute sensor monitors data more swiftly and effectively while requiring less power is becoming more feasible. The temperature of a patient's body is one of the heart rhythms that could be used to evaluate the health status of that patient. Body temperature changes can emerge due to a variety of factors, including infection, joint pain, heart attacks, and electric shocks. Body temperature was previously determined using a thermometer; however, temperature sensors and equivalent electrical and electronic technologies have essentially replaced this method of measurement (Chen, et al., 2017).

Accurate acceleration measurements along a specified axis are required but only within a specific frequency range. Accelerometers are used in this situation. They are also employed to monitor human behavior. They are usually used around wearables and implantables to detect falls, which is their primary application. Accelerometers are available in various configurations, including piezoelectric, impedance, and piezoresistive types, among others. The respiratory rate is another crucial physiological parameter to monitor. The breathing rate tends to rise as a consequence of injury and physical activity, among other things. When it comes to detecting health problems, estimating heart rate can be pretty helpful. Persistent cardiac problems are also a source of concern, and they must be closely monitored. In these settings, ECG sensors could be utilized to diagnose cardiovascular diseases. In

Table 4.2   Sensor with their incidental bio-signals.

Sensor type	Bio-signal type	Data that has been measured
Temperature	The temperature of the body or the skin	The capacity of the body to generate and dissipate heat
Piezoelectric sensor	Rate of respiration	Phonocardiograph Heartbeat measurement with a stethoscope
	Breathing rate per unit of time	
Electrodes on the chest or the skin	ECG	Sound wave recording of the contraction and relaxation phases of the heart cycle
Accelerometer	Body movements	Measurement of acceleration forces in three-dimensional spaces
The oximeter of the pulse	Saturation with oxygen	Oxygen carried by human blood in a given volume

addition to providing information on the consistency and frequency of the heart's pulse, ECG sensors also provide information that can be used to diagnose cardiovascular illness. Several nano sensors and the business derived from them are summarized in Table 4.2, which may be viewed here.

4.4 Involvement of AI in Patient-Monitoring

4.4.1 Mobility aids the living environment

Assisted living in the home (also known as ambient assisted living (AAL)) is a term that refers to the integration of modern technology into a person's regular life to fulfill the increased healthcare cost and backing, later refining the distinct overall peculiarity of life. In certain, AAL systems established on AI are incredibly beneficial to the aged, the disabled, and patients with severe environments. Remote patient monitoring, personality, emergency response services, and in-home quality healthcare are some of the successful healthcare solutions provided by these organizations (Darwish, Senn, Lohr, & Kermarrec, 2014). Examples include remote monitoring with AI-based sensors, which can be used to observe patients' everyday activities and provide constant help to elderly persons under the supervision of careers. In AAL systems (Davis, Owusu, BastaniX, & Marcenaro, 2018), an ANN is a vital component of the system. Several technologies, including the Web of Things and big data, are used to power it. Massive volumes of data can be managed by AAL systems, which can then be interpreted with appropriate models, accurate patterns detected, and predictions generated, which may then be used for decision-making in employing AI approaches such as ML.

The approach uses a variety of ML approaches to recognize different types of mobility patterns, including RF, Bayesian network, *K*-means, and other classifiers. Using the IoT, for example, a variety of sensors, including AI technologies and AAL mechanisms, may be used to monitor in real time this same patient's various activities, such as lying down, lying on the floor, standing, stepping on flat ground, walking upwards, or climbing the stairs (Chetty, White, & Akther, 2015).

Numerous AAL systems practice the detectors on the smartphone to detect activity (accelerometer and gyroscope). Davis-Owusu *et al.* (Davis, Owusu, BastaniX, & Marcenaro, 2018) accustomed a smartphone with integrated sensor detectors to accumulate particulars from senior individuals. It was discovered that they were able to distinguish six specific activities using three ML techniques: SVM, ANN, and an SVM-hidden Markov model (HMM) hybrid approach. Many functionalities have also been conducted in AAL systems by utilizing techniques that are based on DL (Ronao & Cho, 2016). A significant amount of research has been done on applying convolution neural network (CNN) to recognize everyday movements such as exercise, standing up, walking down a staircase, going for a walk, reclining, and standing (De-La-Hoz-Franco, Ariza-Colpas, & Quero, 2018) among other things. It is possible to build AAL systems utilizing a variety of sensors. It is common to see smartphones and wearables being used, and because they are generally unobtrusive and inexpensive, this results in highly profitable AAL systems, similar to e-textiles, building automation, and helping robots (Majumder, Mondal, & Deen, 2017).

Syed and colleagues (Syed, Jabeen, Manimala, & Alsaeedi, Smart healthcare framework for ambient assisted living using IoMT and big data analytics techniques, 2019) introduced an intelligent medical system treating AAL that uses AI and the IoT to measure the aerobic exercise of older people. To evolve this structure, the researchers possessed to position many sensors on various segments of the participant's frame. Sensors placed on the person's ankle, arm, and breast were utilized to collect data and then sent through the IoT. In order to execute multiple tasks on the data, advanced automation technologies such as MapReduce and multivariate regression classifiers are being used to complete the jobs. When it came to remotely forecasting 12 physical actions, the developed algorithm was 98.2% accurate. The interface for this project is a mobile medical app that permits telemedicine through data visualization and reports generation and report generation. As shown in Figure 4.2, a distant specialist app allows specialists to screen their sufferers from a distance and provide records based on categorized everyday activities.

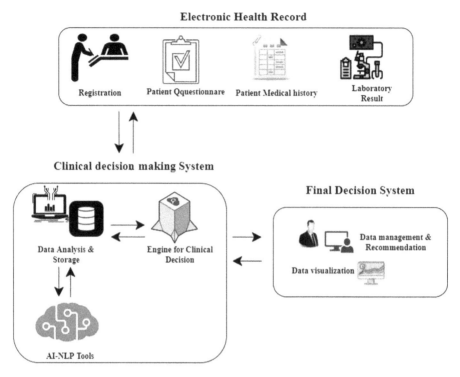

Figure 4.2 Clinical decision-making assistance.

In conformity with the results of bodily activity classification, experts might make appropriate judgments and provide relevant records. For caretakers who get treatment recommendation, communication cautions to assess potential threats and intervene accordingly in the case of a problem such as cardiogenic shock or fall detection. As it can distinguish numerous physical motions and thus remotely monitor the patient's health status, such a framework may be termed a solution for reintroducing AAL to the elderly and crippled.

4.4.2 Clinical decision-making assistance

In recent years, evaluated and clinical findings decision support systems (CDSS) for house telemonitoring of persons with chronic illnesses, including respiratory distress, have been devised and constructed (Sanchez-Morillo, Fernandez-Granero, & Leon-Jimenez, 2016). The CDSS's primary purpose is to monitor the patient's health daily to detect any recurrence or progression of the ailment early on and averted [50]. CDSS is predicated on machine-readable medical experience, evidenced particulars, and encapsulated AI patterns that exploit this data and expertise to offer relevant information to doctors and aid in their decision-making (Mills, 2019). The most common and developed CDSS are those that target drug protection and loyalty. These CDSSs constitute the backbone of electronic prescribing and clinical decision support. In general, sophisticated AI algorithms are not required to assist physicians; to have available data is enough to aid clinicians' decisions. However, because CDSS retain large quantities of sufferer data (Lilly, et al., 2014), AI is emerging in complete agreement with a wide number of rich datasets. Technology and neural networks, for example, have been developed to manage large datasets utilizing developed feature selection approaches (Kindle, Badawi, Celi, & Sturland, 2019). Due to the use of AI to progress huge quantities of data, CDSS could work on various plateaus. The AI algorithms embedded into a CDSS, for example, can use the center's electronic health data to forecast emergent admissions at various levels. It intends to permit the hospice to develop more effective decision-making methods for crisis room administration, resulting in better patient care and lower costs. Furthermore, by enabling the app before the device to monitor patient medical conditions and alert them to potential exacerbations, these solutions can improve quality of life (Iadanza, Mudura, Melillo, & Gherardell, 2020).

Furthermore, the presence of a CDSS does not always guarantee that clinicians will use it in an intended way. According to new findings, nurses rely on clinical information systems while they are first starting their careers; nonetheless, once they expand involvement, they mostly practice the system to "double-check" or even make judgments due to personal experiences (Dowding, Randell, Mitchell, & Foster). The suggestions of these systems, according to some nurses, limit their specialized findings and, when unwell constructed, contradict their judgment (Ernesäter, Holmström, & Engström, 2009). It suggests that although building furthermore assessing each CDSS, specific, precise strategies must be succeeded.

1. The five principal rules should be followed, as described in the literature:

2. Appropriate information (clinical expertise, clinical protocols, and AI algorithms that are acceptable)

3. Appropriate individuals (practitioners, members of interdisciplinary teams, and patients all require data to make decisions)

4. Appropriate format (decision data is presented through alerts, prompting, guidelines, templates, and information buttons)

5.   Appropriate channels (determining the data's time structure and properly incorporating it into the decision-making process (Borum, 2018))

A satisfying medical staff results when these requirements are met, and clinicians acquire valuable data that does not contradict their professional judgment or take more time and effort.

4.4.3 Smartphones, apps, sensors, and devices

AI as well as the IoT, which prevent potential cellphone apps, social networks, also location-tracking expertise, be able to help chronic disease patients receive timely detection, treatment, and self-management (Malasinghe, Ramzan, & Daha, 2019). Passive monitoring, participatory sensing, and muscle strength are the three types of remote healthcare sensing technology. The most common passive sensing device is a smartphone. The built-in sensors on a smartphone (accelerometer, odometer, and magnetometer) provide physics-based features such as determining a user's daily step. They may provide air pressure information, bright sources, voice, and also pressure on the touchscreen. Additionally, the created camera allows for more creative uses of these detectors. Such as turning the phone first into fall detection technique, analysis refers to the process (through detecting sound data gathered via speakers), or a fitness tracker (Coppetti, et al., 2017) as demonstrated in Figure 4.3.

Wristbands, are examples of wearable devices, components, are becoming more common in the modern period and include some of the same

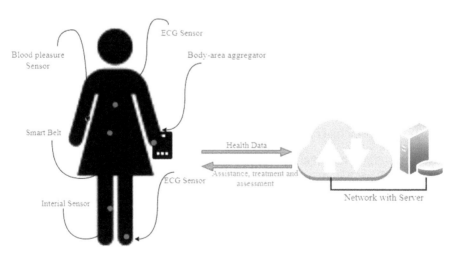

Figure 4.3   Wearable sensors in healthcare.

sensors found in telephones. They may detect vibrations associated with smoke (Saleheen, et al., 2015) as well as seizure occurrence. In 2017, 18% of persons in the United States wear a gadget that could be worn, such as a smartwatch or a wrist-worn gadget. Photoplethysmography sensors, which monitor fluctuations in reflected light caused by perivascular blood flow adjustments with each heartbeat, are typically seen on wrist sensors. It gives details on the characteristics that regulate the cardiac rhythm. Clinicians also use wearable sensors; for example, smartwatch patches can measure physical movement and posture, radio frequency sensors placed over clothing can check respiration rate, and others can track the healing process for a range of neurological disorders (McLaren, Joseph, Baguley, & Taylor, 2016).

A tiny sensor put in a pill delivers signals to a compostable patch when the tablet reaches the stomach in an emerging gadget for medication adherence (Hafezi, et al., 2014). Active sensing requires the observation of a patient, whereas passive sensing collects just observable evidence. As a result, active sensing allows for a subjective perspective of a patient's health condition to be obtained through questionnaires that ask patients to report their sensations during specific time intervals. Patient-reported data is equally as crucial as measured data, and it needs to be factored into AI models for an accurate assessment of results. Active and passive sensing is combined in operational assessment approaches, with the patient's subjective assessment and specific device sensor's highly objective data collecting. It evaluates the functioning of patients' mobile health devices to fulfill maintenance appropriately; furthermore, it assesses cognitive processes like memory and responding speed via smartphone platforms (Sim, 2019).

The bulk of mobile apps used to assess cognitive performance are mental-stimulation-based brain training aids. Many emotional health solutions have improved users' mental health by enhancing cognitive qualities such as velocity, memory, attention, and problem-solving. Apple's Lumosity and Elevation were among the first apps on the market; these apps use fascinating mini-games created by researchers to improve the brain's neuroplasticity. AI can be included through these apps to provide users with a user-friendly monitoring interface for analyzing their cognitive functions in real time. Another significant usage of technology in cognitive health monitoring is chatbot coaches. Bots, Youper, Wysa, and Unmind (Apps, avatars, and robots: The future of mental healthcare, 2019) are all smartphone health apps that use conversational AI chatbots to provide everyday assistance and mental health tracking. Cognitive behavioral therapy approaches created by psychology/mental health researchers are used to teach and evaluate chatbots' AI algorithms.

4.4.4 Processing of text language

The scale and complexity of patient-provided medical data are growing drastically. These enormous datasets contain critical information that could prove to be beneficial in medical decision-making. Unfortunately, due to expressing the personal, unstructured nature (e.g., available clinical records), these data are limited in their use, and almost no local workforce can manage a large number of records because the majority of them is not well-organized (stored data). On the other hand, AI-powered computer software may scan all gathered and save pertinent information for potential use. These AI techniques use natural language processing (NLP) to extract a human-written text's meaning, emotion, or intent. NLP algorithms help computers understand human language, making clinical research easier by automating their extraction of relevant data and evaluating free-text documents. Probabilistic, ML, and DL are the most commonly used methodologies (Senders, et al., 2019; Remote patient monitoring using artificial intelligence, 2020).

4.4.5 Healthcare applications of text processing technology

NLP-based solutions are becoming more common. Hands-free communication is very important in healthcare. These technologies have been implemented in several AI-based applications, the most notable of which were Chatbots for healthcare management via wearable technology. The following are a few examples of NLP uses in healthcare: aside from NLP tools in education, where a variety of mobile apps could be beneficial to students with reports and summaries, specific NLP-based applications can help people with significant points notes during their doctor visit. Using voice recognition algorithms, the integrated language processing technique allows for the summary and extraction of crucial information. Doctors and nurses can also record speech updates using NLP-based mobile apps. This same surgeon, for example, can orally document and communicate with the rest of the team promptly (Medical internet of things and big data in healthcare, 2016).

Secure remote control: NLP technology allows people with disabilities to communicate without having to use their hands. NLP technology in the smartphone app may simulate human conversation and engage consumers via autonomous systems, chatbots. NLP applications within the chatbot industry: m-health apps. Chatbots are innovative conversation programs that mimic a human narrative using powerful AI technology to read and reply to what users say. These virtual assistants will resemble family doctors today and, in the future, will be ready to provide sufferers with immediate medical assistance. In the field of healthcare, several chatbot technologies have shown potential. Babylon health: the healthcare guidance app incorporates many cutting-edge patients who can use AI algorithms to help them remotely consult surgeons and other health professionals. Patients are assigned consultations by appropriate AI algorithms based on their medical background and other health considerations. Users can also carry out real-time video chats with doctors about their health profiles using this program.

This software uses NLP to create a symptom check Chatbot that can provide people with helpful information like solutions and subscriptions to their medical issues. With the assistance of the national health system of the United Kingdom, this app is now giving a limited range of free services to satisfy the medical needs of many individuals. The health of the Buoy: the Buoy health app promotes patients/users in discovering the etiology of their disease, with the main focus on diagnosis. Using built AI applications qualified on vast medical datasets, this application provides a discussion of a real-time chatbot that can aid patients. The chatbot provides information on patients. A reasonably precise diagnosis of their illness based on their sensations, as well as relevant medical advice for their health complications.

4.4.6 Using consumer technology to its full potential

AI is being more widely applied in various fields of monitoring systems due to society's rising technology adoption. According to the most recent data from Pew Research, 81% of people own a smartphone, with far more than 63% of intelligent phones completing heath or medical searches. In addition, a 2018 study on smartphone usage and fitness app use by disadvantaged populations in the United States discovered that 38% of responders use a cell phone app to keep pace with innovation overall health (Vangeepuram, et al., 2018). Significant progress in areas such as data science and computer vision has been made in intelligent machines, which are serving to meet consumer expectation. Consumer demand for healthcare technologies is influenced by factors at the community level, including population increase and aging, rising healthcare costs, healthcare gaps in care provision, and the significant prevalence of diseases like asthma, cardiovascular disease, dementia, and Parkinson's (Gibbons & Shaikh, 2019). This requires periodically assessing one's food, physical exercise, and medical care. There is a growing demand for how AI technologies for monitoring patients turn out, as they are seen as a crucial step toward bettering pain treatment and self-management.

Furthermore, as mentioned in the national healthcare system (Dall, West, Chakrabarti, & Iacobucci, 2015), there is growing concern about personnel shortages. According to the 2010 American association consumer survey on healthcare access, these shortages are more likely to affect underserved customers in regional and interior areas. Additionally, the existing health staff in the United States is aging, resulting in a greater retirement rate. As a result, AI's role in monitoring patients has been broadened to include developing advanced healthcare monitoring systems that can help people cope with increased demands. The most exciting developments would be those that focus on patient stabilization, reducing the length and cost of staying, performing therapy and diagnostics under physician oversight, and monitoring the patients while they will not be in the hospital, and those who are going through the recuperation process.

Furthermore, data suggest that low-income persons have poorer health, have less healthcare coverage, are more prone to being health-conscious, and live in generally adverse environments (Gibbons & Shaikh, 2019). These disparities persist for various cultural, socioeconomic, behavioral, regional, and healthcare system characteristics. As a result, whenever the system is inattentive to or insufficiently sensitive to genuine concerns, significant distrust and issues occur between these individuals and healthcare practitioners (Wickramasinghe, John, George, & Vogel, 2019).

4.4.7 AI's function in diabetes forecasts and management

Diabetes is a widespread condition that affects millions of people worldwide. Healthcare systems worldwide spend a lot of money on prescription drugs (Syed, Jabeen, Manimala, & Alsaeedi , Smart healthcare framework for ambient assisted living using IoMT and big data analytics techniques, 2019). While there are many alternative treatments, there is an outstanding essential for enlightening diabetes patients' eminence of life and illness management to reduce diabetes-related difficulties. Supplementary to only maintaining blood glucose levels like glycated HbA1C, diabetes management also comprises sensation improved and understanding or monitoring the illness (Fagherazzi & Ravaud, 2019). Improvements in the device and new technologies offer an opportunity to rethink the ways diabetes is treated and managed in this context. A substantial number of articles on AI systems for diabetes care have been published in the recent decade, demonstrating the huge interest in the topic and the range of ways to use AI to progress the lives of individuals with diabetes. These systems can be rummage-sale for diabetes screenings, diagnosis, dealing, or administration, for example (Dankwa-Mullan, et al., 2019). We will delve a little deeper into each of these diabetes patient monitoring solutions.

4.4.7.1 Apps and technologies for diabetes monitoring

The Behavior Change Technique (BCT) app, for diabetes, is one of the numerous commercially accessible diabetic treatment methods for patient monitoring. BCT is a phone app that allows diabetics to display their sugar levels, in real-time levels, and estimates potential swings in blood sugar levels ranging from an hour ahead of time. Using CGM, mobile phone sensors, and personal user inputs, diabetes uses ML algorithms to estimate future blood glucose variations. This feature allows patients to take preventative measures to keep their sugar levels within the target range. This is also combined with technologies for historical analysis to effectively monitor and control the condition and achieve improved blood glucose control.

Diabetic is a diabetes care program for people who have insulin resistance and type-2 diabetes. It allows for diabetes management that is tailored to the individual by stimulating the blood sugar system. It also supports the handler to understand the influence of various inducements, such as diets, on the insulin-level tendency to avoid hyperglycemia. To receive continuous glucose monitor data, the app is used with Dexcom, a third-party program that displays glucose levels. It may also be linked to a Fitbit or an Apple Watch to incorporate data like temperament rate, diet, and step score. These extra efforts help to improve forecast accuracy. DiaBits' ML algorithm takes three to seven days to get to know the user's physiology, and as the user's data grows, so does its accuracy at forecasting blood glucose.

SVM and unsupervised-DL techniques are performed relying just on volumes of information and whether it derives personalities or groups of people. Although auto-regressors are useful, they can only forecast blood glucose values 15 minutes in advance [68]. BCT's technology was successfully implemented in the Canada Hospital for Children in Canada, with a prediction accuracy increase of 96%. The diabetic app was tested in eight adolescent patients for 60 days using a blood glucose meter and Fitbit trackers, gathering information about personal activities, breathing rate, and glucose level. This was before the model was used to anticipate the user's near-future glucose level results after 30 days of training just on the effect of activity, heartbeat, and glucose levels. Using statistics from a constant glucometer and a fitness tracker, the researchers discovered that being projected up to 1 hour ahead of time, blood glucose levels can be anticipated. DiaBits' algorithms are currently being integrated into Premier Health's healthcare app.

Premier Health's ecosystem includes 292 clinics and over three million patients. To enhance the performance of overall patient care and health-related decisions, Prestigious Health would continue to incorporate clinical decision tools into its electronic health record (EHR) platform. Pilot tests are being conducted with people who use the Premier mobile applications to track their pulse rate, hypertension, age, sex, and strength. Will it be fascinating to see how the software works when used in a larger sample of patients? To see if they can improve glucose control, most diabetes treatment programs have been tested on a small- to medium-sized patient group. Long-term trials with a significant proportion of people are required to obtain appropriate data for treatment outcomes and tolerability (Cvetković, et al., 2016).

For tasks such as patient self-monitoring, physician decision support, customizable risk assessment, and also the detection of retinal and other delayed concerns, many diabetes monitoring devices are being developed by employing AI-based methodologies. Recent advancements in innovative consumer electronics such as glucose detection, stents, and fitness tracking have resulted in a slew of clinical research, evaluating and seeking to improve glycemic management and prevent hypoglycemic episodes by merging genuine CGM with AI algorithms. A closed-loop insulin pump with automation control and reporting was employed in several types of research (Thabit, et al., 2015).

4.5 AI-Assisted Monitoring of the Heart

AI holds potential encouraging implications in cardiovascular medicine, particularly in diagnostic and therapeutic procedures, and sufferer intensive care. These requests are separated into two groups: virtual and physical investigations. Virtual study topics for healthcare management and automated CDSS include ML and text analysis. In the construction sector, AI enforcements mostly lay robotic surgical procedures.

4.5.1 AI in cardiology with virtual applications

The use of ML in imaging can be broken down into three phases: image processing, renovation, and analysis. The implementation of advanced AI systems on vast volumes of images acquired has eased picture analysis by establishing new organizational patterns for image interpretation. Several AI-based solutions have been developed to automate various image processing tasks. To automate the heartbeat phase and post-analytical frame, Molony *et al.* (Molony, Hosseini, & Samady, 2018) constructed and validated DL models on communication skills for cardio endoscopies of 3900 people. Data mining algorithms are used in several AI frameworks; for example, University of Pennsylvania researchers developed a neural network approach for cardiac ultrasound image segmentation. This method automates the lumen area and plaques weight computation operations in real time. This method yielded promising picture segmentation results and received positive evaluations from field experts. The researchers anticipate that combining data mining and ML methodologies in creative AI systems would enable autonomous monitoring and diagnosis of many image-based disorders soon, in a manner that closely resembles a specialist's role (Sardar, et al., 2019).

4.5.2 Supporting system in clinical decisions

CDSS are also progressing in cardiology toward inner systems that use DL, machine training, and language handling to mimic a human expert's decision-making process. Watson in intellectual health technology, for example, makes substantial use of data from health files (health records, lab, or imaging reports) as well as web information sources (Syeda-Mahmood, 2018). AI is a machine training solution, which are mostly based on data science, and is designed to caricature the decision-making processes of the anthropoid brain. IBM has built an arrangement for medical practitioners and radiologists to use in their cardiovascular medical decision-making. Various cardiac monitoring functions, especially in imaging, are included in IBM's Medical Sieve project, such as automatic detection of coronary vascular disease undergoing angiography (Sardar, et al., 2019).

4.5.3 Augmented reality (AR), virtual reality (VR), and virtual assistants

Among the most important requests of techniques in cardiology is the grouping with AI techniques with augmented worlds and AR technology. When it comes to treatment, the foremost VR frameworks are aimed at postoperative and presence modalities such as cardiovascular treatments and stress management for sufferers. AR platforms also assist interventionists during heart procedures by giving important real-time data through many monitoring displays. Vocal-style automated systems, including Google, Apple's Siri, and Amazon's Alexa, identify speech utilizing extremely advanced AI schemes. Adopting physicians and operators can quickly communicate with virtual assistants, along with speech interface, effectively and simply look through evidence in their EMR arrangement or gain web evidence using only their voice signature credentials (Steinhubl & Topol, 2018).

4.5.4 Automated analysis with data

Databases have become progressively more composite and also difficult to evaluate using outdated statistical approaches as a result of the growth of varied and useful datasets, including such sequence data, social networks, and cardiac imaging. Approaches to AI that use big data methodologies for automatic theory construction have shown analytical and predictive potential (Jiang, et al., 2017). DL algorithms, for example, can sift through incredibly detailed facts regarding specific cardiovascular diseases, identify key risk factors, and conduct accurate analysis. ML established approaches may be used to forecast the danger of a heart attack and mortality associated with various cardiac operations for patient monitoring. In cardiology, ML was used to forecast a year's worth of events. Death in patients with diastolic dysfunction and long-term illness survival in patients with cardiovascular disease (Johnson, et al., 2018).

4.6 Neural Applications Linked to AI and Patient Monitoring

4.6.1 AI for dementia patients

In the latter days, there has been substantial growth in the demographic characteristics of the number of western countries, as well as a corresponding fall in the birth rate, signaling those civilizations are aging. Simultaneously, difficulties are becoming increasingly difficult to handle due to a shortage of medical personnel, notably doctors, nurses, and elderly caretakers. Persons with age-related problems, such as dementia, face a difficult time ensuring high-quality nursing care. Dementia affects 50 million people globally today, with 10 million new cases diagnosed each year (Pharmacotherapy of dementia in Germany: Results from a nationwide claims database, 2015). As a result, enhancing clinical care facilities and providing psychiatric patients with the standards of security, they require to live a standard life and will require a concerted effort.

The majority of dementia patients are successfully preserved, albeit their special assistance needs vary depending on the stage of the infection. These demands may be transitory, but they might be things if the resources and efforts required exceed the capabilities of the careers. As a result, Alzheimer's management necessitates collaboration across professional and sectoral lines. Coordination is less effective in practice due to delays or perhaps a lack of data flow among players. This recognizes the significance of establishing shared support systems that include medical services, family members providing informal care, and distance care. Regularities in the routine behavior of persons with dementia or emergency can be detected automatically using a similar support system, allowing easy synchronization of necessary measures. The German scientific project "SAKI," which aims to overcome dementia-related challenges and meet patients' requirements in their homes, is a comparable support system.

4.6.2 Dementia monitoring

Recent AI-based dementia monitoring services and infrastructure are intended to follow the succession of the complaint about time and help patients maintain their daily activities. The computational orthosis for assisting activities system is a dementia-residence system. This device was intended to assist individuals with dementia in performing daily tasks such as washing their hands. This method is generally based on the field of computer-vision-based, which allows it to monitor the present stage as well as an activity before determining the right spoken or pictorial instructions to offer, whereas dementia sufferers wash their manpower. This research popularized the idea of "zero activity innovations" or technologies that require little to no energy to operate, resulting in the development of multiple successful systems. The primary idea behind these technologies is to gather data, evaluate it, and use the necessary ICT, such as computer vision, algorithms, sensor systems, as well as the IoT, without interfering with people's lives. A recent advanced study has concentrated on embedding AI systems for emotions and personality assessment into those support structures to make them easier to incorporate into users' lifestyles (Robillard & Hoey, 2018).

4.6.3 Supporting dementia patients

With the elaboration of AI with the IoT infrastructures in people's homes, the notion of intelligent housing was formed. ML and IoT developments have broadened intelligent homes by introducing a new supporting system based on prepared and clothing (An assistive technology system that provides personalized dressing support for people living with dementia: capability study, 2018). The Gloucester HomeKit is indeed an early form of a dementia-friendly smart home. It has a digital forum for discussion for visual and voice guidance, immersion and microwave screens, a computerized night light system, an item finder, and a digital conversation forum for visible and auditory directions. This method also used voice hasty to attentive patients when potentially hazardous situations arose, such as when heaters were left on and when midnight wandering happened. This automated remote support system was tested at a care facility as part of such an intelligent installation to see how it affected the lives of people with severe dementia. By evaluating data from monitors and surveys, this technology has promised to boost a sufferer's independence. This intelligent home system assisted the client in recovering from urine incontinence, sleeping two hours longer each night, and cutting night-time explorations in half (Orpwood, Adlam, Evans, Chadd, & Self).

Intelligent machine advancements have improved the cognition of these innovative homes and are shrewd by allowing for more user engagement. Intelligent home elements can learn about their owners' routines and automatically adjust their programming to match their needs thanks to these powerful AI algorithms. Human-like (user action, not physically) robots are currently the topic of investigation; in fact, most smart home gadgets are expected to be equipped with artificially intelligent systems in the future, allowing them to assist people in various ways. In Japan, charming and friendly nurse robots have already been used to aid patient care. These intelligent machines are calculated to assist individuals with sensual activities; however, they can also look after every patient's overall well-being, a crucial component of life for people with disabilities like dementia sufferers. As an illustration, PARO robotic is just a fun and engaging seal toy that soothes and relaxes while also providing mammal therapy.

4.7 AI for Migraine Patients

Migraine is a prevalent neurological disease marked by repeated headaches, vomiting, and light sensitivity (Olesen, 2008). It is the third common disease in the globe and the seventh most devastating neurological disorder. On the other hand, migraine is a highly complex disease that is commonly misinterpreted since its symptoms are similar to those of other conditions, including convulsions and tension headaches. Several studies using the simple flash method have been undertaken in the brand-new few years to tackle the dilemma of recognizing migraine diagnosis, with positive results. The premise behind this method is to employ multichannel EEG to analyze the patient's brain reactions to different speeds of flash stimulation; the results are most accurate when the flash frequency is 4 Hz (Akben, Subasi, & Tuncel, 2012).

Furthermore, diagnostic aids for tension headache and, subsequently, migraine were proposed. When the remembrance rate, declining rate, F-score, and overall accuracy were taken into account, they came up with more accurate results (Yin, Lu, Yu, Chena, & Duan, 2015). Numerous AI-based migraine management efforts are ongoing, intending to manage migraines only through a mobile phone app or connected phone. The collected information from migraine patients is used in AI-based big data approaches (Akben, Tuncel, & Alkan, Classification of multi-channel EEG signals for migraine detection, 2016). To determine migraine gestures, wavelet-based characteristics were employed. The EEG data were fragmented into two associations using neural networks: migraine and onslaught (Akben, Tuncel, & Alkan, Classification of multi-channel EEG signals for migraine detection, 2016). Another study (Krawczyk, Simić, Simić, & Woźniak, 2013) used ML mechanisms to contrive a decision-assistance system for headache categorization.

For various migraines, including nerve pain and migraine headaches, the method described above yielded adequate categorization results. Although there are numerous challenges in the area of cluster headache identification, the most significant is the symptomatic nature of the condition. Furthermore, it has been proved that emerging technologies merging AI and the Internet of Things can begin to produce promising results if a massive amount of data on epilepsy patients is available (Subasi, Ahmed, Aličković, & Hassan, 2019). Migraine researchers have recently focused on relevant quality mobile health apps that notify patients of migraines attacks. To construct effective predictive models, these AI-powered smartphone apps use a range of big data methodologies. As vast datasets with various sorts of data are acquired, these models become increasingly accurate. For example, a mobile medical app called Migraine Alert was developed through collaborators at the Cleveland Clinic, the University of Washington, and Two View Health (a digital health startup). This program uses multivariate ML algorithms to create personalized and predictive forecasting that can forecast migraine attacks daily for an individual.

The AI representations in the indicated app have been skilled on a dataset that comprised migraines sensations, daily inputs, stimuli, and other biological variables (Fitbit) from just the clinical trials of other collaborators. In contrast, Migraines Buddy and Reduce My Pain Pro are two mobile health apps that provide high-quality treatments of migraine reduction using massive data based on ML techniques. Experts created this monitoring tool in psychology and data analytics to help patients track potential everyday stressors and contextual factors that cause migraine attacks. Doctors have found this program an excellent remote monitoring tool because it allows them to propose various alleviation options based on their medical illness.

The former program leverages user-provided data on migraine discomfort to assist users in tracking their symptoms by offering an intelligent tailored visualization of their migraine status in the form of graphs, charts, and calendar views. There are several mobile health programs for migraine effective strategies to address on the same premise, including Migraine Insights and Migraine Monitoring. More apps are not created explicitly for migraine monitoring but instead focus on pain management techniques. Backlight Filter and Night Vision are two examples of apps that reduce exposure to blue light linked to migraines. Other examples are Sam Harris's Relax Tunes:

Sleep Sound or Waking Up, which provide thoughtful remedies through hypnosis and directed meditations employing relaxing melodies.

4.8 Conclusion

Although more people will develop chronic conditions that necessitate behavior and generic drug control, there has been a significant increase in interest in remote monitoring technology that can assist patients and healthcare professionals in disease management. This monitoring technology lets users collect real-time data either in the household or while traveling and then save or discharge it to dental professionals for therapeutic consultation. Several essential requests for medical decision support were addressed in this article. These included glucose measures in favor of diabetes, mobility 24-hour care Parkinson's illness, and also ECG forecasting for focal point problems. However, many more are already in use or development. Despite the enormous potential, there is just a small body of testimony to encourage security and surveillance tactics regarding improved health services and cost savings.

Large-scale employment is mandatory to demonstrate the true assistances of certain therapies crosswise a wide range of sectors and individuals within the supply chain network. Over the past few years, promotions in data science have directed to virtually every problem in surveillance systems in the field of computer vision, with encouraging results. Massive datasets generated by secure web monitoring systems like medical procedures, wearables, and applications can be handled using ML methods. Because it is capable of handling big datasets, hypothesis, classification, and decision mentoring can be improved. Improved health results have been documented in a wide range of vehicle tracking situations, apart from cardiovascular disease, migraines, and diabetes management, when an AI-based system is utilized. Many issues, including barriers to evaluation, medical education, data, security and privacy, and integration into standard healthcare services, need to be addressed in the meantime.

4.9 Acknowledgment

We are thankful to all coauthors for contributing to this chapter.

4.10 Funding

None.

4.11 Conflict of Interest

Authors have no conflicts of interest, financial or other to disclose.

References

[1] Ahad, A., Tahir, M., & Yau, K. (2019). 5G-based smart healthcare network: architecture, taxonomy, challenges and future research directions. *IEEE access, 7*, 100747–100762.

[2] Ahmed, H., & Serener, A. (2016). Effects of external factors in CGM sensor glucose concentration prediction. *Procedia Computer Science, 102*, 623–629.

[3] Akben, S., Subasi, A., & Tuncel, D. (2012). Analysis of repetitive flash stimulation frequencies and record periods to detect migraine using artificial neural network. *Journal of medical systems, 36*(2), 925–931.

[4] Akben, S., Tuncel, D., & Alkan, A. (2016). Classification of multichannel EEG signals for migraine detection. *Biomedical Research, 27*(3), 743–748.

[5] Albahri, O., Zaidan, A., Zaidan, B., Hashim, M., Albahri, A., & Alsalem, M. (2018). Real-time remote health-monitoring Systems in a Medical Centre: A review of the provision of healthcare services-based body sensor information, open challenges and methodological aspects. *Journal of medical systems, 42*(9).

[6] An assistive technology system that provides personalized dressing support for people living with dementia: capability study. (2018). *JMIR medical informatics, 6*(2), e5587.

[7] Apps, avatars, and robots: The future of mental healthcare. (2019). *Issues in mental health nursing, 40*(3), 208–214.

[8] Ashok Vegesna, Melody Tran, Michele Angelaccio, & Steve Arcona. (2017). Remote Patient Monitoring via Non-Invasive Digital Technologies: A Systematic Review. *Telemedicine and e-Health*, 3–17. Retrieved from http://doi.org/10.1089/tmj.2016.0051

[9] Baig, M., GholamHosseini,, H., Moqeem, A., Mirza, F., & Lindén, M. (2017). A systematic review of wearable patient monitoring systems– current challenges and opportunities for clinical adoption. *Journal of medical systems, 41*(7), 1–9.

[10] Bashi, N., Karunanithi, M., Fatehi, F., Ding,, H., & Walters, D. (2017). Remote monitoring of patients with heart failure: an overview of systematic reviews. *Journal of medical Internet research, 19*(1), e18.

[11] Block, V., Pitsch, E., Tahir, P., Cree, B., Allen,, D., & Gelfand, J. (2016). Remote physical activity monitoring in neurological disease: a systematic revie. *PloS one, 11*(1), e0154335.

[12] Borum, C. (2018). Barriers for hospital-based nurse practitioners utilizing clinical decision support systems: A systematic review. *CIN: Computers, Informatics, Nursing, 36*(4), 177–182.

[13] Catherine Klersy, Annalisa De Silvestri, & Gabriella. (2009). A Meta-Analysis of Remote Monitoring of Heart Failure Patients. *Journal of the American College of Cardiology*, 1683–1694. Retrieved from https://doi.org/10.1016/j.jacc.2009.08.017

[14] Charleonnan, A., Fufaung, T., Niyomwong, T., Chokchueypattanakit, W., Suwannawach,, & Ninchawee, N. (2016). Predictive analytics for chronic kidney disease using machine learning techniques. In IEEE (Ed.), *In 2016 management and innovation technology international conference (MITicon)*, (pp. MIT-80).

[15] Chatrati, S., Hossain, G., Goyal, A., Bhan, A., Bhattacharya, S., Gaurav, D., & Tiwari, S. (2020). Smart home health monitoring system for predicting type 2 diabetes and hypertension. *Journal of King Saud University-Computer and Information Sciences.*

[16] Chen, C., Zhao, X., Li, Z., Zhu, Z., Qian, S., & Flewitt, A. (2017). Current and emerging technology for continuous glucose monitoring. *Sensors, 17*(1), 182.

[17] Chen, M., Yang, J., Zhou, J., Hao, Y., & Zhang, J. (2018). 5G-smart diabetes: Toward personalized diabetes diagnosis with healthcare big data clouds. *IEEE Communications Magazine, 56*(4), 16–23.

[18] Chetty, G., White, M., & Akther, F. (2015). Smart phone based data mining for human activity recognition. *Procedia Computer Science, 46*, 1181–1187.

[19] Clifton, L., Clifton, D., Pimentel, M., Watkinson, P., & Tarassenko, L. (2012). Gaussian processes for personalized e-health monitoring with wearable sensors. *IEEE Transactions on Biomedical Engineering, 60*(1), 193–197.

[20] Coppetti, T., Brauchlin, A., Müggler, S., Attinger-Toller, A., Templin, C., Schönrath, F., & Wyss, C. (2017). Accuracy of smartphone apps for heart rate measurement. *European journal of preventive cardiology, 24*(12), 1287–1293.

[21] Cvetković, B., Janko, V., Romero, A., Kafalı, O., Stathis, K., & Luštrek, M. (2016). Activity recognition for diabetic patients using a smartphone. *Journal of medical systems, 40*(12), 1–8.

[22] Dall, T., West, T., Chakrabarti, R., & Iacobucci, W. (2015). The complexities of physician supply and demand: projections from 2013 to 2025. *Washington, DC: Association of American Medical Colleges.*

[23] Dankwa-Mullan, I., Rivo, M., Sepulveda, M., Park, Y., Snowdon, J., & Rhee, K. (2019). Transforming diabetes care through artificial intelligence: the future is here. *Population health management,, 22*(3), 229–242.

[24] Darwish, M., Senn, E., Lohr, C., & Kermarrec, Y. (2014). A comparison between ambient assisted living systems. *In International Conference on Smart Homes and Health Telematics* (pp. 231–237). Cham: Springer.

[25] Davis, K., Owusu, E., BastaniX, V., & Marcenaro, L. (2018). Activity recognition based on inertial sensors for ambient assisted living. In IEEE (Ed.), *In 2016 19th international conference on information fusion (fusion. 6*, pp. 371–378. IEEE.

[26] De-La-Hoz-Franco, E., Ariza-Colpas, P., & Quero, J. (2018). Sensor-based datasets for human activity recognition–a systematic review of literature. *IEEE Access, 6*, 59192–5921.

[27] Deshmukh, S., & Shilaskar, S. (2015). Wearable sensors and patient monitoring system: A Review. *In 2015 International Conference on Pervasive Computing (ICPC)* (pp. 1–3). IEEE.

[28] Dowding, D., Randell, R., Mitchell, N., & Foster, R. (n.d.). Experience and nurses use of computerised decision support systems. *In Connecting Health and Humans* (pp. 506–510). IOS Press.

[29] Ernesäter, A., Holmström, I., & Engström, M. (2009). Telenurses' experiences of working with computerized decision support: supporting, inhibiting and quality improving. *Journal of advanced nursing, 65*(5), 1074–1083.

[30] Fagherazzi, G., & Ravaud, P. (2019). Digital diabetes: Perspectives for diabetes prevention, management and research. *Diabetes & metabolism, 45*(4), 322–32.

[31] Gibbons, M., & Shaikh, Y. (2019). Introduction to consumer health informatics and digital inclusion. *In Consumer Informatics and Digital Health* (pp. 25–41). Springer, Cham.

[32] González-Valenzuela, S., Chen, M., & Leung, V. (2011). Mobility support for health monitoring at home using wearable sensors. *IEEE Transactions on Information Technology in Biomedicine, 15*(4), 539–549.

[33] Gopalsami, N., Osorio, I., Kulikov, S., Buyko, S., Martynov, A., & Raptis, A. (2007). SAW Microsensor Brain Implant for Prediction

and Monitoring of Seizures. *IEEE Sensors Journal, 7*(7), 977–982. doi:10.1109/JSEN.2007.895974

[34] Hafezi, H., Robertson, T., Moon, G., Au-Yeung, K., Zdeblick, M., & Savage, G. (2014). An ingestible sensor for measuring medication adherence. *IEEE Transactions on Biomedical Engineering, 62*(1), 99–109.

[35] Hamine, S., Gerth-Guyette, E., Faulx, D., & Green, B. (Journal of medical Internet research). Hamine, S., Gerth-Guyette, E., Faulx, D., Green, B. B., & Ginsburg, A. S. *Impact of mHealth chronic disease management on treatment adherence and patient outcomes: a systematic review, 17*(2), e3951.

[36] Iadanza, E., Mudura, M., Melillo, P., & Gherardell, M. (2020). An automatic system supporting clinical decision for chronic obstructive pulmonary disease. *Health and Technology, 10*(2), 487–498.

[37] Izonin, L., Trostianchyn, A., Duriagina, Z., Tkachenko, R., Tepla, T., & Lotoshynska, N. (2018). The combined use of the wiener polynomial and SVM for material classification task in medical implants production. *nternational Journal of Intelligent Systems and Applications, 10*(9), 40–47.

[38] Jiang, F., Jiang, Y., Zhi, H., Dong, G., Li,, H., Ma, S., & Wang, Y. (2017). Artificial intelligence in healthcare: past, present and future. *Stroke and vascular neurology, 2*(4).

[39] Johnson, K., Torres Soto, J., Glicksberg, B., Shameer, K., Miotto, R., Ali, M., & Dudley, J. (2018). Artificial intelligence in cardiology. *Journal of the American College of Cardiology, 71*(23), 2668–2679.

[40] Kannadasan, K., Edla, D., & Kuppili, V. (2019). Type 2 diabetes data classification using stacked autoencoders in deep neural networks. *Clinical Epidemiology and Global Health,, 7*(4), 530–535.

[41] Kindle, R., Badawi, O., Celi, L., & Sturland, S. (2019). Intensive care unit telemedicine in the era of big data, artificial intelligence, and computer clinical decision support systems. *Critical care clinics, 35*(3), 483–495.

[42] Krawczyk, B., Simić, D., Simić, S., & Woźniak, M. (2013). Automatic diagnosis of primary headaches by machine learning methods. *Central European Journal of Medicine, 8*(2), 157–165.

[43] Lilly, C., Zubrow, M., Kempner, K., Reynolds, H., Subramanian, S., & Eriksson, E. (2014). Critical care telemedicine: evolution and state of the art. *Critical care medicine, 42*(11), 2429–2436.

[44] Lloret, J., Parra, L., Taha, M., & Tomás, J. (2017). An architecture and protocol for smart continuous eHealth monitoring using 5G. *Computer Networks, 129*, 340–351.

[45] Lonini, L., Dai, A., Shawen,, N., Simuni, T., Poon,, C., Shimanovich, L., & Jayaraman, A. (2018). Wearable sensors for Parkinson's disease: which data are worth collecting for training symptom detection models. *NPJ digital medicine, 1*(1), 1–8.

[46] Majumder, S., Mondal, T., & Deen, M. (2017). Wearable sensors for remote health monitoring. *Sensors, 17*(1), 130.

[47] Malasinghe, L., Ramzan, N., & Daha, K. (2019). Remote patient monitoring: a comprehensive study. *J Ambient Intell Human Computer, 10,* 57–76. Retrieved from https://doi.org/10.1007/s12652–017-0598-x

[48] McLaren, R., Joseph, F., Baguley, C., & Taylor, D. (2016). A review of e-textiles in neurological rehabilitation: How close are we? *Journal of neuroengineering and rehabilitation, 13*(1), 1–13.

[49] Medical internet of things and big data in healthcare. (2016). *Healthcare informatics research, 22*(3), 156–163.

[50] Mills, S. (2019). Electronic health records and use of clinical decision support. *Crit Care Nurs Clin North Am, 31*(2), 125–131.

[51] Molony, D., Hosseini, H., & Samady, H. (2018). TCT-2 Deep IVUS: A machine learning framework for fully automatic IVUS segmentation. *Journal of the American College of Cardiology, 72*(135), B1-B1.

[52] Mukhopadhyay, S. (2014). Wearable sensors for human activity monitoring: A review. *IEEE sensors journal, 15*(3), 1321–1330.

[53] Najm, I., Hamoud, A., Lloret, J., & Bosch, I. (2019). Machine learning prediction approach to enhance congestion control in 5G IoT environment. *Electronics, 8*(6), 607.

[54] Neuman, M. (2010). Measurement of Vital Signs: Temperature [Tutorial]. *IEEE Pulse, 1*(2), 40–49.

[55] Olesen, J. (2008). The international classification of headache disorders. Headache:. *The Journal of Head and Face Pain, 48*(5), 691–693.

[56] Ong, M., Romano, P., Edgington, S., Aronow, H., Auerbach, A., Black, J., & Fonarow, G. (2016). Effectiveness of remote patient monitoring after discharge of hospitalized patients with heart failure: the better effectiveness after transition–heart failure (BEAT-HF) randomized clinical trial. *JAMA internal medicine, 176*(3), 310–318.

[57] Orpwood, R., Adlam, T., Evans, N., Chadd, J., & Self, D. (n.d.). Evaluation of an assisted-living smart home for someone with dementia. *Journal of Assistive Technologies.*

[58] Pantelopoulos, A., & Bourbakis, N. (2009). A survey on wearable sensor-based systems for health monitoring and prognosis. *IEEE Transactions on Systems, Man, and Cybernetics, Part C (Applications and Reviews), 40*(1), 1–12.

[59] Pharmacotherapy of dementia in Germany: Results from a nationwide claims database. (2015). *European neuropsychopharmacology, 25*(12), 2333–2338.

[60] Ramathulasi, T., & Babu, M. (2020). Comprehensive Survey of IoT Communication Technologies. *Emerging Research in Data Engineering Systems and Computer Communications*, (pp. 303–311).

[61] Remote patient monitoring using artificial intelligence. (2020). *In Artificial Intelligence in Healthcare*, 203–234.

[62] Robillard, J., & Hoey, J. (2018). Emotion and motivation in cognitive assistive technologies for dementia. *Computer, 51*(3), 24–34.

[63] Rodbard, D. (2016). Continuous glucose monitoring: a review of successes, challenges, and opportunities. *Diabetes technology & therapeutics, 18*(2).

[64] Ronao, C., & Cho, S. (2016). Human activity recognition with smartphone sensors using deep learning neural networks. *Expert systems with applications, 59*, 235–244.

[65] Saleheen, N., Ali, A., Hossain, S., Sarker, H., Chatterjee, S., Marlin, B., & Kumar, S. (2015). puffMarker: a multi-sensor approach for pinpointing the timing of first lapse in smoking cessation. *In Proceedings of the 2015 ACM International Joint Conference on Pervasive and Ubiquitous Computing*, (pp. 999–1010).

[66] Sanchez-Morillo, D., Fernandez-Granero, M., & Leon-Jimenez, A. (2016). Use of predictive algorithms in-home monitoring of chronic obstructive pulmonary disease and asthma: a systematic review. *Chronic respiratory disease, 13*(3), 264–283.

[67] Sardar, P., Abbott,, J., Kundu, A., Aronow, H., Granada, J., & Giri, J. (2019). Impact of artificial intelligence on interventional cardiology: from decision-making aid to advanced interventional procedure assistance. *Cardiovascular Interventions, 12*(14), 1293–1303.

[68] Savarese, G., & Lund, L. (2017). Global public health burden of heart failure. *Cardiac failure review, 3*(1), 7.

[69] Senders, J., Karhade, A., CoteX, D., Mehrtash, A., Lamba, N., DiRisio, A., & Arnaout, O. (2019). Natural language processing for automated quantification of brain metastases reported in free-text radiology reports. *JCO clinical cancer informatics, 3*, 1–9.

[70] Sim, I. (2019). Mobile devices and health. *New England Journal of Medicine, 381*(10), 956–968.

[71] Steinhubl, S., & Topol, E. (2018). Now we're talking: bringing a voice to digital medicine. *The Lancet, 392*(10148), 627.

[72] Subasi, A., Ahmed, A., Aličković, E., & Hassan, A. (2019). Effect of photic stimulation for migraine detection using random forest and

discrete wavelet transform. *Biomedical signal processing and control, 49*, 231–239.

[73] Syed, L., Jabeen, S., Manimala, S., & Alsaeedi , A. (2019). Smart healthcare framework for ambient assisted living using IoMT and big data analytics techniques. *Future Generation Computer Systems, 101*, 136–15.

[74] Syed, L., Jabeen, S., Manimala, S., & Alsaeedi, A. (2019). Smart healthcare framework for ambient assisted living using IoMT and big data analytics techniques. *Future Generation Computer Systems, 101*, 136–151.

[75] Syeda-Mahmood, T. (2018). Role of big data and machine learning in diagnostic decision support in radiology. *Journal of the American College of Radiology, 15*(3), 569–576.

[76] Teichmann, D., De Matteis, D., Bartelt, T., Walter, M., & Leonhardt, S. (2015). A bendable and wearable cardiorespiratory monitoring device fusing two noncontact sensor principles. *IEEE journal of biomedical and health informatics, 19*(3), 784–793.

[77] Tepla, T., Izonin, I., DuriaginaZ. A, Tkachenko, R., Trostianchyn, A., Lemishka, I., & Kovbasyuk, T. (2018). Alloys selection based on the supervised learning technique for design of biocompatible medical materials. . *Archives of Materials Science and Engineering, 93*(1), 32–40.

[78] Thabit, H., Tauschmann, M., Allen, J., Leelarathna, L., Hartnell, S., Wilinska, M., & Hovorka, R. (2015). Home use of an artificial beta cell in type 1 diabetes. *New England Journal of Medicine, 373*(22), 2129–2140.

[79] Tkachenko, R., Doroshenko,, A., Izonin, I., Tsymbal, Y., & Havrysh, B. (2018). imbalance data classification via neural-like structures of geometric transformations model: Local and global approaches. In Springer (Ed.), *In International conference on computer science, engineering and education applications*, (pp. 112–122). CHAM.

[80] Vangeepuram, N., Mayer, V., Fei, K., V., Hanlen-Rosado, E., Andrade, C., Wright, S., & Horowitz,, C. (2018). Smartphone ownership and perspectives on health apps among a vulnerable population in East Harlem. *Mhealth,, 4*.

[81] Wang, L., Wang, X., Chen, A., Jin, X., & Che, H. (2020). Prediction of type 2 diabetes risk and its effect evaluation based on the XGBoost model. In Healthcare. *Multidisciplinary Digital Publishing Institute, 8*(3), 247.

[82] Wickramasinghe, N. (2019). Essential considerations for successful consumer health informatics solutions. *Yearbook of medical informatics, 28*(01), 158–164.

[83] Wickramasinghe, N., John, B., George, J., & Vogel, D. (2019). Achieving Value-Based Care in Chronic Disease Management: Intervention Study. *JMIR Diabetes, 4*(2). doi:10.2196/10368

[84] Xiao, F., Miao, Q., Xie, X., Sun, L., & Wang, R. (2018). Indoor anti-collision alarm system based on wearable Internet of Things for smart healthcare. *IEEE Communications Magazine, 56*(4), 53–59.

[85] Yan, L., Yoo, J., Kim, B., & Yoo, H. (2010). A 0.5-μ V $_{\rm rms}$ $12-$\mu$ W Wirelessly Powered Patch-Type Healthcare Sensor for Wearable Body Sensor Network. *IEEE journal of solid-state circuits, 11*, 45.

[86] Yin, Z., Lu, X., Yu, S., Chena, X., & Duan, H. (2015). A clinical decision support system for the diagnosis of probable migraine and probable tension-type headache based on case-based reasoning. *The journal of headache and pain, 16*(1), 1–9.

[87] Yoo, E., & Lee, S. (2010). Glucose biosensors: an overview of use in clinical practice. *Sensors, 10*(5), 4558–4576.

[88] Yoo, H., Han, S., & Chung, K. (2020). A frequency pattern mining model based on deep neural network for real-time classification of heart conditions. In Healthcare. *Multidisciplinary Digital Publishing Institute, 8*(3), 234).

5

Artificial Intelligence: A Promising Approach Toward Targeted Drug Therapy in Cancer Treatment

Amrita Shukla[1], Simran Ludhiani[2], Neeraj Kumar[1], Shahid Rja[1], Sudhanshu Mishra[3*], and Subasini Uthirapathy[4]

[1]Dr. M C Saxena College of Pharmacy, India
[2]Department of Pharmacy, Shri Govindram Seksaria Institute of Technology and Science, India
[3]Department of Pharmaceutical Science & Technology, Madan Mohan Malaviya University of Technology, India
[4]Faculty of Pharmacy, Tishak International University, Iraq
*Corresponding Author
Sudhanshu Mishra
Department of Pharmaceutical Science & Technology, Madan Mohan Malaviya University of Technology, India
Email: msudhanshu22@gmail.com

Abstract

Cancer is one of the leading causes of death due to its high morbidity and mortality throughout the world. Various anticancer therapies like chemotherapy, radiation therapy, hormone therapy, surgical approaches, etc., are widely used in treating cancer but have serious adverse effects mostly due to cytotoxic action toward normal cells. AI is a boon in the field of oncology too. Anticancer drug activity is predicted using AI, and AI is used to help in the development of anticancer medicines. Various cancers and medications may react differently, and recent screening tools have repeatedly revealed a connection between cancer cell genetic variety and therapeutic efficacy. Its main features are lesion recognition, target area delineation, three-dimensional tumor localization, clinical and pathological analysis, quantitative tumor

129

analysis, and tumor picture segmentation. The present review will cover the current status of various monoclonal antibodies used in targeted drug therapy of cancer along with the prospects of therapy. This article highlights the application of AI in various facets of the pharmaceutical sector with a focus on cancer treatment.

5.1 Introduction

Artificial intelligence has already become an integral part of our day-to-day lives and proposes to improve it further. Constantly increasing data volumes, improvements in algorithms, and continuous evolution of computer power and storage are some of the major reasons for the popularity of AI. This concept of artificial intelligence was rooted in the early 1950s and was defined as the science of developing intelligent machines by one of the founders of this field, John McCarthy [1]. AI is the capability of machines to learn and simulate the tasks that are often related to human behavior; it can also be described as a set of self-learning techniques [2]. AI is not a single technology but a cluster of various other technologies like machine learning (ML) and deep learning (DL) which are used separately or in a combination for completing the provided tasks. Machine learning is a computational process that requires input data to achieve a desired task [3]. It comprises a lot of theories and algorithms. Algorithms are nothing but a set of rules that create a model. These algorithms can be classified into supervised, unsupervised, and reinforcement learning as shown in Figure 5.1 [4].

If any desired output is described by certain attributes, then machine learning portrays and connects those attributes for achieving final results or desired output.

5.2 AI, Machine Learning, and Deep Learning

Artificial intelligence is a concept that arises from an idea to use technology that will utilize behavior that imitates man (Figure 5.1). This idea gave rise to another recommendation called machine learning, which comprises statistical strategies to understand by being specifically programmed; it can also be understood in cases where programming is unknown [5]. Machine learning incorporates supervised learning (observed and directed planning), unsupervised learning (unobserved and in-directed planning), and reinforcement learning (supportive planning). Supervised learning is the determining method; it encompasses regression methods along with classification methods where prognostic models are framed based on data from the input

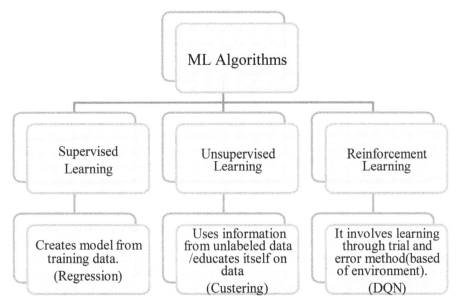

Figure 5.1 Classification of machine learning algorithms.

and the output sources. The classification sub-section output implies illness identification, and the regression sub-section identifies the efficacy of the drug and absorption, distribution, metabolism, excretion, and therapeutics prediction. Unsupervised learning is the un-determined method, which takes into consideration grouping and characterizing approaches based on inputs [6, 7]. As per this concept, the output identification (sub-section to illness) can be identified from the grouped inputs and target identification can be done by the characterized inputs. Reinforcement learning is practiced by the ideology of composing decisions in specified circumstances and enforcing them to augment performance. The output of this is de novo drug design which is a part of making the decision and experimental designs under the enforcement or execution. These all are achievable by modeling and quantum chemistry [8].

Machine learning is further sub-classified into another aspect that utilizes machine-made neural networks which learn from huge experimental data; this is called machine learning. The huge database gives more probability to the chances of discovering a new molecule that can, in turn, be a novel drug for a disease or disorder [9]. As technology is advancing, new methods of data management aid in handling huge data and synchronize the concept of ML. The concept revolves around neural networks and their subtypes like conventional, recurrent, and fully connected feed-forward networks.

This idea will give rise to an era of successful clinical trials with negligible errors and maximum achievable efficiency with the fastest possible speed and economical process [10].

Another important methodology or element of AI is deep learning (DL). DL projects the input data toward the output by utilizing representation learning or feature learning. This conversion process takes place inside a cluster of numerous mathematical processing units also known as neurons. These neurons derive a logical relationship between the input and the output with the help of forming a deep neural network (DNN) [11]. It is evident that the representation of data by deep learning provides better results in better sample generation and better classification modeling and it also enables the automated extraction of depictions from the unsupervised data (shown in Figure 5.2).

In the current scenario, AI has revolutionized almost all the facets of the healthcare industry including drug discovery, diagnosis of complex diseases, patient care and monitoring, assisting experts in decision making, and so on. In the same way, AI is playing a crucial role in oncology too [12]. The way to reduce the mortality rate due to cancer is through early detection and treatment. The utilization of AI complex algorithms can help in accessing the patient's relevant clinical information so that any type of inaccuracy in

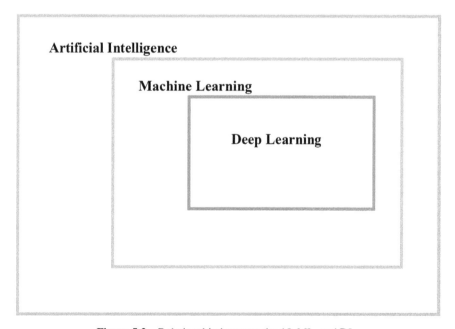

Figure 5.2 Relationship between the AI, ML, and DL.

diagnosis or treatment can be avoided. Neural networks and deep learning also provide genetic analysis and detection data which makes it easier to analyze the treatment outcome. Similarly, the application of AI in radiomics allows easy and accurate diagnosis of complex malignant tumors which otherwise cannot be detected by the human eye [13]. The clinical oncologists obtain the image of tumor sites and, with the help of certain software, outline the tumor for providing the radiotherapy dose; AI helps in marking these spots and margins so that the therapy acts at the site and prevents potential unnecessary side effects. It also allows the experts to identify organs at risk to be more accurate and protect them from side effects. This process is called radiotherapy target delineation or contouring which can be performed precisely with the application of AI technologies [14]. Deep neural network (DNN) can also support the classification of cancer subtypes by using medical images. One of the important elements of cancer detection is the determination of the stage. The stage decides the kind of therapies or treatments to be given to the patients. Gleason score (merger of two scores indicating the presence of tumor at two different locations in the body) is a component that aids determination of stage in the prostate cancer. DNN has proved to give promising results in calculating the Gleason score by using histopathological images of tumors. AI has opened new ways for early diagnosis of cancer via various novel detection techniques like liquid biopsies for circulation tumor DNA (ctDNA). This technique involves minimal invasion in the body (detected by blood samples) and allows tracking of the possible risk of relapse and predicting the appropriate treatment options. Along with early detection as seen in the above examples, the AI also serves other purposes like identification of key mutations by utilizing the histopathological images and detecting the origin of tumors for providing effective chemotherapy to the patients [15].

Cancer research, drug discovery, and development cost excessive money and time, making the cancer treatment expensive; it is very essential to make it affordable and accessible to common people. The involvement of AI plays a key role in making this process more efficient. This is somehow achieved by integrating various sets of data, for example, integration of clinical data and gene expressions, thus inscribing all the components of the drug discovery process. Along with its application in discovery, AI is also applied in drug designing for the generation of new molecules (*in silico*) consisting of specific properties and target affinities, though there are certain problems and difficulties in modeling complex targets and certain specific objectives still help serve the purpose [16]. One of the examples is designing of structural analog of celecoxib and sulfur fewer compounds (Figure 5.3).

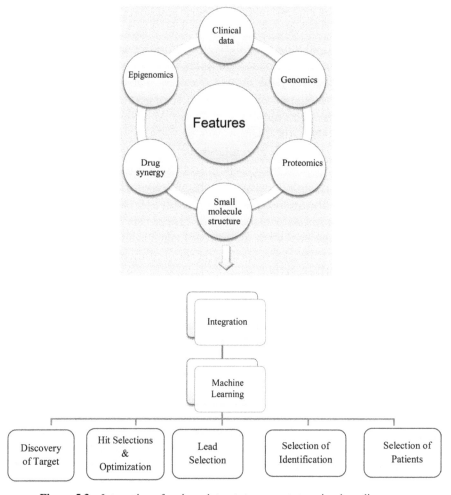

Figure 5.3   Integration of various datasets to support stepwise drug discovery.

5.3 Drug Development Process

The data is available from various substantial sources. The derivation can be from high-degree efficiency compound and fragment screening, computer modeling, and literature sources. These multifactor variables are used to commence the feedback-driven drug development process. The inductive and deductive analysis is a vital part of this process. These are used for the optimization of the identified hits and also lead compounds. The drug development process has several components and their automation leads to a substantial decrease in the unpredictability and probabilities of errors, which, in turn, enhances the efficiency of the process. Design method such

Table 5.1 Implications of AI during multiple stages of cancer therapy.

Phases	Opportunities	Challenges
Discovery	Minimize off-target effect and toxicity and enhance drug exposure.	Identifying optimal targets and properly validating AI-designed drugs.
Development	Optimize drug and dose selection and match patients to therapies and trials.	Improving trial outcome and stratification with the right patient data.
Administration	Sustained dose optimization. Overcoming resistance with game theory.	Moe clinical validation is needed. Use in more cancer types.

as de-novo takes comprehensions of organic chemistry for the synthesis of *in silico* compounds and virtual screening of models [17]. This, in turn, gives a place to efficacy and toxicity analysis of biochemical and biological parameters. These algorithms enable the invention and identification of novel compounds with various anti-disease activities. The fundamental step is to identify the novel compounds that comprise a promising biological activity. Biological activity denotes the interaction of the compound to the organism as a whole or it can even be limited to an enzyme. When a compound exhibits a promising biological activity to a target, it is considered as a "hit." To identify the hits(finding), screening methods include using chemical databases, substances extracted from plants, fungus, bacteria, etc., naturally and computer simulations can be screened to identify the hits. Once the hits are identified, then the subsequent step is the recognition of the lead molecule. Lead can be the new drug for disease treatment. The lead is then optimized based on the chemical structure, and alterations are done to get a compound that gives more efficacy, safety, and therapeutic benefit. The compound is characterized for its safety and efficacy on animal models or cell-based assays [18]. The implications of artificial intelligence during multiple stages of cancer therapy with the challenges are summarized in Table 5.1 [19].

5.3.1 Role of AI in chemotherapy

Artificial Intelligence can be widely applied during chemotherapy, focusing majorly on the patient's response to drugs. There are many milestones of implications of artificial intelligence achieved by researchers in cancer treatment. As of now, researchers claim that AI can be successfully used to optimize and manage overall chemotherapy and drug tolerance. Chen *et al.* used stacked RBMs deep learning method to predict the synergy of drug regimen or drug in combinations carried in by patients during the chemotherapy. Out of many effective ways, the synergistic effect caused by the combination of a drug is one of the most advantageous properties known for the treatment of cancer. However, the desirable synergistic effect was challenged by the prediction of drug combination, which is effective. Drug synergy was predicted using gene expression, pathway, and ontology fingerprints, which are the literature-derived ontological profile of the genes, a method that is novel in the scope of chemotherapy and is based on a deep belief network [20]. Using deep belief networks, we can create a fairly decent framework to capture predictions despite comprehending the underlying mechanisms if we have enough, well-annotated data for training. This enables us to extend the present model to include more types of data, such as mutation, methylation, and proteomics data. Because of its scalability, we can improve the sustainability of our technique by including data collected in the future.

Levine *et al*. applied AI to the electronic health record (EHR) to maintain the well-timed data of the effect of chemotherapy on patient's outcomes. The EHR helps the doctors and clinicians to have insight into the real-world patient experience – good, bad, or some troublesome ADR. The AI was applied to the extracted data from the EHR after it was imported into the IBM cloud [21]. Data were retrieved from the EHR of patients with stage III breast cancer who presented between 2013 and 2015, de-identified, and put further into the IBM Cloud. Medical concepts were extracted from unstructured clinical literature and transformed into structured attributes using specialized natural language processing (NLP) annotators. These annotators were tested on 19 more patients with stage III breast cancer within the same period during the validation phase. For nine critical indications, the generated data was compared to that in the medical chart (gold standard). To examine the patient journey, data from the EHR can be extracted, read, and combined. The degree of correlation between NLP and the gold standard was found to be high and significant, indicating that it was legitimate.

Pantuck *et al*. used an AI platform named CURATE. AI for the optimization of combination chemotherapy thus applies synergism as a curative and effective way for the treatment of cancer. The two drugs ZEN-3694 and enzalutamide were used in combination when the study.CURATE.AI was found to identify significant dose modifications for ZEN-3694 and enzalutamide, improving the efficacy of treatment and tolerance. It additionally indicates that ZEN-3694's involvement in the regimen is responsible for the patient's long-term response. The patient was likely to progress with the combined treatment because of the CURATE.AI's improved safety and effectiveness, leading to a sustained response and no tumor progression based on CURATE. PSA levels were kept under control by AI, and the size of the lesion was reduced. The introduction of technology platforms like CURATE.AI has made it possible to modulate combination therapy doses to improve therapeutic efficacy and maintain patient tolerance. These qualities could help to enhance crucial clinical trial results like objective response rates and overall survival, among other things. CURATE.AI was able to individualize ZEN-3694 and enzalutamide delivery in combination in this study due to dose adjustments in a patient, receiving combination therapy for mCRPC [22].

5.3.2 Role of AI in radiotherapy

In radiotherapy for the treatment of cancer, the role of artificial intelligence is quite specific. Radiotherapy includes mapping out the target regions to cure them via radiations. Artificial intelligence was found to assist radiotherapy, from targeting the affected region to defining the specific radiation for the same [23]. In general, radiotherapy for the treatment of cancer consists of seven different stages: imaging, treatment planning (TP), simulation, radiotherapy accessories, radiation delivery, radiotherapy verification, and patient monitoring [24].

The first step in radiotherapy is imaging, which is the diagnostic stage for the presence of a tumor. The detection of the tumor leads to a collection of information. The process of imaging delivers the gross volume of the tumor, its location size, and information about its vicinity. Due to the implication of AI, presently, there are many models for the imaging of tumors, like positron emission tomography (PET), single-photon emission tomography (SPECT), and computed tomography (CT). The TP and simulation process is intended to obtain the data of patients under recovery which mainly often includes a mass of the tumor, patient's body weight, height, BMI, and pre-exposures to any treatment. All the details are noted and calculated to obtain the best outcomes. As the name itself suggests, treatment planning (TP) includes risk and failure estimation, optimization of treatment planning, beam intensity shaping, etc. [25].

To immobilize the patient under treatment, radiotherapy accessories are used and then the radiations are delivered. The main target of the ionizing radiations is to destroy the tumor cells while safeguarding the healthy cells. The current radiotherapy modalities include stereotactic body radiotherapy (SBRT), proton therapy, electron therapy, etc. After the successful delivery of radiotherapy, the patient is followed for the period of months to years, to make sure his well-being after the exposure of the therapy, i.e., patient is monitored.

Lin *et al.* studied and constructed a deep learning tool for the contouring of primary gross tumor volume in patients with nasopharyngeal carcinoma (NPC). The use of artificial intelligence in the treatment of patients under study made the treatment accurate and precise; this could be seen to have a fruitful impact in controlling and reducing the tumor and in the survival of the patient. Babier *et al.* used deep learning to develop a software that offers time reduction in the course of radiation therapy from days to just a few hours [26, 27].

The switch from traditional AI to deep learning algorithms of modern AI requires as much data as possible. As the novel AI is helping out to produce automated regimens for the cure, data sharing becomes necessary with the guarantee of patient privacy. This need became the root of growing IT infrastructure promoting data sharing [28, 29].

The manual delineation of the targeted area before AI takes 4–5 hours, while automated delineation after the implication of AI in radiotherapy takes 15–20 minutes; thus, the use of AI mainly focuses on targeting cancer-affected areas and the formation of an automatic radiotherapy plan. The AI plays its role effectively without the hustle of manual image extraction, registration, and interpolation. In manual radiotherapy, treatment of some organs requires the doctor to manually change the location after the result is generated. Also as the AI offers treatment planning (TP), it becomes easier for doctors to take a follow-up, and, thus, AI accelerates the overall treatment duration [30].

5.3.3 Role of AI in cancer drug development

The application of AI is not limited to diagnosis, but these technologies are actively applied in anticancer drug development too. One of the most important factors in drug development is determining the interaction between drugs and the cancer cell genome. Many scientists have worked in this direction of utilizing machine learning for identifying accurate interactions. Lind *et al.* amalgamated machine learning technology with the screening data, which resulted in a forest model for forecasting the action of the antitumor drugs as per the mutation state of the cancer cell genome [31]. Similarly, Wang and friends created a machine-learning-based prediction model that was known as an elastic regression model that successfully forecasted the sensitivity of the patients suffering from ovarian cancer, gastric cancer, and endometrial cancer who were treated with tamoxifen, 5FU, and paclitaxel, respectively.

AI can also easily help in assessing how the tumor cells acquire resistance toward cancer drugs by analyzing the large datasets [32, 33].

The development of a drug is a tedious process that requires a lot of time and a huge sum of money. Many of the developed molecules are rejected in clinical trials due to certain toxicity-related problems or other issues. AI in various forms can reduce the intensity of these factors and make the drug development process less tedious and economic. Virtual screening of molecules is a very promising process that involves the identification of the potential molecules from millions of different compounds. The association of machine learning and high throughput screening may easily reduce the cases of false predictions [34]. In technologies, the researchers adopt the most complex and effective algorithms to perform this screening, such as SVM, Bayesian, deep neural network, RF, etc. Xie *et al.* used SVM along with docking-based method, while Meslamani *et al.* described the use of PROFILER for determining which of the ligands have the highest probability for combining with bioactive compounds [35, 36]. Precision medicine is a rapidly evolving strategy for disease management. This allows the experts to create personalized and more accurate treatment plans for the patients by analyzing their genetic profiles, type of tumor, and other medical records. Assessment of such a huge dataset and drug discovery is again supported by AI technologies. These are the various approaches where AI supports the development of cancer drugs, making the process more efficient and economic.

5.3.4 Role of AI in immunotherapy

Immunotherapy is one of the most critical therapies of all the treatments adopted for the treatment of cancer. It involves curing cancer by acting patient's immune system or defense system by utilizing the substances either made by the body or in the laboratory. This therapy is proved to be effective in treating different types of cancer; yet, there are certain limitations of this therapy including high cost and frequent adverse effects (autoimmune disorders) in patients. Application of the AI can make this therapy more efficient by elevating diagnosis accuracy, reducing human resource costs, predicting the outcome of the treatment with the aid of medical imaging, immune signatures, and histological analysis. It is evident that AI enhances the success ratio of immunotherapy by forecasting the outcome of therapy in patients with the help of immune predictive scores like immunophenoscore and immunoscores. The identification of major histocompatibility complex (MHC) by AI technologies renders 99.66% accuracy in recognition of patterns related to immune response. So, in a nutshell, it is evident that the combination of

AI algorithms and clinicians' interpretations may lead to better results for patients [37, 38].

The advancements in immunotherapy are often subject to the identification of targets that are connected to the development of resistance against tumors or tumor-causing factors; hence, AI is of great use in broadening the applicability of immunotherapy in treating cancer. A large number of samples and assays are required for effectively analyzing the interaction of tumor cells and immune cells along with patient's response to this interaction; these assays, in return, generate a huge amount of datasets that are quite difficult to scrutinize manually. AI supports easy and quick investigation of large datasets. The presentation of peptides that binds with the MHC is important for the development of cancer vaccine; thus, machine learning has been implemented in recognition of neoantigens presented by the solid tumors [39]. The recognition of neoantigens properly requires screening of a large number of synthetic peptides and difficult to acquire clinical specimens or human leukocyte antigen (HLA). Sullivan *et al.* evolved an AI method involving deep learning that uses tumor HLA peptide mass spectrometry datasets for enhancing neoantigen recognition [40].

5.4 Monoclonal Antibodies (mAbs) used in Cancer Treatment

Monoclonal antibodies in the treatment of cancer have been established as a milestone around various pre-existing therapeutic strategies. Comes under immunotherapy monoclonal antibodies are now considered as one of the most effective elements for cancer treatment [41].

Monoclonal antibodies are featured to have a specific and common antigen-binding site in all the antibodies produced homogeneously from a single cell line. Thus, all the antibodies produced are identical in their protein sequence and have the same affinity and biological interactions [42].

Antibodies are potent enough to elicit the later immune response, by first recognizing the foreign antigen and then neutralizing them. Structurally, antibodies are glycoproteins and belong to Ig (immunoglobin) superfamily.

In the structure of an antibody, the fragment-antigen binding (Fab) region, as its name suggests, is for the identification of the specific antigen. Another region that is located downside the Y structure of the antibody, namely the fragment crystallizable (Fc) region, is responsible for the interaction between antibody and other elements of the immune system. These Fc regions are identified by Fc receptors (FcRs) present on the immune cells. Based on the heavy chain, there are five types of antibodies, namely IgM,

IgG, IgA, IgD, and IgE. Among them, the IgG is the most common antibody that is used in immunotherapy and antibody therapy [43, 44].

5.5 MOA of mAbs

The monoclonal antibodies can effectively cause cancerous cells death by various known mechanisms. The very first and head-on mechanism known is the blocking of growth factor receptor (GFR) signaling. When mAbs binds to the target GFR while controlling their activation and ligand binding state, then, eventually, the growth of the tumor is unsettled. One very fine example of a mAb drug that follows this MOA is Cetuximab, which is an anti-epidermal growth factor receptor (anti-EGFR) monoclonal antibody. The overexpression of EGFR in cancer cells eventually leads to tumor cells multiplication and migration. Cetuximab initiates apoptosis in tumor cells by blocking the ligand-binding site and dimerization of the growth factor [45].

Another mechanism involving the growth factor follows internalization, a type of endocytosis because the growth factor has no ligand, and, thus, they follow heterodimerization for their activation. One such growth factor is the human epidermal growth factor receptor 2 (HER2) is a tyrosine kinase receptor that is overexpressed in breast and ovarian cancerous cells. Monoclonal antibodies treat such cancerous cells by inhibiting heterodimerization and internalization. The first FDA-approved mAb was Trastuzumab, an anti-HER2 that remains an effective treatment for breast and ovarian cancer.

Apart from the two mentioned direct mechanisms, various indirect mechanisms also exist, which need the element of the host immune system to function, namely complement-dependent cytotoxicity (CDC), antibody-dependent cellular phagocytosis (ADCP), and antibody-dependent cell-mediated cytotoxicity (ADCC). Table 5.2 summarizes FDA-approved mAbs [46].

Even though monoclonal antibody treatment has had some remarkable clinical achievements, therapeutic resistance still poses a major barrier. Additional studies should concentrate on examining the mode of action of mAbs to find novel ways to improve clinical efficacy.

5.6 Future Prospects

Currently, the applications of AI in oncology are vast spread. Though the major challenges and questions in oncology including analysis of a large amount of data, lack of early diagnostic techniques, difficulty in implementing patient treatment plans, drug development, etc., are efficiently combated and answered by AI technologies to a large extent, there are still many more

Table 5.2 Monoclonal antibodies approved by FDA for cancer treatment.

Name	Antigen
Atezolizumab	PD-L1
Avelumab	PD-L1
Bevacizumab	VEGF
Cemiplimab	PD-1
Cetuximab	EGFR
Daratumumab	CD38
Dinutuximab	GD2
Durvalumab	PD-L1
Elotuzumab	SLAMF7
Ipilimumab	CTLA-4
Isatuximab	CD38
Mogamulizumab	CCR4
Necitumumab	EGFR
Nivolumab	PD-1
Obinutuzumab	CD20
Ofatumumab	CD20
Olaratumab	PDGFRα
Panitumumab	EGFR
Pembrolizumab	PD-1
Pertuzumab	HER2
Ramucirumab	VEGFR2
Rituximab	CD20
Trastuzumab	HER2
Gemtuzumabozogamicin	CD33
Brentuximab vedotin	CD30
Trastuzumab emtansine	HER2
Inotuzumabozogamicin	CD22
Polatuzumabvedotin	CD79B
Enfortumabvedotin	Nectin-4
Trastuzumab deruxtecan	HER2

obstacles at the ground implementation level that are needed to be managed yet. The certain areas that require to be addressed to avoid pitfalls include the building of cancer AI research communities, access to quality cancer data, black box problem (lack of rationale in predictions made by machines), etc. Also to get complete benefits of AI, it is very important to fill the knowledge gaps. Today, the clinicians are least informed about data science and technology, and in a similar way, the tech experts are least informed about the field of oncology; bridging this gap will lead to utilization of AI to its maximum potential. It is evident from many studies that in the coming years, AI will be incorporated in the clinical decision making, care, and diagnosis of cancer patients in a much more advanced manner. Cancer diagnostics is a traditional starting point for developing effective therapeutic methods and management of diseases; its AI-based refinement is a significant success. Moreover, future AI innovations should take into account undiscovered but critical boundaries in this scenario, such as medication discovery, therapy administration, and follow-up tactics. Indeed, the expansion of AI, as per our viewpoint, needs to follow through and integrative patterns to determine a substantial improvement in the diagnosis treatment of cancer patients. This is among the most significant benefits of AI, as it will allow for the proper interaction and amalgamation of domains related to cancer on a single patient, enabling the difficult goals of personalized therapy. The ability to combine various and composite data produced from multi-omics techniques to oncologic patients is among the most promising AI expectations. AI's potential tools may be the only ones capable of handling large amounts of data from many sorts of analysis, such as information collected from DNA and RNA fingerprinting. In this vein, the recent publication of the American College of Medical Genetics' criteria and recommendations for the interpretation of sequence variants has sparked a new generation of AI development, with new possibilities in precision oncology.

At present, work is needed to ensure the consistent implementation of AI in medical institutions and hospitals. Many experts believe that AI technologies hold the massive potential to take oncology, cancer research, and patient care to another level, and combating challenges by continuous research and study will boost this potential further.

5.7 Conclusion

AI has certainly made some significant contributions as far as cancer research and drug development are concerned. We cannot deny the fact that the human brain is restricted in many ways, making it difficult to discover and formulate the most appropriate treatment along with the identification of minute details. This may deprive the patients of getting the best possible treatment and care. AI plays the key role here; it provides the experts and clinicians with a perception that otherwise would be very difficult to obtain. AI technologies have enabled us to make cancer research and drug development more efficient and economic. It has also contributed to speeding up the cancer drug discovery process, and precision drug discovery making patient care more effective. However, there are still many challenges needed to be addressed for more unprecedented advancements in the field. More research and studies are required to take complete advantage of these AI technologies, but it is very certain that the integration of AI will be the driving force for future advancements in cancer research and will bring about promising changes in the existing technologies. Talking about integration, the most significant challenges for completing the "AI-revolution" in oncology are the creation of integrative and interdisciplinary research formative beliefs, the prompt understanding of the relevance of all malignancies, including rare tumors, and the continuous support for ensuring its growth. As discussed in this chapter, AI is having an increasing impact on every domain of oncology. The initial steps in establishing new development strategies with practical implications are to understand AI's historical background and current successes. AI is currently being used in oncologic clinical practice, but continued and increased efforts are required to allow AI to reach its full potential.

5.8 Acknowledgment

I would like to thank my co-authors for contributing their knowledge and time and giving their support in compiling the work.

5.9 Funding

None.

5.10 Conflicts of Interest

The authors declare no conflict of interest.

References

[1] McCarthy, John. "What is artificial intelligence?." (1998).

[2] February 15, 2019 Benjamin H. Kann, MD, Reid Thompson, MD, PhD, Charles R. Thomas, Jr, MD, Adam Dicker, MD, PhD, Oncology, Oncology Vol 33 No 2, Volume 33, Issue 2.

[3] El Naqa I., Murphy M.J. (2015) What Is Machine Learning?. In: El Naqa I., Li R., Murphy M. (eds) Machine Learning in Radiation Oncology.

[4] Bonetto, R., &Latzko, V. (2021). *Machine learning. Computing in Communication Networks,* 135–167.

[5] Cunningham P., Cord M., Delany S.J. (2008) Supervised Learning. In: Cord M., Cunningham P. (eds) Machine Learning Techniques for Multimedia. Cognitive Technologies.

[6] Sutton R.S. (1992) Introduction: The Challenge of Reinforcement Learning. In: Sutton R.S. (eds) Reinforcement Learning. The Springer

International Series in Engineering and Computer Science (Knowledge Representation, Learning and Expert Systems), vol 173.

[7] Cohen, S. (2021). *The basics of machine learning: strategies and techniques. Artificial Intelligence and Deep Learning in Pathology,* 13–40.

[8] VoPham, T. et al. (2018) Emerging trends in geospatial artificial intelligence (geoAI): potential applications for environmental epidemiology. Environ. Heal. 17, 40

[9] *Chen, H. et al. (2018) The rise of deep learning in drug discovery. Drug Discov. Today 23,* 1241–1250

[10] Jiang, F. et al. (2017) Artificial intelligence in healthcare: past, present and future. Stroke Vasc. Neurol. 2, 230–243

[11] Georgevici, A.I., Terblanche, M. Neural networks and deep learning: a brief introduction. *Intensive Care Med* 45, 712–714 (2019).

[12] Najafabadi, M.M., Villanustre, F., Khoshgoftaar, T.M. *et al.* Deep learning applications and challenges in big data analytics. *Journal of Big Data* 2, 1 (2015).

[13] Zodwa Dlamini, Flavia Zita Francies, Rodney Hull, RahabaMarima,Artificial intelligence (AI) and big data in cancer and precision oncology,Computational and Structural Biotechnology Journal,Volume 18, 2020, Pages 2300–2311.

[14] Boon, Ian S et al. "Assessing the Role of Artificial Intelligence (AI) in Clinical Oncology: Utility of Machine Learning in Radiotherapy Target Volume Delineation." *Medicines (Basel, Switzerland)* vol. 5,4 131. 11 Dec. 2018, doi:10.3390/medicines5040131

[15] Bhinder B, Gilvary C, Madhukar NS, Elemento O. Artificial Intelligence in Cancer Research and Precision Medicine. Cancer Discov. 2021 Apr;11(4):900–915.

[16] Ho, D. (2020). Artificial intelligence in cancer therapy. *Science, 367*(6481), 982–983.

[17] Yuan, Y. et al. (2011) LigBuilder 2: a practical de novo drug design approach. J. Chem. Inf. Model. 51, 1083–1091.

[18] Zhu, T. et al. (2013) Hit identification and optimization in virtual screening: practical recommendations based on a critical literature analysis. J. Med. Chem. 56, 6560–6572

[19] Chen, G., Tsoi, A., Xu, H., & Zheng, W. J. (2018). Predict effective drug combination by deep belief network and ontology fingerprints. *Journal of biomedical informatics*, *85*, 149–154.

[20] Pantuck, A. J., Lee, D. K., Kee, T., Wang, P., Lakhotia, S., Silverman, M. H., ... & Ho, D. (2018). Modulating BET Bromodomain inhibitor ZEN-3694 and enzalutamide combination dosing in a metastatic prostate

cancer patient using CURATE. AI, an artificial intelligence platform. *Advanced Therapeutics*, *1*(6), 1800104.

[21] Levine, M. N., Alexander, G., Sathiyapalan, A., Agrawal, A., & Pond, G. (2019). Learning health system for breast cancer: pilot project experience. *JCO clinical cancer informatics*, *3*, 1–11.

[22] Hussein, M., Heijmen, B. J., Verellen, D., & Nisbet, A. (2018). Automation in intensity modulated radiotherapy treatment planning—a review of recent innovations. *The British journal of radiology*, *91*(1092), 20180270.

[23] Khoo VS. Radiotherapeutic Techniques for Prostate Cancer, Dose Escalation and Brachytherapy. Clin Oncol. 2005;17(7):560–571. doi:10.1016/J.CLON.2005.07.006

[24] Deshmukh P, Levy MS. Effective radiation dose in coronary imaging modalities: Back to Basics. Catheter Cardiovasc Interv. 2015;85(7):1182–1183. doi:10.1002/ccd.26013

[25] Siddique, Sarkar; Chow, James C.L. (2020). *Artificial intelligence in radiotherapy. Reports of Practical Oncology & Radiotherapy, (), S1507136720300444– .doi:10.1016/j.rpor.2020.03.015.*

[26] Lin, Li; Dou, Qi; Jin, Yue-Ming; Zhou, Guan-Qun; Tang, Yi-Qiang; Chen, Wei-Lin; Su, Bao-An; Liu, Feng; Tao, Chang-Juan; Jiang, Ning; Li, Jun-Yun; Tang, Ling-Long; Xie, Chuan-Miao; Huang, Shao-Min; Ma, Jun; Heng, Pheng-Ann; Wee, Joseph T. S.; Chua, Melvin L. K.; Chen, Hao; Sun, Ying (2019). *Deep Learning for Automated Contouring of Primary Tumor Volumes by MRI for Nasopharyngeal Carcinoma. Radiology, (), 182012*

[27] Babier, A., Boutilier, J. J., McNiven, A. L., & Chan, T. C. (2018). Knowledge-based automated planning for oropharyngeal cancer. *Medical physics*, *45*(7), 2875–2883.

[28] Lambin, P., Van Stiphout, R. G., Starmans, M. H., Rios-Velazquez, E., Nalbantov, G., Aerts, H. J., ... & Dekker, A. (2013). Predicting outcomes in radiation oncology— multifactorial decision support systems. *Nature reviews Clinical oncology*, *10*(1), 27–40

[29] Liang, G., Fan, W., Luo, H., & Zhu, X. (2020). The emerging roles of artificial intelligence in cancer drug development and precision therapy. *Biomedicine & Pharmacotherapy*, *128*, 110255.

[30] Lind AP, Anderson PC (2019) Predicting drug activity against cancer cells by random forest models based on minimal genomic information and chemical properties. PLoS ONE 14(7): e0219774.

[31] Wang, Y., Wang, Z., Xu, J. *et al.* Systematic identification of non-coding pharmacogenomic landscape in cancer. *Nat Commun* 9, 3192 (2018).

[32] Yanagisawa K, Toratani M, Asai A, Konno M, Niioka H, Mizushima T, Satoh T, Miyake J, Ogawa K, Vecchione A, Doki Y, Eguchi H, Ishii H. Convolutional Neural Network Can Recognize Drug Resistance of Single Cancer Cells. Int J Mol Sci. 2020 Apr 30;21(9):3166.

[33] Nagasundaram Nagarajan, Edward K. Y. Yapp, Nguyen Quoc Khanh Le, Balu Kamaraj, Abeer Mohammed Al-Subaie, Hui-Yuan Yeh, "Application of Computational Biology and Artificial Intelligence Technologies in Cancer Precision Drug Discovery", *BioMed Research International*, vol. 2019, Article ID 8427042, 15 pages, 2019.

[34] Q.-Q. Xie, L. Zhong, Y.-L. Pan et al., "Combined SVM-based and docking-based virtual screening for retrieving novel inhibitors of c-met," *European Journal of Medicinal Chemistry*, vol. 46, no. 9, pp. 3675–3680, 2011.

[35] J. Meslamani, R. Bhajun, F. Martz, and D. Rognan, "Computational profiling of bioactive compounds using a target-dependent composite workflow," *Journal of Chemical Information and Modeling*, vol. 53, no. 9, pp. 2322–2333, 2013.

[36] Zhijie Xu, Xiang Wang, Shuangshuang Zeng, Xinxin Ren, Yuanliang Yan, ZhichengGong,Applying artificial intelligence for cancer immunotherapy,Acta Pharmaceutica Sinica B,2021.

[37] Esfahani, K et al. "A review of cancer immunotherapy: from the past, to the present, to the future." *Current oncology (Toronto, Ont.)* vol. 27,Suppl 2 (2020): S87–S97.

[38] Zhou, X., Qu, M., Tebon, P., Jiang, X., Wang, C., Xue, Y., Khademhosseini, A. (2020). *Screening Cancer Immunotherapy: When Engineering Approaches Meet Artificial Intelligence.*

[39] Bulik-Sullivan, B., Busby, J., Palmer, C. *et al.* Deep learning using tumor HLA peptide mass spectrometry datasets improves neoantigen identification. *Nat Biotechnol* 37, 55–63 (2019).

[40] Zahavi, D., & Weiner, L. (2020). Monoclonal antibodies in cancer therapy. *Antibodies*, 9(3), 34.

[41] Murphy, K., & Weaver, C. (2016). *Janeway's immunobiology*. Garland science.

[42] Weiner, L. M., Surana, R., & Wang, S. (2010). Antibodies and cancer therapy: versatile platforms for cancer immunotherapy. *Nature reviews. Immunology*, 10(5), 317.

[43] Li, S., Schmitz, K. R., Jeffrey, P. D., Wiltzius, J. J., Kussie, P., & Ferguson, K. M. (2005). Structural basis for inhibition of the epidermal growth factor receptor by cetuximab. *Cancer cell*, 7(4), 301–311.

[44] Patel, D., Bassi, R., Hooper, A., Prewett, M., Hicklin, D. J., & Kang, X. (2009). Anti- epidermal growth factor receptor monoclonal antibody cetuximab inhibits EGFR/HER-2 heterodimerization and activation. *International journal of oncology*, *34*(1), 25–32.

[45] Slamon, D. J., Godolphin, W., Jones, L. A., Holt, J. A., Wong, S. G., Keith, D. E., ... & Press, M. F. (1989). Studies of the HER-2/neu proto-oncogene in human breast and ovarian cancer. *science*, *244*(4905), 707–712.

[46] Chen, J. S., Lan, K., & Hung, M. C. (2003). Strategies to target HER2/ neu overexpression for cancer therapy. *Drug resistance updates*, *6*(3), 129–136.

6

Artificial-Intelligence-Based Cloud Computing Techniques for Patient Data Management

Akanksha Sharma[1*], Ashish Verma[2], Rishabha Malviya[3], and Mahendran Sekar[4]

[1]Monad College of Pharmacy, Monad University, India
[2]School of Pharmacy, Monad University, India
[3]Department of Pharmacy, School of Medical and Allied Sciences, Galgotias University, India
[4]Department of Pharmaceutical Chemistry, Faculty of Pharmacy and Health Sciences, Royal College of Medicine Perak, Universiti Kuala Lumpur, Malaysia
[*]**Corresponding Author**
Akanksha Sharma
Monad College of Pharmacy, Monad University, India
Email: akankshasona012@gmail.com

Abstract

This chapter describes the role of artificial-intelligence-based cloud computing techniques in inpatient data management. The practice of installing a remote server accessed by the internet to manage, store, and process healthcare data is called cloud computing in healthcare. In contrast, setting up an onsite data center or data hosting on a personal computer with servers are both options. The manuscript describes the role of cloud computing research in health management like telemedicine or teleconsultation, patients' self-management and public health, management of a hospital, therapy, and secondary utilization of data. Cloud computing is a low-cost alternative that helps in storing vast volumes of data; it is also accessible via telehealth which increases the patient experience. However, it has some challenges of security which will get resolved in future and provide more efficient healthcare services.

6.1 Introduction of Artificial Intelligence Based Cloud Computing Techniques

"Cloud" is frequently utilized as a symbol for the internet (which is frequently shown in ICT courses as cloud illustrations). Some individuals attribute the name to Google's CEO, Eric Schmidt, who is reported to have coined the term "cloud computing" in a 2006 conference. Financial pressures, management of many stakeholders for service delivery, and aging populations are all the issues that are faced by the healthcare business. Surged utilization of information and communications technology (ICT) can aid the health industry in solving these difficulties. ICT advancements, along with the need to make healthcare more efficient, have increased health ICT applications. ICT has been used to help healthcare professionals; it has greater access to patient records and also helps to make better decisions. ICT has much more potential to assist the healthcare sector in lowering costs and improving service outcomes [1, 2].

Cloud computing increases the availability of IT services at all times and from any location. It is not a novel technique but, rather, a novel method of providing computing resources. Google Docs and Microsoft Office 365 are instances of cloud-based nonmedical platforms, whereas Google Health and Microsoft HealthVault are examples of medical-related applications. In comparison to traditional computing, the cloud computing approach has three key advantages, i.e., on-demand access to powerful computing resources, supply of services without requiring clients to commit upfront, and availability for short time utilization. Several industries have been affected by the cloud model, and in upcomig few years, it is expected that about 80% of existing industries would have adopted cloud computing. Furthermore, firms that lack the resources and infrastructure to set up on-premises apps might use cloud computing [2].

Health information technology (HIT) and related computer-based information or data can increase the quality and productivity of health-related facilities and are, thus, believed as an important part of the health-care industry's achievement. Traditional IT techniques for health, in which health organizations build or buy hardware infrastructures and in-house software applications are frequently inadequate to meet the ever-changing and growing needs in the health sector. Healthcare institutions, especially those in rural locations, sometimes face a shortage of IT sources like computation and capacity of storage [3, 4].

To stay cost-efficient, effective, and timely while providing high-quality facilities, healthcare requires continual and systematic innovation. Cloud computing, according to many managers and experts, will enhance healthcare facilities, promote research in healthcare, and reshape the information technology (IT) sector. Cloud computing was thought to have lower initial costs for electronic health records (EHRs), like hardware, networking, software, staff, license fees, and, hence, boost adoption. The biomedical infromatics community is a group which exchange data and applications which benefit in new computing paradigms like cloud computing [5].

Both healthcare organizations and cloud service providers must take proper precautions to ensure the patient data's secure handling to safeguard the security and privacy of healthcare data. The majority of the problems occur due to the storage of personal information and medical data on cloud servers, a virtual environment from where data can be readily drudged. As a result, before introducing any cloud-based healthcare facilities, strict security procedures and guarantees must be implemented. Government regulations and rules must be followed to confirm that facility providers of the cloud respect the law and take all important precautions to preserve the security and privacy of patient information. Whether or not cloud-based systems were used, the Health Insurance Portability and Accountability Act (HIPAA) controls and governs the security and privacy of patient information. Protected health information (PHI) is a privacy provision established by HIPAA that assures that patient information cannot be utilized without an order of a court or the patient's agreement and approval [6].

6.2 Cloud Computing: A New Economic Computing Model

Traditional health IT approaches can be enhanced by cloud computing's (CC) unique IT service philosophy. Cloud computing has three paradigmatic models in terms of services: software, platform, and infrastructure. As a result, CC can provide resources of IT (via IaaS), IT programs with programming devices, libraries, and languages for software deployment or development (via PaaS), or software applications which are ready to use and run on the infrastructure of cloud (via SaaS) to healthcare industries [3, 4]. Figure 6.1 shows the cloud computing paradigmatic models that help to provide the services.

6.2.1 Infrastructure as a service (IaaS)

Infrastructure as a service is a model of computing in which customers rent storage, processing, networks, and additional computer resources over that they can deploy and execute software such as applications and operating systems. The equipment is owned by the supplier, who is also responsible for

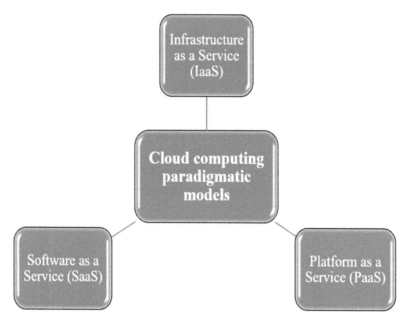

Figure 6.1   A schematic diagram that shows the cloud computing paradigmatic models that help to provide the services.

operating, housing, and repairing. Typically, the customer pays on a basis of per use.

6.2.2 Platform as a service (PaaS)

It is a service model in that a service provider facilitates access to a cloud-based system to customers, in which they can grow and deploy applications while the provider manages the underlying infrastructure. The development of tools (for example, operating systems) is organized by a cloud that is accessed via using a browser. Developers may use PaaS to create web applications without implementing any devices on computers and then use them without needing any particular administration abilities.

6.2.3 Software as a service (SaaS)

Consumers use provider applications through a cloud application via the program interface or a customer interface such as a web browser in software as a service (SaaS). A cloud facility provider hosts the programs (for example, EHRs) and makes them available to consumers over a network, usually the

Table 6.1 List of main cloud computing service providers [8].

S. No.	Type of Service	Providers	Example
a.	(IaaS) Infrastructure as a service	GoGrid	Cloud Hosting
		Savvis	Colocation Hosting
		Amazon	Amazon Services
b.	(PaaS) Platform as a service	Salesforce	Force.com
		Microsoft	Azure
		Google	Google Apps
c.	(SaaS) Software as a service	Apple	iCloud
		Netsuite	NetSuite CRM+
		Salesforce.com	Sale Cloud
		Google	Google Docs

internet [7]. Table 6.1 summarizes the list of major cloud computing service providers.

6.3 The US National Institute of Standards and Technology (NIST) has Identified four Models for Cloud Computing Deployment

6.3.1 Public cloud

On a pay-as-you-go basis, a cloud service provider prepares resources (storage and application) that get available to the common public using the internet. Users can borrow a virtual computer from Amazon Elastic Compute Cloud (EC2) to run their apps. EC2 is a cloud computing service that runs on Amazon's network infrastructure and data centers, allowing users to spend only for the services they utilize, with no minimum cost.

6.3.2 Community cloud

Various firms utilize cloud infrastructure and they all have similar challenges (for example, mission, requirements of security, compliance, and policy). The Google GovCloud facilitates a secure data environment to the Los Angeles City Council which helps in the storage of information and application which are only manageable by city interventions.

6.3.3 Private cloud

A cloud infrastructure works only for the benefit of a specific company. The proprietary network or information center provides facilities for a particular people's group. With the help of Microsoft Azure or Dynamic Data Center Toolkit, clients can make the base of private cloud infrastructure by utilizing the Windows Server.

6.3.4 Hybrid cloud

There are two or more clouds in the cloud infrastructure (public, community, and private). In this infrastructure, a business manages and delivers few resources in its data center while outsourcing another. For instance, IBM and Juniper Networks cooperate to supply a hybrid cloud infrastructure to associations that allow them to expand their private clouds toward the remote servers in a safe public cloud [8].

6.4 Cloud Computing from the Perspective of Management, Security, Technology, and Legality

6.4.1 Management aspect

The huge increase in average life expectancy has resulted in the rapid aging of the population. As a result, there was an increasing requirement for more resources and a wider range of medical services. To help healthcare professionals, it is necessary to find more productive ways to handle this worldwide crisis, which needs innovative and cost-effective techniques. Cloud computing can provide practical answers, as demonstrated by IBM and Active Health Management in the novel clinical data management system, "Collaborative Care Solution," which was released in November 2010. Cloud-computing-based systems have the goal to allow healthcare and medical professionals to quickly access healthcare information from a different source such as electronic health records (EHRs). Patients with chronic diseases are benefited by connecting with their doctors and following up on their recommended drugs. With the expanding volume of information and data of patients available on personal and electronic health records, data management became more efficient. This may be seen from the standpoint of information storage and various servers are required to handle such massive volumes of data.

Furthermore, EHR systems include built-in security features that assure secure environment availability for patient data management. Patients are involved in the handling of their health information and data process in a secure manner. The addition of proper authentication mechanisms to the role-based system helps to provide a more secure environment for storing and managing patient data. There are security considerations to be taken when transferring healthcare information to a cloud system, including data encryption. A public key infrastructure (PKI) is utilized to manage and maintain public-key encryption, which is an expensive technology [6, 9, 10].

6.4.2 Technology aspect

Cloud computing is a modern computational technique that may be able to assist in the resolution of various issues. Many storages, computation, hardware, networking, and software resources are available as facilities (without the need of considerable configuration) and on-demand with cloud computing (when required). Microsoft Azure, AWS (Amazon Web Services), and Google Cloud Platform (GCP) are a few of the commercially available cloud platforms available today [11].

Because it is simple to utilize, has network resources on-demand access, takes minimum administrative effort, and is of lower price, cloud computing is gaining a lot of traction among IT firms, academia, and individual users. This novel cloud computing technique has taken off in the industry, with an enhancement in the number of businesses adopting it. It captivates cloud consumers by offering low-cost services, a pay-for-use policy, distributed nature, and a quick supply of computing resources, as well as storage of data center with endless space and high computing capability for managing and storing the data. Despite potential benefits of cloud computing, organizations are less interested in implementing it because of some drawbacks like account hijacking, data cleaning, low control over the process, data loss, cloud service providers (CSPs) insider attacks, lack of migration or portability from one service provider to other, lack of legal aspects, less reliability, lack of suitability, and lower quality of service (QoS). There are various obstacles in cloud computing adoption like virtualization, resource scheduling, interoperability, load balancing, multi-tenancy, and security [12].

6.4.3 Security aspect

There are various security difficulties and problems in cloud computing since it covers multiple techniques like databases, networks, operating systems, resource scheduling, virtualization, management of transactions, memory management, and concurrent control. This is critical since the cloud service provider must confirm that consumers do not face catastrophic issues such as data loss or theft, which could result in significant losses depending on the sensitivity of the data kept in the cloud. A malevolent client may impersonate normal users to infect the cloud. Data theft is a relatively typical problem that cloud service companies face nowadays. Furthermore, due to the cost flexibility and effectiveness, few cloud service providers do not even offer their servers. There are also occurrences of loss of data that can be a severe issue for customers.

For instance, the server may be unexpectedly closed down, resulting in a loss of data for customers. Moreover, data may be damaged or corrupted as an outcome of a natural disaster. Furthermore, a major securities problem in cloud applications is the physical availability of data. Cloud computing security should be addressed on both the supplier and user sides. Users should not tamper with the data of other users; so the cloud service provider should facilitate a good layer of security for them. Cloud computing is a better way to cut costs and offer additional storage only if both the provider and user take care of security.

According to the report, regulatory reform is necessary to safeguard the important data in the cloud because a major difficult aspect of cloud computing is ensuring the consumer's trust toward the privacy and security of their data. For maintaining data security available in the cloud application, the design of the cloud application environment is critical. The consumers must comprehend the notion of the cloud service provider's data storage restrictions. Some of the best choices include cloud service providers that provide security solutions that comply with requirements such as PCI DSS, HIPAA, and EU rules of data protection [13].

6.4.4 Legal aspect

Privacy and data protection are important factors for making the trust of the customer which is required by a cloud computing system to attain its proper potential. Customers would be capable to estimate the problems they face if providers do not adopt better and clearer policies and practices. The various main providers, fortunately, have made promises to adopt the best policies and procedures to safeguard consumers' privacy and data. In addition to service providers' promises toward this protection, certain groups like Cloud Security Alliance have published a detailed roadmap to address privacy and security concerns. The Trusted Computing Group, a non-profit organization, has proposed a set of software and hardware solutions for building trusted platforms. Governments have an important role in promoting generally agreed-on laws for both users and suppliers [14].

Physical storage on the cloud could be dispersed across numerous jurisdictions, each with its security and intellectual property, own set of rules governing data privacy, and usage. For instance, the US Health Insurance Portability and Accountability Act (HIPAA) prohibits organizations by revealing personal health information to unaffiliated third parties, and the Providing Appropriate Tools Required to Intercept and Obstruct Terrorism Act (PATRIOT) provides the authority to the government of US to demand information if conditions are declared a necessary or emergency to the security of homeland [15].

Likewise, the Personal Information Protection and Electronic Documents Act (PIPEDA) in Canada restricts an organization's ability to acquire, utilize, or disclose personal information in the course of business. However, a provider may shift customers' data from one jurisdiction to another without notifying the user [16].

6.5 Cloud Computing Strategic Planning

When a healthcare company decides to go to the cloud applications, it requires a model for strategic planning to assure perfect goal for the work, tangible steps are taken to reach the goal, and all positive and bad aspects of the effort are identified and handled. There is a model for cloud computing strategic planning that a health industry can use to define its strategy, resource, and direction allocation for transitioning traditional information technology (IT) infrastructure to clouds. Identification, action, evaluation, and follow-up are four stages of the model [17].

6.5.1 Stage I – Identification

The first stage of the approach is to assess the present state of the association's service process and determine the primary goal of quality improvement (QI) by listening to the voice of the patients or the voice of the customer (VOC). The root cause analysis (RCA) technique is used to investigate existing service process issues. To serve patients more effectively and efficiently, the identification objective and its scope must be cleared. The strategic planning team must develop and explain the quality indicators of healthcare services as well as their objective and application. More importantly, the description of the performance indicators and the procedures for evaluating them must be agreed upon and approved by all parties concerned. This model gives the strategic planning to team a clear picture of the challenge through which they are dealing [17, 18].

6.5.2 Stage II – Evaluation

The model's second stage is related to the benefits of assessment and drawbacks of cloud-based computing adoption. ENISA, the Cloud Security

Alliance, and the National Institute of Standards and Technology have created thorough directions to assess the advantages and hazards of cloud-based computing adoption. To assess the feasibility of the cloud-based method, a capable user can do a "strengths, weaknesses, opportunities, and threats" (SWOT) analysis.

Cloud Security Alliance identifies 12 cloud computing security domains. Governance and operations are the two basic groups that make up the domains. For each domain, there are also proposals for solutions. Many significant cloud security and privacy challenges, as well as the associated preventative guidelines for organizations to take while developing or launching a public cloud service outsourcing arrangement, are listed in NIST Guidelines on Privacy and Security in Public Based Cloud Computing [19].

6.5.3 Stage III – Action

The organization will be able to decide whether or not to embrace the new computing paradigm after evaluating it. If the response is positive, an implementation plan must be developed. There is the four-step plan as follows.

6.5.3.1 Step 1: Determination of cloud service and deployment model

Cloud-based computing can relate with a variety of services (PaaS, IaaS, and SaaS) as well as deployment models (public, private, hybrid, and community cloud). Every model of service or deployment has its own set of advantages and drawbacks. As a result, when contracting for various types of deployment or services patterns, the primary factors should differ [20].

6.5.3.2 Step 2: Obtain confirmation from a chosen cloud provider

The company needs proof that the chosen provider will deliver a high-quality facility while adhering to solid privacy, security, and legal policies and legislation. Pay-per-use, on-demand access, quick flexibility, on-time technical support, and operational openness are among the quality-of-service guarantees. Data integrity, availability, confidentiality, authenticity, nonrepudiation, and authorization all are covered by the privacy and security assurances. In addition, the provider should ensure that data including any backups is held exclusively in places allowed by contract, regulation, and service level agreement [21].

6.5.3.3 Step 3: Take consideration in migration of future data

The association may have to migrate information and services to other providers or return to an in-house IT atmosphere because the provider stops the service or business operations (like the latest suspension of Google Health), has an undesirable reduction in quality of service, or has a contract dispute. Data portability should be factored into the strategy from the start [22].

6.5.3.4 Step 4: Start of implementation of pilot

Several previous strategic planning approaches recommend that a company with no prior cloud expertise begin with a trial project. The pilot should be sufficient to demonstrate the advantages of cloud computing to the enterprise [23].

6.5.4 Stage IV – Follow-up

The final step is to set up cloud-based computing infrastructure and create a follow-up strategy. The strategy specifies how and when service developments will be measured. To assess the size of the improvement, reasonable targets are defined ahead of time and outcomes of novel facilities are determined against the performance indicators or particular targets. If the novel facility condition is not met, the health industry must examine what factors affect the attainment of a goal. If the cloud provider is the primary source of unsatisfactory service, the business will consult and negotiate ways to enhance service with the provider, or it may consider shifting of services and data toward other providers and back to its in-house IT system [24].

6.6 Cloud Computing Research Utilization in Healthcare

Figure 6.2 shows the various application of cloud computing research in healthcare.

6.6.1 Cloud computing in telemedicine/teleconsultation

Cloud computing provides transparent service, flexibility and scalability, remuneration service support, omni-accessibility, and other benefits. This paradigm not only allows customers to take advantage of convenient, diverse, and efficient services but also frees them from upkeep. The telemedicine cloud, which is attached to smart mobile devices, is a promising strategy to provide cost-effective and pervasive healthcare. Although many telemedicine systems take advantage of powerful cloud computing characteristics, the telemedicine

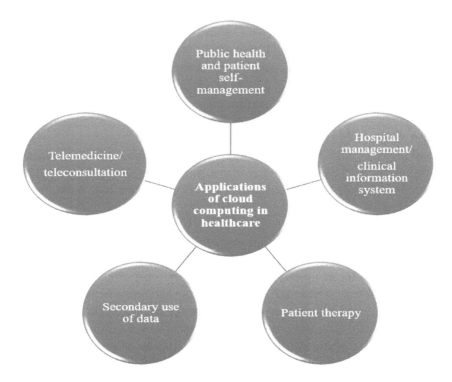

Figure 6.2   A schematic diagram that shows various cloud computing research applications in healthcare.

cloud is still in its infancy. The promise for cloud computing in telemedicine is evident, but there are still a lot of questions to be answered [25].

Countries throughout the world are grappling with healthcare concerns such as equity, access, cost-effectiveness, and quality. Modern information and communication technologies (ICTs) like computers, networks, and mobile devices, which are transforming how people convey with one another, also show a lot of promise for tackling the world's increasingly serious and diversified health issues. Telemedicine is the most important ICT-enabled service and it has improved and promoted novel ICTs which have become more accessible.

Though no single explanation of telemedicine exists, the World Health Organization (WHO) states it as "long-distance healthcare delivery uses the communication and information techniques for sharing of information for injuries and disease therapy, diagnosis and prevention." Patient meetings through video conferencing, medical picture transmission and storage, remote vital sign monitoring, e-health services (including patient portals), continuing education, nursing contact centers, and consumer-focused wireless applications are all examples. Recent breakthroughs and the widespread adoption of wireless mobile technology have fueled a surge in interest in telemedicine as a means of delivering healthcare [26].

According to a survey, cloud-based healthcare applications are divided into five groups: the Health Cloud eXchange (HCX), emergency medical system (EMS), health ATM kiosks, @HealthCloud, and DICOM-based system.

EMS stands for emergency medical services, and it is a system that allows patients' personal health records to be accessed to give prompt care. EMS application, personal health record (PHR) stage, and a portal to access the former are three key components. A medical recorder and user interface make up the PHR platform. Patients can access their medical history data through the user interface, while authorized healthcare providers can access certain aspects of the data. The EMS application, on the other hand, stores emergency medical information as well as application software. A variety of web services are included in the application software, but they are only accessible by an approved person in the ambulance and emergency department.

A distributed web interactive system called HCX offers private cloud-based information exchanging service that enables dynamic innovation of different records of health and healthcare services. HCX enables the health information to be shared between multiple electronic health record (EHR) systems. It responds to modifications done in the cloud system automatically [27]. Patients can use health ATM kiosks to handle their health-related data. It allows patients and care providers to get fast access to relevant health data. Individuals can manage their care online by reviewing personal account information and doing transactions.

The DICOM-based system was created to cope with the huge number of images of medical and diagnostic imaging procedures that traditional healthcare data has to deal with. The internal hospital network is used to store and manage image archives. A firewall keeps the network safe. HealthCloud is a cloud-based and Android-based mobile healthcare information management solution. It uses Amazon Simple Storage Service to update, retrieve, and store healthcare information [27, 28].

Teleconsultation is a viable and promising approach for making healthcare more inexpensive and of greater quality. It offers better patient outcomes and a higher quality of life as well as lower national healthcare costs by decreasing unneeded assessments, in-person appointments, and transfer of the patient. Various types of data must be retrieved during telemedicine consultations by using heterogeneous customer devices in various communication settings to provide elevated quality ongoing hygienic care whenever and wherever it is required. In the context of clinical information sharing, video conferencing and electronic monitoring also possess significant issues for data privacy and interoperability. Clinical data of patients are still dispersed, inaccessible, and fragmented.

Isolated in a variety of medical facilities, when patients want teleconsultation, consulting doctors must manually collect clinical data from them. In some cases, a lack of clinical data can lead to information loss, misdiagnoses, and repeat medicine prescriptions, among other things. This results in excessive costs and inefficiency as well as medical security issues. The telemedicine consulation, compatibility, accessibility, availability and portability is get improved by decereasing the price. For remote medical consultation, a prototype of a cloud-based mobile telemedicine consultation method is used. Many diverse technologies are necessary to establish a worldwide infrastructure for offering collaborative diagnosis, remote teleconsulting, and emergency scenario management. It is made easier by cloud computing, which provides a solution that allows for interoperability and data interchange at the time of consultation. In a cloud setting, web services are employed to solve clinical information interchange issues automatically rather than manually. When used as a mobile telemedicine device, the iPad is easy and portable [29].

6.6.2 Cloud computing in public health and patient self-management

Individual residents and patients, as well as big demographic groupings, are the focus of public health (epidemiology) [30]. In home-based healthcare, especially for long-term disease treatment, patient self-management is promoted. Patient self-management could be improved by sharing health information. In comparison to previous technologies, cloud computing can give a long time of sustainable elucidation. A hybrid cloud has been recognized as a viable option for allowing patients to share health information to improve chronic disease treatment.

Chronic diseases have a long-term impact on sufferers. Most chronic diseases are caused by physical activities such as lifestyle, nutrition, and metabolism. Chronic disease treatment is mainly reliant on the patients' daily habits. As a result, most chronic disease symptoms can be relieved by modifying daily habits such as quitting drinking and smoking, implementing and maintaining a nutritious diet, or enhancing physical activity [31, 32].

Care recipients may be able to live independently at home with the help of home-based healthcare. Healthcare providers may keep an eye on the patients based on their daily health information, which is shared, facilitate medical recommendations, and also provide feedback via medical examination reports. In addition, more people are urged to help with home-based healthcare, like family members and other patients having similar symptoms. The patient will be encouraged to take an active role in their care as a result of patient-centered home-based healthcare, which will include few self-management actions and the sharing of vital health data with their physicians or other research institutes.

The advancement of ICT has resulted in the widespread usage of wireless personal devices such as cellphones, PCs, and other self-monitoring equipment. This could be a viable option for home-based healthcare. HealthVault and Apple Health, for example, are commercial or research-based elucidations for self-management of healthcare.

Cloud computing "may deliver distributed, swiftly provisioned and adjustable computing sources (like servers, applications, storage, networks, and other facilities) that are on-demand, elastic and measurable with network connections" according to Wikipedia. The obvious benefits of cloud computing, like large scalability, security, and availability, have made it a trend in the eHealth sector in recent years. Because patient daily health data is so large, cloud storage appears to be a better option than sharing this information among healthcare groups, hospitals, or third-party institutions of research. Remote services like various IT systems of hospitals and other healthcare institutions, and also some third-party online care providers, can process huge healthcare data in the cloud. These benefits add to the case for using cloud computing to share health data [32].

The majority of elderly adults have chronic illnesses that can lead to morbidity and mortality. Cardiovascular diseases (stroke, coronary heart disease, and other cerebrovascular illness), diseases of the respiratory system (trachea, lung, bronchus cancers, pulmonary hypertension, lower respiratory infections, obstructive sleep apnea syndrome, asthma, bronchiectasis, and chronic obstructive pulmonary disease), and arterial hypertension are general clinical problems encountered in each clinical practices among the elder patients also in the general population.

However, it has been discovered that there is a link between the different chronic illnesses and total healthcare spending. Furthermore, there appears to be a link between concomitant chronic illnesses, the occurrence of ambulatory care sensitive hospitalizations, and the prevalence of care problems. The major issue is that healthcare providers recently have little evidence or guidance on how to approach treatment decisions for these patients. As a result, knowing how to effectively care for individuals with co-morbidities may lead to enhancements in life quality, healthcare utilization, safety, morbidity, and mortality.

In research, it was found that congestive heart failure is frequently associated with diabetes, hypotension, atrial fibrillation, hypertension, myocardial infarction, cognitive impairment, dementia, depression, or chronic obstructive pulmonary disease, which is often followed by osteoporosis, glaucoma, ischemic heart disease, osteopenia, cachexia, peripheral muscle dysfunction, cancer, ventricular arrhythmias, malnutrition, and other conditions.

It will be incredibly beneficial if telemonitoring programs could be implemented to cover a variety of conditions. Diabetes and failure of heart and chronic obstructive pulmonary diseases (COPD) are examples of such pairings. As the impacts of chronic illness, the number of chronic illnesses and particular difficulties like comorbidities continues to rise, resulting in polypharmacy (multidrug) and a deterioration in life quality. They can also contribute to an increase in long time care and costs of therapy (mostly in drug administration).

Telemonitoring offers various solutions for the management of patients' health, ranging from non-invasive monitoring of important parameters like using wearable sensor networks and signal processing technologies to making a proper analysis, identifying right therapies, and intervening at the correct time, to overall treatment regulation. Sensory systems should be designed in such a way that they can be implanted within the skin or housed in wearable fabrics. Low bandwidth networks based on improved low-power wireless technologies can be utilized to send and receive data. Wearable sensors in context with detection algorithms help to activate a message at a specific time, which can be delivered to the appropriate location via mobile technique [33].

Hybrid cloud was found to be an appropriate paradigm for sharing well-being data in this application based on outcomes of literature research and evaluation. The National Institute of Standards and Technology explains hybrid cloud as a "composite of two or more independent cloud services and infrastructures which are separate entities but get linked together via methodologies." By combining private cloud and public cloud, a hybrid cloud is generated, which benefits in the public cloud's cost savings and scalability without putting key apps and data on a third-party public cloud. It appears to be the best cloud utilization approach for simultaneously sharing health information created by the patient to the hospital and third parties. Security, effective scalability, customization, expandability, and relatively low cost are the advantages of installing a hybrid cloud [34].

6.6.3 Cloud computing in hospital management/ clinical information systems

Medical records have increased considerably as a result of fast social and economic growth. To meet the requirements of medical records, hospitals and organizations must use advanced science and technology innovation to continuously raise the level of modern records administration. Various issues come with the traditional medical management system, such as quality and storage capacity of electronic files are limited by hardware tools; as the volume of data grows, the system's speed will reduce due to channel access mode and bandwidth; the current backup mode is insufficient for the long-term storage and security of medical files and records. The standard storage model of data does not allow resource sharing.

Cloud computing is the technique that is used to create virtual data centers that give software applications with a container environment. Dynamically deployable compute and storage resources, as well as dynamic tuning and recovery, are all possible. Virtualization and parallel computing, based on cloud techniques can not only meet the basic requirements for archival information retrieval and storage but also importantly enhance the storage of system, computing, transportation, and other resource utilization of existing electronic records information management solutions [35].

The term "hospital information system" means a computer system that is used to manage clinical data and conduct online activities. It includes all services and operations of hospitals. Radiology information system (RIS), laboratory information system (LIS), and ultrasound information system (UIS) are all examples of general HIS. Information sharing is recognized as critical to lessen the complexity of HIS maintenance and construction. Due to a lack of appropriate technology for establishing an efficient and secure connection, clinics have developed their own HIS system. Apart from expensive management and construction costs, it is unable to provide enough information sharing. Although the province is beginning to develop a consistent medical archive, it is still difficult to expand and impossible to resolve the repetitive construction. All of these issues are expected to be resolved, which will need creativity and improvements to the overall structure of HIS.

Cloud computing, which has been proposed in recent years and is being promoted by Google, Yahoo, IBM, Amazon, Microsoft, and other significant corporations, is a new network paradigm. Narrow cloud computing is a distribution and usage pattern for IT infrastructure that allows users to claim sources with increased needs and capabilities via a network. This network concept is extended to any service in a broader meaning of cloud computing.

Simple Storage Service (S3) and Amazon's Elastic Compute Cloud (EC2) offer facilities primarily for enterprises; IBM's "Blue Cloud" computing facilitates an open cloud computing system for consumers with buy-to-use cloud policy, and many large organizations, including Microsoft, are actively increasing the utilization of cloud computing. The cloud computing concept offers a novel approach to addressing the existing challenges. Recently, certain research works have been conducted in an attempt to use the cloud approach in clinical practices.

The paper discusses the challenges that medical imaging analysis (MIA) researchers and doctors faced and proposes a novel paradigm for cancer imaging research based on the cloud computing concept. Though it is not specifically about hospitals, the problems in processing clinical data and the importance of cloud computing are discussed in depth. Furthermore, the article suggests a cloud-based mechanism for collecting patient data. It can automate everything from data collecting at the bedside to data delivery and remote access for clinical personnel. This research demonstrates how cloud computing is used in clinical operations and makes recommendations for future HIS development [36].

Both in-hospital and pre-hospital services are used in emergency medical procedures. From administrative to medical care, there are a variety of associated operations. As a result, proper information sharing and communication between emergency medical services (EMS) and healthcare organizations are required. The obtainability of patient data has an impact on the emergency care plan as well as the mode of transportation (like ambulance or helicopter) and the accurate relaying of the patient's state. Furthermore, it leads to more precise case prioritization based on severity, the avoidance of superfluous testing, and complete development in the quality of emergency care. Emergency responders' communication is just as crucial as information transmission between healthcare companies and EMS. The Emergency Data Exchange Language (EDXL) was created by the Organization for the Advancement of Structured Information Standards (OASIS) Emergency Management Technical Committee to allow EMS to transmit important information between national and regional organizations. Integrating cloud computing technologies with emergency care services systems allows for faster data recovery from the personal healthcare record (PHR) or electronic health record (EHR) of patients. This data will assist paramedics in making the best decision and managing patients' conditions depending on their medical histories.

In addition, an emergency personal healthcare record (EPHR) is created to assist paramedics in gaining access to data that is constantly updated from multiple resources. This data is divided into three categories: emergency documents, patient documents (comprise the medical history of the patient), and resource documents (that include instructional sources). Through the use of cloud computing services, EPHR assists EMS workers in accessing patient-oriented data that can be delivered in either Clinical Document Architecture (CDA) or Health Level 7 (HL7) standard [37].

6.6.4 Cloud computing in therapy

A cloud computing expert system was created and installed to assist diabetics in managing their condition and providing advice on what they should do. This method can also benefit care providers and physicians by assisting them in speeding up the decision-making process for which diabetic therapy is most appropriate. Personal information about the patient, such as age, gender, weight, and diabetes types, such as gestational and type-1 and type-2 diabetes, are input to the system. In addition, the system uses three blood sugar tests as input: breakfast, lunch, and dinner.

The HBA1C (A1C) test is an additional input. A1C is a blood glucose index that is calculated for patients with diabetes over the past three to four months. It indicates the fraction of hemoglobin that has glucose clinging to it. The lower the amount of glucose in the bloodstream, the better the diabetic's health will remain. A1C levels in adults without diabetes are typically between 4% and 6%. After the patient has entered all of the necessary information, the system calculates the average of the three tests and compares the result to the normal blood sugar range of 80–140; if it is more than 140, it is high, and if it is lower than 80, it is low.

The system determines the right treatment based on the patient's diabetes type and the results of two tests (average blood sugar test and A1C test). If type-1, the therapy will consist of a calculated diet, scheduled physical exercise, numerous daily insulin injections, and multiple daily home blood glucose tests. Insulin doses could be increased, lowered, or kept the same based on an average blood sugar test. Type-2 diabetes is treated with a combination of diet, exercise, home blood glucose testing, and oral medications.

The patient must take insulin injections if the A1C test is over 10% or over 7.5% plus the average blood test glucose is over 324. If the patient wants to keep a record or not, there are two options in the system. He must log in if he wants a record; as a result, the system keeps his information and allows him to retrieve it. If the patient does not want a record kept, he can just submit all of the information and the system will handle him appropriately [38].

Cloud computing is crucial for the implementation of assisted living environments. In addition, patients can be monitored via cloud computing services. In various poor countries, where clinical facilities and knowledge are rare, the aged and others with disabilities live in inaccessible or remote towns and villages.

Women suffering from cancer of the breast are frequently left undiagnosed in rural locations, and by the time they reach the doctors in developed cities, it is often too late. Doctors can diagnose patients who are unable to reach them due to financial constraints using cloud technology. They can utilize cloud-based applications for eHealth and telemedicine, which comprises the medical data transmission from remote areas to specialized physicians and major hospitals in other geographical regions. During an emergency, cloud-based applications also allow critical services like a speedy search engine for organ and blood donations [39].

6.6.5 Cloud computing in secondary use of data

Cloud computing can be used to make clinical data available for secondary use, such as mining of text, analysis of data,, and clinical research. To achieve a HIPAA-compliant atmosphere, cloud computing can be utilized to store and exchange research health information and data from electronic health records in a cloud structure. For them, cloud computing has the benefit of offering vast computing resources to researchers. Security of data can be gained by allowing researchers to establish their virtual servers and customized networks by using proprietary cloud technologies.

A method for enabling cloud-based services with great scalability and data security that complies with HIPAA. A cloud based software allows the extraction, processing, management and comparison of medical data of several hospitals. However, it is unclear how the security of data should be gained because, at this time, the information in the cloud is not anonymous and will only be available to the specific data provider.

While the stated project makes possible the use of a community cloud, the OpenNebula-based application shows its utilization in both private and public environments. The key advantages of cloud computing are reduced costs of data processing (no upfront investment, pay as you go) and managed services that allow data providers to use complicated, computation-intensive services [40]. Data mining models and findings may be shared between different clinics via a cloud-based server [41, 42].

6.7 Conclusion

Cloud computing helps in the management of the healthcare of patients. Cloud computing comprises three archetypal models: platform, software, and infrastructure. The manuscript describes the cloud computing opportunities from the various aspects of technology, security, management, and legality. It describes the cloud computing strategic planning which includes identification, action, evaluation, and follow-up. The manuscript also enclosed detailed information on cloud computing categorization for research in healthcare such as telemedicine or teleconsultation, patient self-management, and public health, hospital management or medical information systems, secondary use of data, and therapy. It also focuses on the challenges regarding privacy and security of cloud-based computing which creates problems in data handling. This issue will be resolved in the future and help to provide better healthcare services.

6.8 Acknowledgment

We are thankful to all co-authors for their contribution.

6.9 Funding

None.

6.10 Conflict of Interest

There is no conflict of interest.

References

[1] Sultan, N., 2014. Making use of cloud computing for healthcare provision: Opportunities and challenges. *International Journal of Information Management*, *34*(2), pp.177–184.

[2] Ali, O., Shrestha, A., Soar, J. and Wamba, S.F., 2018. Cloud computing-enabled healthcare opportunities, issues, and applications: A systematic review. *International Journal of Information Management*, *43*, pp.146–158.

[3] Gao, F. and Sunyaev, A., 2019. Context matters: A review of the determinant factors in the decision to adopt cloud computing in healthcare. *International Journal of Information Management*, *48*, pp.120–138.

[4] Ermakova, T., Huenges, J., Erek, K. and Zarnekow, R., 2013. Cloud computing in healthcare–a literature review on current state of research.

[5] Rosenthal, A., Mork, P., Li, M.H., Stanford, J., Koester, D. and Reynolds, P., 2010. Cloud computing: a new business paradigm for bio-medical information sharing. *Journal of biomedical informatics*, *43*(2), pp.342–353.

[6] Aziz, H.A. and Guled, A., 2016. Cloud computing and healthcare services.

[7] Ogwel, B., Odhiambo-Otieno, G., Otieno, G., Abila, J. and Omore, R., 2022. Leveraging cloud computing for improved health service delivery: Findings from public health facilities in Kisumu County, Western Kenya 2019. *Learning health systems*, *6*(1), p.e10276.

[8] http://www.jecr.org/sites/default/files/12_4_p01.pdf

[9] Rodrigues, J.J., de la Torre, I., Fernández, G. and López-Coronado, M., 2013. Analysis of the security and privacy requirements of cloud-based electronic health records systems. *Journal of medical Internet research*, *15*(8), p.e2494.

[10] Alshehri, S., Radziszowski, S.P. and Raj, R.K., 2012, April. Secure access for healthcare data in the cloud using ciphertext-policy attribute-based encryption. In *2012 IEEE 28th international conference on data engineering workshops* (pp. 143–146). IEEE.

[11] Mrozek, D., 2020. A review of Cloud computing technologies for comprehensive microRNA analyses. *Computational biology and chemistry*, *88*, p.107365.

[12] Birje, M.N., Challagidad, P.S., Goudar, R.H. and Tapale, M.T., 2017. Cloud computing review: concepts, technology, challenges and security. *International Journal of Cloud Computing*, *6*(1), pp.32–57.

[13] An, Y.Z., Zaaba, Z.F. and Samsudin, N.F., 2016, November. Reviews on security issues and challenges in cloud computing. In *IOP Conference Series: Materials Science and Engineering* (Vol. 160, No. 1, p. 012106). IOP Publishing.

[14] Act, A., 1996. Health insurance portability and accountability act of 1996.*Public law*, *104*, p.191.

[15] Alexander, Y. and Brenner, E.H., 2002. Uniting and Strengthening America by Providing Appropriate Tools Required to Intercept and Obstruct Terrorism Act of 2001 (USA Patriot Act): HR 3162. In *US Federal Legal Responses to Terrorism* (pp. 309–496). Brill Nijhoff.

[16] https://laws-lois.justice.gc.ca/pdf/p-8.6.pdf

[17] Mu-Hsing, K., 2012. A healthcare cloud computing strategic planning model. In *Computer Science and Convergence* (pp. 769–775). Springer, Dordrecht.

[18] Lee, T.S. and Kuo, M.H., 2009. Toyota A3 report: a tool for process improvement in healthcare. *Stud Health Technol Inform*, *143*, pp. 235–240.

[19] Kuo, M.H., 2011. Opportunities and challenges of cloud computing to improve health care services. *Journal of medical Internet research*, *13*(3), p.e1867.

[20] Zhang, R. and Liu, L., 2010, July. Security models and requirements for healthcare application clouds. In *2010 IEEE 3rd International Conference on cloud Computing* (pp. 268–275). IEEE.

[21] ENISA, C.C., 2009. Benefits, risks and recommendations for information security. *European Network and Information Security*, *23*.

[22] Gagliardi, F. and Muscella, S., 2010. Cloud computing–data confidentiality and interoperability challenges. In *Cloud Computing* (pp. 257–270). Springer, London.

[23] Stanoevska-Slabeva, K., Wozniak, T. and Hoyer, V., 2010. Practical guidelines for evolving IT infrastructure towards Grids and Clouds. In *Grid and Cloud Computing* (pp. 225–243). Springer, Berlin, Heidelberg.

[24] Lee, T.S. and Kuo, M.H., 2009. Toyota A3 report: a tool for process improvement in healthcare. *Stud Health Technol Inform*, *143*, pp. 235–240.

[25] Ahmed, S. and Abdullah, A., 2011, March. Telemedicine in a cloud—A review. In *2011 IEEE Symposium on Computers & Informatics* (pp. 776–781). IEEE.

[26] Jin, Z. and Chen, Y., 2015. Telemedicine in the cloud era: Prospects and challenges. *IEEE Pervasive Computing*, *14*(1), pp.54–61.

[27] Wooten, R., Klink, R., Sinek, F., Bai, Y. and Sharma, M., 2012, May. Design and implementation of a secure healthcare social cloud system. In *2012 12th IEEE/ACM International Symposium on Cluster, Cloud and Grid Computing (ccgrid 2012)* (pp. 805–810). IEEE.

[28] Matlani, P. and Londhe, N.D., 2013, January. A cloud computing based telemedicine service. In *2013 IEEE Point-of-Care Healthcare Technologies (PHT)* (pp. 326–330). IEEE.

[29] Wang, H., Li, T.H. and Wu, F., 2014. A Cloud-Based Mobile Telemedicine Consultation System based on iPad. In *Applied Mechanics and Materials* (Vol. 644, pp. 3208–3211). Trans Tech Publications Ltd.

[30] Griebel, L., Prokosch, H.U., Köpcke, F., Toddenroth, D., Christoph, J., Leb, I., Engel, I. and Sedlmayr, M., 2015. A scoping review of cloud

computing in healthcare. *BMC medical informatics and decision making*, *15*(1), pp.1–16.

[31] Danaei, G., Ding, E.L., Mozaffarian, D., Taylor, B., Rehm, J., Murray, C.J. and Ezzati, M., 2009. The preventable causes of death in the United States: comparative risk assessment of dietary, lifestyle, and metabolic risk factors. *PLoS medicine*, *6*(4), p.e1000058.

[32] Hu, Y., Peng, C. and Bai, G., 2015, June. Sharing health data through hybrid cloud for self-management. In *2015 IEEE International Conference on Multimedia & Expo Workshops (ICMEW)* (pp. 1–6). IEEE.

[33] Arif, M.J., El Emary, I.M. and Koutsouris, D.D., 2014. A review on the technologies and services used in the self-management of health and independent living of elderly. *Technology and Health Care*, *22*(5), pp. 677–687.

[34] Ma, J., Peng, C. and Chen, Q., 2014, November. Health Information Exchange for Home-Based Chronic Disease Self-Management—A Hybrid Cloud Approach. In *2014 5th International Conference on Digital Home* (pp. 246–251). IEEE.

[35] Guo, L., Chen, F., Chen, L. and Tang, X., 2010, April. The building of cloud computing environment for e-health. In *2010 International Conference on E-Health Networking Digital Ecosystems and Technologies (EDT)* (Vol. 1, pp. 89–92). IEEE.

[36] He, C., Jin, X., Zhao, Z. and Xiang, T., 2010, October. A cloud computing solution for hospital information system. In *2010 IEEE International Conference on Intelligent Computing and Intelligent Systems* (Vol. 2, pp. 517–520). IEEE.

[37] Aziz, H.A. and Guled, A., 2016. Cloud computing and healthcare services.

[38] Al-Ghamdi, D.A.A.M., Wazzan, M.A., Mujallid, F.M. and Bakhsh, N.K., 2011. An expert system of determining diabetes treatment based on cloud computing platforms. *IJCSIT) International Journal of Computer Science and Information Technologies*, *2*(5), pp.1982–1987.

[39] Lahoura, V., Singh, H., Aggarwal, A., Sharma, B., Mohammed, M.A., Damaševičius, R., Kadry, S. and Cengiz, K., 2021. Cloud computing-based framework for breast cancer diagnosis using extreme learning machine. *Diagnostics*, *11*(2), p.241.

[40] Griebel, L., Prokosch, H.U., Köpcke, F., Toddenroth, D., Christoph, J., Leb, I., Engel, I. and Sedlmayr, M., 2015. A scoping review of cloud computing in healthcare. *BMC medical informatics and decision making*, *15*(1), pp.1–16.

[41] Rea, S., Pathak, J., Savova, G., Oniki, T.A., Westberg, L., Beebe, C.E., Tao, C., Parker, C.G., Haug, P.J., Huff, S.M. and Chute, C.G., 2012. Building a robust, scalable and standards-driven infrastructure for secondary use of EHR data: the SHARPn project. *Journal of biomedical informatics*, *45*(4), pp.763–771.

7

Role of Artificial Intelligence and Robotics in Healthcare

Shilpa Rawat[1], Shilpa Singh*[1], Rishabha Malviya[1], Sunita Dahiya[2], and Sonali Sundram[1]

[1]Department of Pharmacy, School of Medical and Allied Sciences, Galgotias University, India
[2]Department of Pharmaceutical Sciences, School of Pharmacy, University of Pureto Rico, Medical Sciences Campus, USA
*Corresponding Author: Shilpa singh
Research Scholar, Department of pharmacy, School of medical and allied science , Galgotias University, India
Email: shilpasingh43206@gmail.com

Abstract

Artificial intelligence (AI) is increasingly being used in healthcare due to the increasing complexity and volume of data. Additionally, AI is currently being used by pharmaceutical and biotechnology companies as well as payers and healthcare providers. Furthermore, AI assists thousands of individuals in diagnosing and treating illnesses, along with keeping patients on track with their treatment plans. Although AI is one of the most exciting robotics topics, it has only recently entered the healthcare industry. It is most commonly used in "deep or non-deep machine learning," the process of making computers smarter through new research. On the other hand, previous models concentrated on a specific function or type of AI rather than supporting patients and physicians during surgery. Based on studies, the broad coverage of AI in medical applications led to the construction of a graphical representation conceptual model. Moreover, blockchain technology is used to connect patients and professionals before, during, and after surgery. Interestingly, in some circumstances, AI can help arrange a doctor's appointment or choose a suitable time for it. Subsequently, AI assists doctors in developing future treatments and patients in preparing for surgery. Besides, the third sort of learning is deep learning. In addition, by analyzing medical data and prescribing therapy, AI can assist post-operative patients.

7.1 Introduction

7.1.1 History of artificial intelligence

The health specialist system is based on Bayesian statistics and decision theory, which were first used in medical research in the 1970s to identify and propose treatments for glaucoma and infectious disease. Advances in Bayesian networks (BN), artificial neural networks (ANN), and hybrid intelligence (HI) accelerated bioinformatics research throughout the late 1990s, culminating in increased acceptance of medical artificial intelligence (MAI). Interestingly, medical installations are expected to save $150 billion by 2026, which will lead to a $6.6 billion investment in MAIs around the world by 2021 [1]. Besides, health information technology (HIT) has had a significant impact on the field of health information management (HIM). The HIM team put their efforts to make medical data accessible, accurate, safe, and secure for healthcare professionals. Furthermore, the digitalization of medical information had a considerable impact on the duties and activities of HIM professionals, pushing many people to take on more technical roles in the collection, storage, and use of medical information. The digitalization of medical information, as well as new data processing and storage technology, has enabled the creation of complicated algorithms in the same way as AI does [2]. John McCarthy invented the term "artificial intelligence" which he defines as the "science and engineering of making intelligent machines." According to him, the word "simple" refers to a machine's ability to perform tasks that humans consider intelligent [3]. In addition, AI applications can be divided into two categories:

1. an attempt to duplicate a person's mental abilities;

2. the advancement of technologies that can perform tasks that previously required human interaction.

Furthermore, AI is divided into several sub-disciplines, each focusing on different subjects, including vision, problem-solving, speech recognition, and learning [4]. Indeed, AI is the replication of human intelligence in systems such as computers or robotics, which are designed to duplicate cognitive processes such as training and problem-solving abilities that people associate with other human brains. AI, machine learning (ML), and deep learning (DL) are buzzwords that everyone seems to be using nowadays because

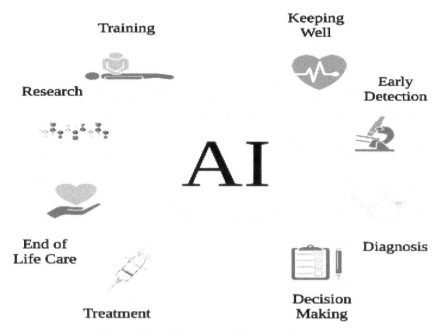

Figure 7.1 AI in various fields of medical study.

they depend on technologies that are becoming increasingly widespread in everyday life. Additionally, the adoption of AI technologies will become a prerequisite for every firm [5]. AI is one of the most exciting and rapidly increasing topics of medical study in the 21st century. Its diverse applications are represented in Figure 7.1 [6].

Before AI technologies can be used in the medical field, they must be "trained" with data generated by medical operations such as testing, assessment, and therapy assignments so that they can recognize comparable subjects and connections between subject characteristics as well as desired outcomes. In this system, the medical evidence includes demographics and data from clinical equipment, diagnostic tests, medical lab tests, and photos. These pieces of evidence could also be used to show how people spend their time [7]. Furthermore, medical care is becoming increasingly important in the age of AI healthcare when everyone is trying to be smart and live a better and longer life by predicting their illness prognosis. It is based on the premise that "caution is the mother of safety," which means that it is preferable to prevent anything from happening in the future than to treat something that has already happened. Interestingly, AI researchers in medicine can now work and prosper [8]. The technique, self-regulating formation, and application of a device (robot) have previously been tested by humans. Besides, many parts of robotics have AI. Robots may be able to think, sense, and know how hot or cold things are. Also, people are capable of making decisions; thus, robotics research is now focused on developing autonomous robots with decision-making abilities. Because today's industrial robots are not human-like, the robot in the human pattern is designed to assist an android [9]. In addition to robotic systems, which were first developed in the 1950s to streamline unsanitary, tedious, and demanding situations, today's modern health and social machines are programmed for entirely different purposes [10]. The improvement center, the theater, and the family compartment are all accessible to individuals who engage with humans in a surgical setting. In addition, commercial and research interests are shared by medical centers and pharmaceutical companies. Intriguingly, the use of robotics in healthcare has increased dramatically in recent years. Consequently, robot-assisted technological advancements are speeding up the development of modern therapy methods by improving patient outcomes and lowering healthcare costs, while also giving patients a different level of care [11].

Furthermore, the need for robotics in healthcare has been identified as one of the primary drivers of improved healthcare education, lower healthcare costs, and shorter recovery times, thereby increasing the number of people who have access to wellness services. New therapeutic methods, improved patient outcomes, and less stress and workload for hospital staff are just a few of the many positive outcomes that could occur with robotics in healthcare.

Healthcare robots can be divided into three categories:

- A medical device that combines robotic surgery, detection, and diagnostic tools.

- Aided robotics include wearable robots and rehabilitative equipment.

- Robots that look like the human body include implants, artificial organs, and body part simulators [12].

Robotics is becoming a vital part of today's healthcare practices. The method for a neurosurgical biopsy has gone over without a glitch since 1985, when a robotic dubbed "PUMA" initially assisted in surgeries [13]. In medicine, robotic surgery refers to the use of machines to perform surgical procedures that are guided by a clinician. The term "robot" was first used in the play Rossum's Universal Robots (RUR) by the Czech novelist and dramatist Karel Capek in 1921. The term is derived from the Czech word for "forced labor." Even though robotic surgery relieves doctors from some repetitious tasks, it still necessitates a high level of surgical experience. Additionally, in the early 21st century, scientists reported on the benefits of robotic surgery. People with fewer complications and fatalities also have lower prices, well-functioning robotic medical tools, and do not report when they do not work. The first medical robots appeared in the 1980s, providing surgical assistance with robotic arm technology. Over the years, AI-enabled computer vision and data analytics have revolutionized health robots, expanding their capabilities into many other areas of healthcare [14]. Medicare robotics aids in surgery by speeding up clinical administration and allowing employees to provide more immediate attention to patients. Besides, robots are revolutionizing resection, easing stock and decontamination, and freeing up time for contributors to interact with patients [15].

7.1.2 The need for AI

Initially, the only way to obtain actual medical data was to use books and periodicals that presented professional suggestions and contained an expert connection. In addition, doctors used to gain expertise by guiding and monitoring patients' diagnoses. Integration of such a large amount of information and expertise to provide specialized healthcare services was a significant challenge. Fortunately, AI has stepped forward to assist by combining a massive amount of healthcare data to improve and expand doctor efficacy. The right questions can help AI in finding relevant information hidden in large datasets, which could play a big role in healthcare decisions [16].

Furthermore, AI has played an essential and expanding role around the world in recent years. The majority of users are unaware of the various ways AI can manifest itself in their daily lives. AI methods, among other things, are utilized to improve efficiency when logging into email accounts, purchasing on digital sites, and looking for vehicle transportation services. Moreover, healthcare is the most important area in which AI is rapidly progressing, particularly in the areas of therapy regulation and diagnostics. As a result, there is concern that AI will outperform humans in both goals and abilities. Numerous studies have shown that AI will improve social judgment, therapeutic choices, and efficiency in the future [17]. Figure 7.2 depicts the broad applications of computational methods in analytics in healthcare.

AI is composed of numerous components, including intelligent machines and natural language processing [3]. Improved diagnostic procedures hold a lot of potential for the betterment of the healthcare sector. ML is a branch of AI that focuses on the development of algorithms both instinctively and by exposing them to vast amounts of "trained" information [18]. Medical care costs are rising all over the world. Rising longevity, an increase in the prevalence of chronic diseases, and the ongoing development of costly new treatments all contribute to this pattern. As a result, it is no surprise that academics predict a bleak future for medical care technology around the

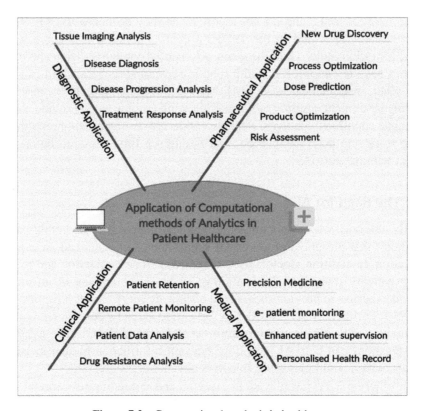

Figure 7.2   Computational methods in healthcare.

world. By improving and lowering the cost of healthcare delivery, AI may be able to mitigate the effects of these changes [19].

7.1.3 How did AI change the way medicine was practiced in the past?

Initially, researchers concentrated on developing robotics that could aid physicians in therapies such as surgery. This effort was a failure, but you already believe that while these machines cannot replace humans, they can certainly assist physicians in anticipating illness and suggesting treatments. Doctors have to remember each person's history, previous prescriptions, ailments, and other details, so that medications can be prescribed based on their case history [20]. The rise of AI is dependent on medical technologies and is accompanied by a flood of clinical data. Electronic healthcare records (EHRs), smart sensors and diagnostic devices, genetic testing, and a variety of other sites provide scientists and doctors access to a massive amount of data. However, the amount of data available outweighs people's ability to cope with and utilize it. Advanced AI technologies, such as ML and prediction algorithms, may help people make sense of the avalanche of information and identify trends that will allow them to make better decisions for patients [21]. When used in conjunction with clinicians, AI techniques and procedures have the potential to improve treatment delivery beyond what they can achieve on their own. Consider precise drugs, which aim to tailor clinical therapy to a patient's specific characteristics. Similarly, AI is expected to change the way doctors provide care by allowing them to find the best medicine doses and determine whether genetic abnormalities cause specific disease. There are a lot of medical, genomic, and diagnostic data at the heart of personalized medicine, which helps doctors be more effective, diagnose correctly, and give the right treatment to the right person [22]. AI and ML are important to making this happen.

7.1.4 Types of AI

AI is divided into three categories as follows:

- **Artificial narrow intelligence (ANI):** It is a type of insensible intelligent machine that is usually targeted at a certain job (weak AI).

- **Artificial general intelligence (AGI: fictional scenario):** A computer that can use intellect on every issue instead of just one, often referred to as "at least as clever as a regular person."

- **Artificial superintelligence (ASI: a fictional scenario):** It is an AI that outperforms even the most brilliant and capable person's brain. The notion is that robots can outperform humans in terms of technology, interpersonal skills, and technical understanding in a variety of fields [23].

7.2 AI in Healthcare

An international health epidemic, such as the coronavirus, brought the medical industry to the forefront for all stakeholders fighting on the front lines. In other nations, including India, the outbreak has been termed a fundamental shift in computerized medical care. Most people believe that it is a perfect time for India to restart its medical system and encourage medical IT companies to fill gaps in the existing system. Besides, in India, many medical care sectors require robotics for a variety of activities [24]. Humans and robots have different benefits and drawbacks, but they can work together to provide and improve medical care. The American Medical Association recently defined AI's role in clinical care as "intelligence amplification," implying that AI would be developed and used to supplement rather than completely replace human intellect [25].

In the field of medicine, AI has several advantages over human intelligence. For instance, AI could learn more effectively than physicians through large datasets, such as inaccessible walls of unstructured data within a digital medical database. Effective AI can quickly retrieve important facts from disconnected or actual data to help improve organizational performance and assist physicians in researching and teaching actual choices. Furthermore, AI is more precise in carrying out current tasks. AI can work indefinitely without losing effectiveness, and, unlike humans, it does not get tired. Moreover, AI tool has the potential to revolutionize the way complicated surgeries are performed [26]. Besides, AI is well suited to medical care delivery. In reality, the use of AI in therapeutic settings has skyrocketed in the last few decades. Subsequently, AI technologies with data mining and data modeling capabilities could be useful because modern drugs have a huge difficulty in obtaining, interpreting, and using structured and unstructured information to cure or control illnesses. Clinical AI is primarily concerned with the development of AI algorithms to aid in disease prediction, diagnosis, therapy, and administration [27]. Interestingly, AI and robots are rapidly evolving in medical services, particularly in recent detection and treatment applications. At the same time, AI is becoming more powerful. AI completes the task that humans are not able to do, often more quickly and efficiently, and at a lower cost [28].

7.2.1 AI tool

AI technologies, according to the preceding explanation, are classified into two types: classical and advanced. The first category includes ML approaches that examine properly programmed data such as genomic and EP datasets. ML algorithms are used in clinical settings to group individuals' characteristics or to predict the likelihood of negative outcomes. The second group includes natural language processing (NLP) approaches that retrieve data from non-fundamental information such as medical documents and medical reviews to augment and enhance organized clinical data. NLP methods aim to convert words into machine-readable information that can then be analyzed using ML methods [7]. Furthermore, when it comes to creating interesting AI systems, big data and data mining are steps in the right direction [29].

7.2.2 Natural language processing

By extracting vital data from patients' vast clinical statistics, NLP technology may aid in improving diagnostic and therapy recommendations. Subsequently, the ability of robots to rapidly absorb massive amounts of visual and textual information using ML and NLP would allow physicians to make rapid diagnostic and treatment recommendations. This technology has a significant impact on healthcare delivery, particularly on how patients are treated [30]. Implementing NLP in the analysis of chest radiography data, for example, can enable antibiotic support technology to alert clinicians to the possibility of anti-infectious treatment being required. Also, NLP is used to manually track difficult impacts in research laboratories [31].

NLP could be useful in a variety of healthcare applications, as given below:

1. **Effective invoicing**: Getting data from doctor records and providing clinical codes for such payment procedures are two of the most obvious uses.

2. **Authorized permission:** Prevent delays and administrative errors while using data from doctors' records.

3. **Clinical decision support system (CDSS)**: Assists employees of the healthcare group in making important decisions, such as forecasting the diagnosis and consequences of patients.

4. **Evaluation of healthcare strategy:** Gathering medical advisors and developing appropriate treatment recommendations [32].

7.2.3 Machine learning

ML, a major subfield of AI, uses large datasets to discover interaction design patterns between parameters [33]. ML programs are becoming more popular due to faster response rates and cloud computing. It can detect abnormalities in images beyond what the human can detect, helping in the diagnosis and treatment of illness. In the coming years, ML will continue to transform the healthcare system. ML has been used in research to create scans for diabetic eye disease and to predict when breast cancer will recur based on clinical data and images [34]. There are three ways to learn ML: unsupervised (the ability to recognize patterns), supervised (the ability to classify and predict methods based on previous instances), and reinforcement learning (the use of reward and penalty systems to make plans for operating in a specific issue area) [35].

1. **Supervised learning**: Throughout this section, the receiver gets labeled with inputs. Depending on such identifiers, the program must generate out-turn.

2. **Unsupervised learning**: The input is not categorized and labeled throughout this class. To effectively generate out-turn, the machine does not have to identify the exact out-turn; instead, it draws its very own conclusions.

3. **Reinforcement learning:** A device that includes a reinforcement learning method generates behavior that depends on its interactions with the environment and attempts to maximize payoff [36].

Furthermore, ML offers a way to reduce rising healthcare expenditures while simultaneously improving the patient–doctor relationship [37]. The findings from ML can aid doctors in deciding which medicines and treatments are best for their patients. They could also help people if they need to come back for follow-up visits. ML approaches, such as artificial neural networks (ANN), represent a completely different approach to AI. Computer systems that use the ANN method build decision-making systems of artificial "neurons" that function in the same way as the human central nervous system does [38].

7.2.4 Algorithms

Some of the well-known ML algorithms are logistic regression (LR), naive Bayesian classification (NBC), *k*-nearest neighbor (KNN), multiple linear regression (MLR), support vector machine (SVM), probabilistic neural network (PNN), binary kernel discrimination (BKD), linear discriminant analysis (LDA), random forest (RF), ANN, partial least-squares (PLS), and principal component analysis (PCA) [39]. In many clinical studies, ANN and SVM are used to analyze scans. ML has high potential and implications in a few areas of medicine. Illness prognosis and diagnosis are fields of application for medicine's efficacy and outcome prediction, repurposing, and innovation [33].

7.2.5 Artificial neural network

An ANN is an ML system that is based on the network organization of the human brain. As the name implies, it is an algorithm based on the neural

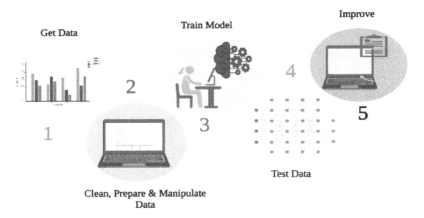

Figure 7.3 Development stages of ANN.

network of lifeforms, specifically the human vision cortical area. Surprisingly, the foundation of AI is the ANN, which is based on a human brain arrangement [40]. ANNs were developed with the goal of better understanding and shaping the functionalities of the nervous system, including computing aspects, while it performs cerebral activities such as sensory sensing, idea classification, idea association, and learning. Nowadays, however, a great effort is put into developing neural networks for purposes such as data analysis and organization, as well as compressed data optimization [41]. Figure 7.3 depicts the various stages of ANN development.

The four main areas of pharmacy in which neural networks may be used are modeling, bio-signal, diagnosis, and prognosis. In a range of clinical applications, ANNs can assist physicians in analyzing, modeling, and interpreting the logic of complex medical investigations. Additionally, categorization is the goal of many ANN medical applications, which indicates that the goal is to place the client into one of a few categories, based on measurable features [42]. ANNs can also be used to help with traditional CDSSs that are based on ancient computer science. They can help predict the prognosis of many diseases, such as cancer, heart problems, and abnormal blood sugar levels. Besides, they can also be used for radiology and histopathology detection [43].

7.2.6 Support vector machine

SVM is an ML-based multi-class approach for data classification that maximizes predicted performance while avoiding modeling error. In imagery analysis, it is more precise than the advanced neural nets [44]. In ML algorithms, each statistic is represented by an equation in the n-sphere, where "n"

denotes the total number of characteristics and each character denotes the quantity of a specific position. This method looks at vectorized data and finds a hyperplane that can tell the two signals apart [45].

7.2.7 Deep learning

DL, also known as deep learning, is an ML method in which an algorithm learns from its flaws and adapts accordingly. DL activities become much more sophisticated as they develop to replicate human minds by tackling challenges through many levels of neuronal networks and then shaping predicted decisions using input information [46]. Such a network, which is influenced by the human mind, is made up of millions or even billions of neurons. DL is frequently applied to and adapts to significantly larger volumes of data than ML [47]. An ANN is a collection of algorithms which uses a technique that simulates the individual mind to identify the hidden correlations in a set of data. On top of that, ANNs are a subset of ML techniques that form the foundation of DL techniques [48]. In medical care, ML is a new topic, as its applications to heart problems are limited when compared to other professions. One of the first commercial applications of DL was ML or the computerized analysis of imagery. Additionally, computer vision has been central to many of AI's early biological applications [49]. Interestingly, DL learns through information with the overall aim of studying processes rather than the feature engineering used in traditional neural networks. Furthermore, DL uses both supervised and unsupervised learning [50].

As the use of robotic systems for spinal fusion procedures becomes more frequent, a substantial source of reference has evolved, focusing solely on the software's correctness, a reduction in intraoperative radiation treatment (IORT), and surgery efficacy. According to research involving 379 orthopedic cases, Mazor-robotics (AI) supported robotic systems minimize surgery problems by five times when compared to manual doctors. The initial effective experiment of robot-assisted eye surgery was done at the University of Oxford. During general anesthesia, 12 people who needed retinal separation were arbitrarily allocated to robot-assisted or hand operations. Even though AI-supported operations have been around for a long time, surgical outcomes have been extremely good in both robotics and human operation groups [51].

7.3 Integrating AI into Healthcare Delivery

The use of AI provides a wide range of capabilities that can assist clinicians in making better decisions, improving care delivery procedures and health

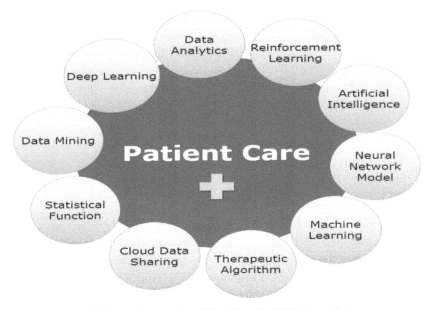

Figure 7.4 Implementation of digital medical fields in patient care.

outcomes, and lowering healthcare costs. The purpose of such research is to locate and summarize current papers on the use of AI in medical care. It reveals the necessary and remarkable outcomes [52].

7.3.1 Patient monitoring

Healthcare is critical in hospitals, operating rooms, intensive care units, and coronary care units, where patient care is defined in seconds, as shown in Figure 7.4. Regular observing tools generate a massive amount of data in these high acuity units (HAU) scenarios, presenting a good opportunity for AI-assisted devices [53]. AI methods can now be used to look at patients in a new way because of the widespread adoption of digital medical fields and the extensive usage of mobile phones and activity trackers. Doctors can now see people's sleep habits, hypertension, pulse, and other vital signs in ways they have never seen before [43].

7.3.2 Disease diagnostics and prediction

The most pressing need for AI in conventional medicine is disease diagnosis. Interestingly, there have been numerous notable developments in this field.

Now, doctors can use AI to diagnose diseases more quickly and accurately. One of the most prevalent methods of diagnostics is inside-body diagnosis using bio-sensors or microchips. Furthermore, the main research tool and genetic makeup could be investigated using ML, which employs AI to identify and discover problems in microarray databases [54]. In the next few years, AI may be able to make a real diagnosis in "visible" health fields like radiology and pathology, dermatology, and ophthalmology [55].

7.3.3 Precision medicine

Personalized or precision medicine (PM) aims to use personal genetic information rather than community genetic information at every stage of a patient's health journey. It entails gathering information from patients, such as genomic data, health check records, or electronic medical records (EMR), and personalizing the therapy using cutting-edge software [32]. Rather than treating the entire population, PM focuses on a proportion of patients. It is useful because a single patient population can have a lot of variations. High efficacy in a few, often insignificant, patient subcategories could influence the observed mean efficiency, while another patient subcategory is ignored [56].

7.3.4 Drug discovery

Using AI and computer-aided drug discovery (CADD) tools, investigators can select a small number of therapeutically successful candidates from a large number of compounds in a fraction of the time, whereas conventional methods can take a very long time. AI also includes programs that help people to determine which documents in a group of documents are healthy and which are unhealthy [57].

7.3.5 Dermatology

Due to its huge clinician dreamscape and dermatopathology picture databases, a dermatologist has taken the lead in implementing AI in the health sector, as represented in Figure 7.5. However, designing and interpreting clinical studies in this field will require a fundamental knowledge of AI. As a result, it is critical to think about AI's possible involvement in dermatological practices. Melanoma, dermatitis, and plaque psoriasis are just a few of the dermatological conditions where AI is slowly gaining attraction. Furthermore, scientists are looking into how AI can be used to improve and enhance existing skin-cancer diagnostic methods [58].

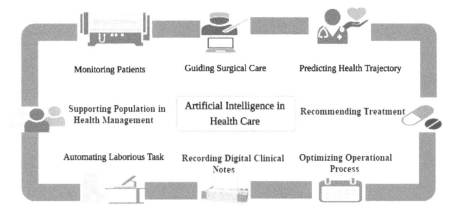

Figure 7.5   AI in healthcare.

7.3.6 Coronavirus

AI automated support robotics and human robots are known as "Cloud Ginger" XR-1 which are used in health centers in Wuhan, China. Its primary purpose is to assist medical personnel in transporting snacks and medications to patients. Besides, its secondary purpose is to keep the patients entertained during their confinement [59].

AI has the potential to have a significant impact as follows:

1.   The contagion should be detected and diagnosed as soon as possible.
2.   Therapy is being monitored.
3.   Tracking of links.
4.   Case studies and mortality estimations in the future.
5.   Medicines and vaccinations are being developed.
6.   Reducing medical staff workload.
7.   Disease prevention [60].

7.3.7 AI in ophthalmology

As ophthalmology is an intensive field that depends mainly on imaging, AI can be utilized to help prevent medication errors [61].

7.3.8 Design of the treatment

AI is enabling breakthroughs in medical procedures such as better medication strategy and data analysis, allowing for the best therapeutic strategies and therapeutic objectives. AI can more quickly and accurately detect disease indications and signs in clinical images such as MRIs, computed tomography (CT) scans, ultrasounds, and X-rays. Diagnostics may now be completed fast, cutting treatment time for patients from days to hours and allowing for more therapy intervention [62]. In addition, AI has been used in a variety of clinical applications, including therapy, diagnosis, rehabilitation, surgery, and prognostic treatments. Another important area of medicine where AI is having an impact is clinical management and medical diagnostics [29].

7.4 The Present State of AI and Its Future

AI techniques are not yet thinking machines, which means they cannot think in the same way that normal practitioners do, relying on "rational thinking" or "medical instinct and expertise." On the other hand, AI functions as a data converter, converting trends from databases into themes. Medical businesses are starting to use AI to automate time-consuming and complex ongoing procedures. Many people have demonstrated that AI can be used to find mistakes in things like diabetic retinopathy and radiotherapy planning [63].

7.4.1 Benefits

1. An AI structure deployed in the healthcare setting can assist clinicians in providing the latest healthcare data from a variety of academic articles. Similarly, AI could aid in the reduction of pathogenic and restorative errors that are not avoidable in in-person facility operations [64].

2. Robot-assisted operations can replace costly operations, which will be economical and useful in psychiatric care.

3. AI will shorten the time it takes to diagnose and cure patients [65].

4. The use of AI in computer models for disease propagation and inception may be measured by applying standard information from the internet, social media, and other media sources [66].

5. Intense healthcare costs may reduce with AI technologies [67].

6. Data from research labs is utilized to track patients' data in real time to diagnose diseases ahead of time [68]. The ease in the process of hospitalization by implementing AI is depicted in Figure 7.6.

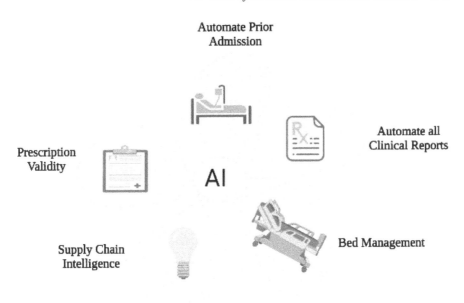

Figure 7.6 AI in hospitalization.

7.4.2 Difficulties of AI in healthcare

- The origin of statistics
- Confidentiality
- It requires comprehensive and impartial data
- Classification schemes are analyzed
- Excessive interpretation
- The long-term consequences of actual entity decisions
- Discriminatory practices
- Premium depending on individual characteristics
- Biases are reinforced [69]

7.5 Role of Robotics in Modern Healthcare

During the COVID-19 pandemic, robotics assisted in reducing pathogen contact. It has become obvious that healthcare robotics can assist in a variety of situations. They can be more efficient in reducing the risk of accidents.

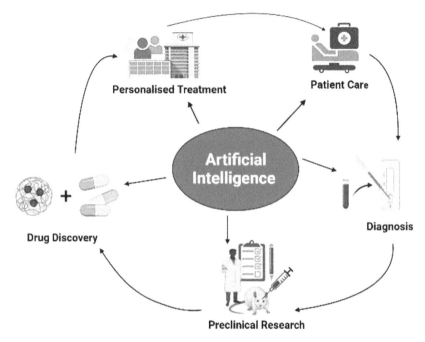

Figure 7.7 AI in various fields of the pharmaceutical industries.

Automation, for example, could clean and prepare people's beds on their own, reducing the amount of person-to-person contact in infectious situations. It is now used in clinical settings to assist health personnel and improve patient care, in addition to operating rooms. During infectious disease cases, hospitals and clinics began using robots for a much broader range of jobs. Also, in hospitals, a machine with AI-enabled medication assists patients. As technology grows, robotics will become more autonomous, potentially executing activities fully on their own. Therefore, doctors, nurses, and other health professionals should help to treat patients with more empathy [15]. The implementation of AI in various fields is indicated in Figure 7.7.

7.5.1 Drug research and development

Robots are used at every stage of the drug supply chain, from basic research to drug manufacturing, quality assurance, and packaging. Robotics aids in the development of critical therapies, allowing for rapid diagnostic procedures for patients and aiding healthcare organizations in meeting stringent drug manufacturing laws while improving drug manufacturing processes. Computers and robots have their origins in the industrial sector; therefore, there is no surprise that they have not made it into the manufacturing process of the drug development lifecycle. The capabilities required to carry out routine tasks with great precision and accuracy are now in place. Robots are most useful during the filing, assembling, and packaging stages of the manufacturing process [70].

7.5.2 Dispensing in pharmacies

Hospital dispensing is a novel robotics application with a promising future. Robots improve the efficiency and accuracy of drug dispensing in pharmacies. Additionally, they are used in both clinical and community drugstores to fulfill digitally placed pharmaceutical orders [70].

7.5.3 Logistics at the hospital

In a typical 300-bed facility, equipment and garbage are transported slightly less than 400 miles per week. It is expected that the use of automation machines will reduce shipping costs by 85%. It also reduces the average daily distance walked by nurses by 3–4 miles. These robots can reduce the number of injuries caused by healthcare lifting and provide caregivers more time to focus on what they do best to provide excellent care [70].

7.6 Robotic Healthcare Is More Advance than Conventional Dispensing

Drug dispensing by hand, preparation of dosages, and organization of medications according to the needs of the patient are all aspects of conventional medical practice. It has been discovered that chemists spend a significant time on those tasks. The growing popularity of robotic pharmacy in the field of medication dispensing has led to the use of robots and automation to perform all of these tasks with the least amount of human interaction, including mixing, classifying, and wrapping drugs. Following are just a few of the numerous benefits.

7.6.1 Increased effectiveness

In addition to a regular human pharmacist, a pharmacy robot can fill and administer a wide range of medications without the need for operator interaction and chances of error. This means that chemists will have more time for higher-value work while also contributing to the smooth operation of the company.

7.6.2 Medical dispensing in an error-free environment

When robots are used, human errors may be reduced to five per lakh drugs. "After you've trained the robot to do the right thing, it will do it again and again, without making errors. Only people contribute to those errors all along the way," says the administrator of the UCSF (Mission Bay hospital pharmacy).

7.6.3 Pharmaceutical operations efficiency

One of the challenges for pharmaceutical companies is that human pharmacists end up advising drug dosage and completing the appropriate patient details on the prescriptions based on patient demand. It could have serious consequences as well as legal and regulatory ramifications.

7.6.4 Confidentiality

In a robotic pharmacy, medicine is safely encrypted in dispensing machines. Due to strict hospital administration policies, these systems can only be managed by someone with network access. With this level of security, the chances of pharmaceutical theft or negligence are significantly reduced. On top of that, the computer software monitors and tracks everything that is distributed.

7.6.5 A toxin-free and secure setting

Infections caused by microbial exposure and improper prescription medication result in a variety of disorders. Interestingly, a robotic drugstore provides a clean atmosphere, which overcomes these problems. Following a dramatic shift from social methods, the pharmaceutical industry has resurrected its operational activities by incorporating automation and robotics. With automated processes, healthcare and clinics can improve operational economies, medication filling levels, count reliability, pharmaceutical errors, people's safety, medical dose consistency, and distribution networks. It may also assist businesses in reducing inventory and waste, as well as expenses because firms will not have to hire additional staff to deal with the increased workload during busy periods. It will be interesting to see how smart healthcare facilities react to this new trend [71].

7.6.6 Advantages of robotics in healthcare systems

1. According to current projections, only 5% of the demand for general surgery is expected to be met. However, the wall has been breached by

robotics. The processor does not use 95% of the available options. Even as surgical procedures become more popular, many surgical professionals accept the robotics. Furthermore, many clinics have expressed an interest in collaborating with AI companies. The next era will be dominated by robotics surgical systems that address the operational and financial problems that current autonomous robots have had [72].

2. Smart technologies must be able to show that they can help people who face a lot of problems like high healthcare costs in complex settings and limitations of hepatic sensing and judgment [73].

3. Because of its accuracy, repetition, and identification ability, robotic digitalization is transforming the medical industry including advances in diagnostics, post-operative planning, post-operative assessment, acute rehabilitation, and chronic assistant devices [74].

4. Machines are not always beneficial to patients, but they can provide a bit of relaxation for doctors and surgeons. As surgeries are not long or stressful with the use of a machine, doctors spend less time in the operating room and are not forced to stand in awkward positions [75].

5. The foremost benefit appears to be that robotic systems are to promote economic growth. Consider how much time and effort doctors would save if they did not have to worry about identifying, administering, or monitoring patients, which results in a successful outcome of 90% of the time. Instead, they could use this time to treat patients, perform life-saving procedures, or even develop innovative healthcare technologies. Another advantage is that robots have unrivaled access to the body; imagine performing surgery in small, difficult-to-reach areas that would have been impossible to reach with only a prosthetic arm [76].

7.7 Acceptance and Implementation of Robots in the Healthcare Business

According to current forecasts, just 5% of the need for general surgery will be met. Robotics, on the other hand, has burst through the barrier. The CPU ignores 95% of the parameters. Even though surgical procedures are becoming more popular, many surgical professionals accept robotics. Many clinics have expressed a desire to work with robotics. Robotics will dominate the next era surgical systems that address current autonomous robots' operational and financial issues [72]. Smart technologies must be able to demonstrate their ability to assist people who face numerous challenges, such as high healthcare costs in complex settings, the expense, and limitations of hepatic sensing, and judgment. These individuals deserve to be recognized and utilized in intelligent ways [73]. This is because robots and digitalization are improving everything from diagnostics to post-operative planning to rehabilitation to chronic assistant devices [74], and acute rehabilitation [75]. Machines are not always beneficial to patients, but they can aid in the relaxation of doctors and surgeons. Doctors spend less time in the operating room and are not forced to stand in awkward positions because surgeries are not as long or as stressful without the use of a machine. The first apparent advantage appears to be that robotic systems promote economic growth. Consider how much time and effort doctors would save if they did not have to worry about identifying, administering, or monitoring patients, which results in 90% of the time having successful outcomes. They could instead use this time to treat patients, perform life-saving procedures, or even develop innovative healthcare technologies. Using robots to do surgery in hard-to-reach places would have been impossible with a prosthetic arm alone [76].

In the medical industry, robotic applications range from surgical procedures to the living room. Robots are used in a variety of settings to assist people with repetitive tasks that are high-risk or require a high level of precision. Robots have also been used in medical care in Europe successfully. "Paro," the seal, a psychological responsibility automaton, was sold all over the world. Around 1200 robots have been sold in Japan, 100 in Denmark, and 100 in the United States. Even though the results of robotics use in different parts of the world and the United States have shown promising and beneficial results in the medical industry, it has been used infrequently in the United States. Currently, research is being conducted on assisted robots. There is mounting evidence that rehabilitative assistance robots can assist people [80–84].

7.8 AI Robotics Emerging Together to Transform the Healthcare System

AI is becoming more capable of doing what humans do, but more effectively and quickly. In addition, AI and robotics both have enormous potential in the medical field. Furthermore, AI and robotics are becoming increasingly important components of our medical environmental friendliness, just as they are in our daily lives. Machine intelligence is improving its ability to perform tasks that humans do much more accurately and quickly, and at a lower cost. By discovering novel genetic code connections and performing machine surgery, AI assists patients, clinicians, and healthcare workers [82]. It is replacing the current medical system with machines that can imagine, comprehend, remember, and respond in ways that humans cannot. Furthermore, in the field of medicine, AI has enormous potential. Both innovations are gradual, and they will become an immediate part of healthcare, just as they are in our daily lives. Many healthcare organizations rely on computers to make decisions with minimal human input. However, it is expected that, in the long run, similar computerized systems will make difficult decisions based on their knowledge without the need for human intervention [81]. We have selected a few examples of how this transformation is currently taking place, which is as follows.

7.8.1 Error rates reduction

Hospitals can use AI and robotics to reduce misdiagnosis and medical error rates, which will result in a lower death toll worldwide each year. In 2015, misdiagnosed illnesses and medical errors were responsible for 10% of all deaths worldwide. By examining individuals' health histories, AI could be useful in predicting and detecting illnesses at a rapid rate. Aside from that, it is designed to assist doctors in implementing a much more comprehensive illness management strategy and in the coordination of care plans. It also helps patients to manage and adhere to lengthy clinical guidelines.

7.8.2 Improving the health of patients

On the other hand, AI empowers individuals to manage their health and well-being, allowing healthcare practitioners to have a better understanding of the problems they confront on a daily basis. Individuals who are attempting to develop self-expression or who are deteriorating due to lifestyle, environment, cell biology, or even other factors can be classified using the modeling approach. AI may also be useful in identifying patients who are developing self-expression or deteriorating due to lifestyle, environment, bioinformatics, or other factors and taking effective action to treat them. Individuals are encouraged to adopt healthier habits through the use of digital software and apps that aid in the proactive maintenance of healthy lifestyles. It also gives customers access to healthcare and well-being services [84]. Furthermore, AI improves healthcare providers' ability to comprehend the day-to-day behaviors and needs of the people they manage, allowing them to provide better advice, direction, and assistance to stay fit.

7.8.3 Early detection

AI has already been used to diagnose diseases such as cancer more precisely and earlier. According to the cancer society, a significant percentage of mammograms produce false results, resulting in one out of every two healthy people being diagnosed with cancer. Mammography is being reviewed and translated at a rate of 30 times faster and with 99% accuracy, reducing the need for unnecessary biopsies. Consumption wearables and other medical equipment, along with AI, have been used to identify and detect serious episodes of high cholesterol levels at an early stage [85]. This allows doctors and other care providers to better supervise and identify deteriorating conditions at this early stage.

7.8.4 Improving decision-making

To improve the healthcare system, health information must be combined with accurate and effective judgments, and prescriptive analytics can assist health professionals in managing administrative duties while providing recommendations on behavior and choices. Watson Healthcare, an IBM division dedicated to data-driven health technology, is said to have made progress in providing clinical decision-making for clinical care. Watson Healthcare is also collaborating with health centers to assist them in utilizing cognition technology to disseminate a large amount of health information and more accurate diagnoses. By combining huge volumes of health data with relevant and appropriate judgments, analytics can help with healthcare decisions and activities, as well as everyday activities, which are high on the agenda for improving treatment. Another area in healthcare where AI is gaining traction is the use of recognized data to identify people who are at high risk of contracting, developing, or worsening disease as a result of their behavior, environment, heredity, or other variables.

7.8.5 Research and development

According to the Californian Medical Research Association, the journey from the laboratory to the person is lengthy and costly, with medicine taking at least 12 years to reach the user. Only 5 out of every 5000 medications that begin in clinical trials make it to animal trials, and only 1 of those 5 is approved for human use. Furthermore, it costs an average of $54.5 million for a manufacturer to develop a drug/medicine for human use. Pharmaceutical research and development is one of the most recent applications of healthcare analytics. There is a way to use the most recent advances in AI to accelerate the development of new drugs and recycling processes [86]. As a result of ongoing technological advancements and innovations, biopharmaceutical companies are paying attention to AI's productivity, dependability, and knowledge. One of the most significant AI discoveries in drug discovery occurred in 2007 when a group of academics tasked a robot named Adam with investigating yeast activities. The robot then sifted through a large amount of data in readable codes to make predictions about the activities of the fungus's 19 chromosomes, and it made nine correct predictions.

7.8.6 Treatment

The technology could indeed assist medical providers in taking a more systematic approach to disease management, best-coordinating wellness programs, and assisting patients in managing and continuing to comply with their lengthy drug treatments, as well as continuing to search medical files to assist suppliers in identifying severely mentally ill people who are already at an increased risk of adverse events. Robots have been used in the medical field for over 30 years. They range in severity from small laboratory robots to overly complicated robotic surgery that can either assist a traditional surgeon or perform surgeries on their own. They are used in hospitals and clinics to do very time-consuming things, like therapy, physiotherapy, care to people who have long-term illnesses, as well as surgery [87].

7.8.7 End-of-life care regeneration

In recent years, robotics has proven to be an excellent component of medical care. They have the potential to transform life care, ranging from simple to advanced experimental robotic systems to extremely difficult tasks such as robotic surgery, which can either assist a conventional physician or perform procedures on their own. In the future, robots will be able to go one step further and have social conversations with humans, which will keep them healthy and their brains sharp as they get older [87, 88].

7.9 Categories of AI Robotic Systems used in the Healthcare System

7.9.1 Assistants to surgeons

Intelligent robots could be developed to assist doctors with a variety of tasks. These things have been getting better and better, and the most recent 3D technology gives surgeons the positional references they need for very complex surgery. These robots have improved natural stereo visualization as well as augmented reality. Using computer vision, these machines are also taught different types of learning data to help them understand the situation and take the right steps [89]. Major companies are expanding their logistics processes in the field of robotic surgical systems. The majority of the market is now controlled by intuitive surgical; however, the situation remains volatile. The entry of large automakers such as Johnson & Johnson and Medtronic has benefited the MED-tech surgical robotics sector. Furthermore, the number of robotic surgical procedures performed in the medical field is steadily increasing. This expansion can be linked to an increase in the global market adoption of robotic surgical equipment. Between 2017 and 2018, the innovative surgery industry saw a 32% increase in global surgeries [90].

7.9.2 Pharmabiotics

It is critical in the medical field to be able to provide precise numbers or medicine, as well as move quickly so that people can recover faster and receive an accurate diagnosis. In addition, another improved pharmacy automation technique has been implemented, allowing robotics to manage powdered, fluid, and extremely sticky ingredients with significantly greater efficiency and agility [89]. Whether at hospitals or healthcare institutions, autonomous robots can offer a wide range of products. In February, Omnicell unveiled the next generation of their Medimat dispensing robots, which are aimed at medical centers. According to Omnicell, Medimat can help a pharmaceutical club in saving time and preventing medication errors by analyzing factors such as repackaging drugs, searching for the exact medication, and staying on top of inventory management. With this type of sub-system, the Medimat system functions as both a small processing facility and a distribution center [91].

7.9.3 Telehealthcare

Telehealth makes use of telemedicine robotics as well as records or displays depicting personal clinical procedures and discussions. Maintaining individual confidentiality is another issue in telehealthcare. When images or voice captures are generated, it appears that a moral quandary exists. Experts at the medical center recommend phone lines, the internet, and other methods for telehealth. Many people with the same illness or a counseling service can see it [92]. As a result, treatment, such as gripping someone's arm, could immediately signal how they were feeling, allowing confidence to be created more easily with such a healthcare professional than with a web camera or indirect communication channels, which could leave all parties guessing about the effectiveness of the communication [93].

7.9.4 Robotics with exoskeletons

Exoskeletons were initially developed to assist handicapped people, quadriplegics, and others in standing. The platform's inventions begin to play a significant role in people's lives. Ekso Bionics, one of the leading players in the industry recently received Federal Drug Administration approval for a new exoskeleton. Individuals with heart and nervous system damage can get their lives back on track faster with this innovative exoskeleton [94]. Exoskeletons or motorized exoskeletons are modular smart machines that use processor systems to control a network of actuators, pneumatic cylinders, valves, or hydraulic systems to recover locomotion [95, 96]. This topic of artificial limbs appears to be current, given the volume of mobility technologies researched and acquired by healthcare institutions and individuals for home use [97, 98]. According to the World Health Organization, regular exercise is any muscular movement that occurs through muscular activities and causes an increase in energy consumption. Without the use of muscles, prosthetic limbs allow for the voluntary contraction of the upper extremities. This appears to be the case in exoskeleton bipedal locomotion when there is an insufficient intake of extra oxygen and nutritional content. As a result, combining functional brain impulses with robot conditioning can be an effective way to combat the problem by increasing muscle activity or energy intake [99–101].

7.9.5 Robots for cleaning and decontamination

Machines are the most effective approach to eliminating or cleaning up polluted areas, preventing individuals from falling ill from viruses such as COVID-19. It can quickly and efficiently remove chemicals and clean a large area. However, robots and AI may become contaminated if used to clean and decontaminate hazardous areas. When people are exposed to infectious environments, they become ill or suffer from a variety of health problems, such as infectious diseases caused by organisms like bacteria, viruses, fungus, or parasites. Unfortunately, healthcare facilities are notoriously unpleasant places to work. Thus, anyone going for therapy may get different illnesses. Bacterial resistance can be found in hospitals since many of the most efficient antibacterial drugs are available there. Sophisticated cleaning robots can be deployed in apartments, which employ high-powered ultraviolet radiation for a few seconds to eradicate any lingering bacteria [102].

7.10 Robotic Programming

When AI-based robots are used in the real world, they are capable of performing a wide range of tasks without the need for human intervention. In reality, autonomous robots are built with ML capabilities that enable them to distinguish between different parts and comprehend complex situations. Computational models are used to distinguish unique characters and properly anticipate the consequences learned from these sets of data, which is accomplished through the use of object identification software. Systems such as robots, which use AI, provide tagged photos of goods that allow computers to detect elements in a wide range of situations and dimensions. Currently, analytics is a well-known brand that offers cognitive computing methods and an attribute selection service. Furthermore, image recognition and smart automation datasets could be used in this case, allowing comparable systems to perform with the highest level of accuracy and speed [89].

Automation allows equations to be generated automatically by analyzing large amounts of data, which speeds up research and simplifies the development of complex systems such as ML tools. Even though massive amounts of data are required for accurate computer learning to function, the data used to train ML models must be precise and of the highest quality. Currently, most mass-produced assistive devices that assist people with daily tasks are significantly less motivated to embrace the design of such a production line with mechanical arms or even the automated personal assistants that were initially anticipated during the space program period. After two centuries, robot brains that are not needed for limbs are scarce in every field. Pattern recognition uses ML levels to aid in the simple use of information. This idea was imagined 20–30 times but never gained popularity due to the limitations of processing capabilities at the time. Because of the groundbreaking capabilities and technological advancements of fully convolutional techniques, ML is finally getting its due. If you have the processing power and the right knowledge, machine understanding can be extremely powerful. It is critical to distinguish between preparation and comprehension when it comes to intelligence processing of such information. Preparation includes situations in which all of the parameters are provided and the robot only needs to figure out how quickly each component should operate to complete a goal, such as collecting an element. During training, on the other hand, the machine has to deal with a lot of different outputs in a very chaotic and demanding way [103].

7.11 Conclusion

Robotic surgery has advanced to the point where it will be capable of transforming human healthcare and communities. The surgical team is now a complicated mixture of segments and subs, video conferencing screens, laptop procedures, and technologically improved operating room procedures. Scanners used by doctors are an excellent complement to robots, providing them with additional options with the least amount of unwanted intervention. Methods are being used, and access to target areas is being expanded. They have also accelerated the delivery of freight and pharmaceuticals throughout the facility. Many more are on the way, each with its unique value proposition. This entails receiving the best medical care possible at the most affordable price. Robotic systems have resulted in the development of critical treatment, rehabilitation, and infant and geriatric care equipment. Furthermore, clinical experience in geriatric care is preferred. The da Vinci surgical system is a field of study that aims to improve people's lives by providing new skills. It will undoubtedly continue to be in high demand in the healthcare industry. Moreover, robotic surgery has a good chance of becoming more popular in the coming years.

7.12 Acknowledgment

We are thankful to our co-authors for their contribution to this chapter.

7.13 Funding

None.

7.14 Conflict of Interest

There is no conflict of interest.

References

[1] Owoyemi, A., Owoyemi, J., Osiyemi, A., & Boyd, A. (2020*). Artificial intelligence for healthcare in Africa*. Frontiers in Digital Health, 2, 6.

[2] Stanfill, M. H., & Marc, D. T. (2019). Health information management: implications of artificial intelligence on healthcare data and information management. *Yearbook of medical informatics*, *28*(01), 056–064.

[3] Schönberger, D. (2019). Artificial intelligence in healthcare: a critical analysis of the legal and ethical implications. *International Journal of Law and Information Technology*, *27*(2), 171–203.

[4] Laï, M. C., Brian, M., & Mamzer, M. F. (2020). Perceptions of artificial intelligence in healthcare: findings from a qualitative survey study among actors in France. *Journal of translational medicine*, *18*(1), 1–13.

[5] Lee, D., & Yoon, S. N. (2021). Application of artificial intelligence-based technologies in the healthcare industry: Opportunities and challenges. *International Journal of Environmental Research and Public Health*, *18*(1), 271.

[6] Ibrahim, H., Liu, X., & Denniston, A. K. (2021). Reporting guidelines for artificial intelligence in healthcare research. *Clinical & experimental ophthalmology*, *49*(5), 470–476.

[7] Jiang, F., Jiang, Y., Zhi, H., Dong, Y., Li, H., Ma, S., & Wang, Y. (2017). Artificial intelligence in healthcare: past, present and future. *Stroke and vascular neurology*, *2*(4).

[8] Mohanta, B., Das, P., & Patnaik, S. (2019, May). Healthcare 5.0: A paradigm shift in digital healthcare system using Artificial Intelligence, IOT and 5G Communication. In *2019 International Conference on Applied Machine Learning (ICAML)* (pp. 191–196). IEEE.

[9] Hans Moravec, Shai Agass; Robotics technology; Britannica https://www.britannica.com/technology/robot-technology .

[10] Okamura, A. M., Matarić, M. J., & Christensen, H. I. (2010). Medical and health-care robotics. *IEEE Robotics & Automation Magazine*, *17*(3), 26–37.

[11] Abbott, D. J., Becke, C., Rothstein, R. I., & Peine, W. J. (2007, October). Design of an endoluminal NOTES robotic system. In *2007 IEEE/RSJ International Conference on Intelligent Robots and Systems* (pp. 410–416). IEEE.

[12] Taylor, R. H. (1997, March). Robots as surgical assistants: where we are, wither we are tending, and how to get there. In *Conference on Artificial Intelligence in Medicine in Europe* (pp. 1–11). Springer, Berlin, Heidelberg.

[13] Kayla Mathews; The growing emergence of robots in health care: key opportunities and benefits;2019; hit consultant.net.

[14] Eric J. Moore; Robotics surgery medical technology; medical technology Britannia.com

[15] https://www.intel.in/content/www/in/en/healthcare-it/robotics-in-healthcare.html

[16] Tekkeşin, A. (2019). Artificial intelligence in healthcare: Past, present and future. *Anatol J Cardiol*, *22*(Suppl 2), 8–9.

[17] Khanna, D. (2018). Use of artificial intelligence in healthcare and medicine. *International Journal of innovations in Engineering Research and technology*.vol 5 issue 12.

[18] Macrae, C. (2019). Governing the safety of artificial intelligence in healthcare. *BMJ quality & safety*, *28*(6), 495–498.

[19] Amann, J., Blasimme, A., Vayena, E., Frey, D., & Madai, V. I. (2020). Explainability for artificial intelligence in healthcare: a multidisciplinary perspective. *BMC Medical Informatics and Decision Making*, 20(1), 1–9.

[20] Arora, A., Batra, N., Grover, P., & Rishik, A. (2021). Role of Artificial Intelligence in Healthcare. *Available at SSRN 3884373*.

[21] https://inside.battelle.org/blog-details/the-algorithm-is-in-5-ways-ai-is-transforming-medicine.

[22] Ahuja, A. S. (2019). The impact of artificial intelligence in medicine on the future role of the physician. *PeerJ*, 7, e7702.

[23] Bajpai, N., & Wadhwa, M. (2021). Artificial Intelligence and Healthcare in India.

[24] Vijai, C., & Wisetsri, W. (2021). Rise of Artificial Intelligence in Healthcare Startups in India. *Advances In Management*, 14(1), 48–52.

[25] Rekha, K. P. (2021). Artificial Intelligence In Healthcare. *KYAMC Journal*, 11(4), 164–165.

[26] Asan, O., Bayrak, A. E., & Choudhury, A. (2020). Artificial intelligence and human trust in healthcare: focus on clinicians. *Journal of medical Internet research*, 22(6), e15154.

[27] Reddy, S. (2018). Use of artificial intelligence in healthcare delivery. In *eHealth-Making Health Care Smarter*. IntechOpen.

[28] Sunarti, S., Rahman, F. F., Naufal, M., Risky, M., Febriyanto, K., & Masnina, R. (2021). Artificial intelligence in healthcare: opportunities and risk for future. *Gaceta Sanitaria*, 35, S67–S70.

[29] Secinaro, S., Calandra, D., Secinaro, A., Muthurangu, V., & Biancone, P. (2021). The role of artificial intelligence in healthcare: a structured literature review. *BMC Medical Informatics and Decision Making*, 21(1), 1–23.

[30] Chen, M., & Decary, M. (2020, January). Artificial intelligence in healthcare: An essential guide for health leaders. In *Healthcare management forum* (Vol. 33, No. 1, pp. 10–18). Sage CA: Los Angeles, CA: SAGE Publications.

[31] Datta, S., Barua, R., & Das, J. (2020). Application of artificial intelligence in modern healthcare system. *Alginates—recent uses of this natural polymer*.

[32] Bohr, A., & Memarzadeh, K. (2020). The rise of artificial intelligence in healthcare applications. In *Artificial Intelligence in healthcare* (pp. 25–60). Academic Press.

[33] Noorbakhsh-Sabet, N., Zand, R., Zhang, Y., & Abedi, V. (2019). Artificial intelligence transforms the future of health care. *The American journal of medicine*, 132(7), 795–801.

[34] https://healthinformatics.uic.edu/blog/machine-learning-in-healthcare/.
[35] Hamet, P., & Tremblay, J. (2017). Artificial intelligence in medicine. *Metabolism*, *69*, S36–S40.
[36] Pillai, R., Oza, P., & Sharma, P. (2020). Review of machine learning techniques in health care. In *Proceedings of ICRIC 2019* (pp. 103–111). Springer, Cham.
[37] Shailaja, K., Seetharamulu, B., & Jabbar, M. A. (2018, March). Machine learning in healthcare: A review. In *2018 Second international conference on electronics, communication and aerospace technology (ICECA)* (pp. 910–914). IEEE.
[38] Paton, C., & Kobayashi, S. (2019). An open science approach to artificial intelligence in healthcare. *Yearbook of medical informatics*, *28*(01), 047–051.
[39] Zhong, F., Xing, J., Li, X., Liu, X., Fu, Z., Xiong, Z., ... & Jiang, H. (2018). Artificial intelligence in drug design. *Science China Life Sciences*, *61*(10), 1191–1204.
[40] Kim, S. K., & Huh, J. H. (2020). Artificial Neural Network Blockchain Techniques for Healthcare System: Focusing on the Personal Health Records. *Electronics*, *9*(5), 763.
[41] Sordo, M. (2002). Introduction to neural networks in healthcare. *Open Clinical: Knowledge Management for Medical Care*.
[42] Kumar, S. S., & Kumar, K. A. (2013). Neural networks in medical and healthcare. *International Journal of Innovation and Research Development*, *2*(8), 241–44.
[43] Reddy, S., Fox, J., & Purohit, M. P. (2019). Artificial intelligence-enabled healthcare delivery. *Journal of the Royal Society of Medicine*, *112*(1), 22–28.
[44] Pai, V. V., & Pai, R. B. (2021). Artificial intelligence in dermatology and healthcare: An overview. *Indian Journal of Dermatology, Venereology & Leprology*, *87*(4).
[45] Dhall, D., Kaur, R., & Juneja, M. (2020). Machine learning: a review of the algorithms and its applications. *Proceedings of ICRIC 2019*, 47–63.
[46] Le Nguyen, T., & Do, T. T. H. (2019, August). Artificial intelligence in healthcare: A new technology benefit for both patients and doctors. In *2019 Portland International Conference on Management of Engineering and Technology (PICMET)* (pp. 1–15). IEEE.
[47] El Kafhali, S., & Lazaar, M. (2021). Artificial Intelligence for Healthcare: Roles, Challenges, and Applications. In *Intelligent Systems in Big Data, Semantic Web and Machine Learning* (pp. 141–156). Springer, Cham.

[48] Manne, R., & Kantheti, S. C. (2021). Application of artificial intelligence in healthcare: chances and challenges. *Current Journal of Applied Science and Technology*, *40*(6), 78–89.

[49] Kipp W. Johnson, Jessica Torres Soto, Benjamin S. Glicksberg, Khader Shameer, Riccardo Miotto, Mohsin Ali, Euan Ashley,Joel T. Dudley. (2018) Journal of the American College of Cardiology.71, (23) 2676.

[50] Wani, I. M., & Arora, S. (2020). Deep neural networks for diagnosis of osteoporosis: a review. *Proceedings of ICRIC 2019*, 65–78.

[51] Abhinav, G. V. K. S., & Subrahmanyam, S. N. (2019). Artificial intelligence in healthcare. *Journal of Drug Delivery and Therapeutics*, *9*(5-s), 164–166.

[52] Yin, J., Ngiam, K. Y., & Teo, H. H. (2021). Role of artificial intelligence applications in real-life clinical practice: systematic review. *Journal of medical Internet research*, *23*(4), e25759.

[53] Yu, K. H., Beam, A. L., & Kohane, I. S. (2018). Artificial intelligence in healthcare. *Nature biomedical engineering*, *2*(10), 719–731.

[54] Rong, G., Mendez, A., Assi, E. B., Zhao, B., & Sawan, M. (2020). Artificial intelligence in healthcare: review and prediction case studies. *Engineering*, *6*(3), 291–301.

[55] Diprose, W., & Buist, N. (2016). Artificial intelligence in medicine: humans need not apply?. *The New Zealand Medical Journal (Online)*, *129*(1434), 73.

[56] Madai, V. I., & Higgins, D. C. (2021). Artificial Intelligence in Healthcare: Lost In Translation?. *arXiv preprint arXiv:2107.13454*.

[57] Kumar, A., Gadag, S., & Nayak, U. Y. (2021). The Beginning of a New Era: Artificial Intelligence in Healthcare. *Advanced Pharmaceutical Bulletin*, *11*(3), 414.

[58] De, A., Sarda, A., Gupta, S., & Das, S. (2020). Use of artificial intelligence in dermatology. *Indian journal of dermatology*, *65*(5), 352.

[59] Dananjayan, S., & Raj, G. M. (2020). Artificial Intelligence during a pandemic: The COVID-19 example. *The International Journal of Health Planning and Management*.

[60] Mohapatra, I., Giri, P. (2020) Artificial Intelligence in Healthcare and Application in the Fight against Current Pandemic - COVID-19. National Journal of Research in Community Medicine. 9, (2)82.

[61] Kapoor, R., Walters, S. P., & Al-Aswad, L. A. (2019). The current state of artificial intelligence in ophthalmology. *Survey of ophthalmology*, *64*(2), 233–240.

[62] N. Murali, N. Sivakumaran. (2018) International Journal of Modern Computation, Information and Communication Technology. 1(6), 105.

[63] Bajwa, J., Munir, U., Nori, A., & Williams, B. (2021). Artificial intelligence in healthcare: transforming the practice of medicine. *Future Healthcare Journal*, 8(2), e188.

[64] Gadde,S.S., Kalli,V.D. (2021) Artificial Intelligence at Healthcare Industry. International Journal for Research in Applied Science & Engineering Technology (IJRASET). 9(2),313.

[65] Sharma, A. (2021) Artificial Intelligence in Health Care. International Journal of Humanities, Arts, Medicine and Science 5(1):106–109

[66] Sharma, A. (2021) . Artificial Intelligence in the Healthcare Industry. International Journal of Research in Humanities, Arts and Science, 5 (1), 2456–5571.

[67] Yu, K. H., Beam, A. L., & Kohane, I. S. (2018). Artificial intelligence in healthcare. *Nature biomedical engineering*, 2(10), 719–731.

[68] Chen, L. K. (2018). Artificial intelligence in medicine and healthcare. *Journal of Clinical Gerontology and Geriatrics*, 9(3), 77–78.

[69] Thesmar, D., Sraer, D., Pinheiro, L., Dadson, N., Veliche, R., & Greenberg, P. (2019). Combining the power of artificial intelligence with the richness of healthcare claims data: Opportunities and challenges. *PharmacoEconomics*, 37(6), 745–752.

[70] Sostero, M. (2020). Automation and robots in services: review of data and taxonomy.

[71] Sd globatech.com/blog/robotic pharmacy -the latest innovation -in medicine depensing

[72] Wall, J., Chandra, V., & Krummel, T. (2008). Robotics in General Surgery, Medical Robotics. *Tech Publications, p 12. European Commission, Information Society.*

[73] Taylor, R. H., Menciassi, A., Fichtinger, G., Fiorini, P., & Dario, P. (2016). Medical robotics and computer-integrated surgery. In *Springer handbook of robotics* (pp. 1657–1684). Springer, Cham.

[74] Taylor, R. H. (1997, March). Robots as surgical assistants: where we are, wither we are tending, and how to get there. In *Conference on Artificial Intelligence in Medicine in Europe* (pp. 1–11). Springer, Berlin, Heidelberg.

[75] Yang, X., She, H., Lu, H., Fukuda, T., & Shen, Y. (2017). State of the art: bipedal robots for lower limb rehabilitation. *Applied Sciences*, 7(11), 1182.

[76] https://themsag.com/blogs/nhs-hot-topics/nhs-hot-topics-robotics-in-healthcare

[77] Riek, L. D. (2017). Healthcare robotics. *Communications of the ACM*, 60(11), 68–78.

[78] Balasubramanian, S., Klein, J., & Burdet, E. (2010). Robot-assisted rehabilitation of hand function. *Current opinion in neurology*, *23*(6), 661–670.

[79] Lluch, M. (2011). Healthcare professionals' organisational barriers to health information technologies—A literature review. *International journal of medical informatics*, *80*(12), 849–862.

[80] Shibata, T., Wada, K., Saito, T., & Tanie, K. (2005, April). Human interactive robot for psychological enrichment and therapy. In *Proc. AISB* (Vol. 5, pp. 98–109).

[81] Sabelli, A. M., Kanda, T., & Hagita, N. (2011, March). A conversational robot in an elderly care center: an ethnographic study. In *2011 6th ACM/ IEEE international conference on human-robot interaction (HRI)* (pp. 37–44). IEEE.

[82] Broekens, J., Heerink, M., & Rosendal, H. (2009). Assistive social robots in elderly care: a review. *Gerontechnology*, *8*(2), 94–103.

[83] Harmo, P., Taipalus, T., Knuuttila, J., Vallet, J., & Halme, A. (2005, August). Needs and solutions-home automation and service robots for the elderly and disabled. In *2005 IEEE/RSJ international conference on intelligent robots and systems* (pp. 3201–3206). IEEE.

[84] Flandorfer, P. (2012). Population ageing and socially assistive robots for elderly persons: the importance of sociodemographic factors for user acceptance. *International Journal of Population Research*, *2012*.

[85] Wired (2016). http://www.wired.co.uk/article/cancer-risk-ai-mammograms (last accessed on 15 january 2022)

[86] California Biomedical Research Association. New Drug Development Process. http://www.ca-biomed.org/pdf/media-kit/fact-sheets/CBRA DrugDevelop.pdf last accessed on 22February 2022)

[87] https://www.pwc.com/gx/en/industries/healthcare/publications/ai-robotics-new-health/transforming-healthcare.html (last accessed on 20 january 2022)

[88] https://www.analyticsinsight.net/how-ai-and-robotics-in-healthcare-are-excelling-beyond-science-fiction/ last accessed on 22 january 2022)

[89] https://www.anolytics.ai/blog/how-ai-robots-used-in-medical-health-care/ (last accessed on 20 February 2021)

[90] https://www.medicaldevice-network.com/comment/what-are-the-main-types-of-robots-used-in-healthcare/ last accessed on 30January 2022)

[91] https://www.pharmaceutical-technology.com/features/robotic-drug-dispensing-digital-pharmacy/ last accessed on 30 january 2022)

[92] Dean, E. (2009, Sept). Robots designed to grin and bear it. Nursing Standard, 24(1), 11. Demiris, G., Doorenbos, A., & Towle, C. (2009). InjRehabil. 2017; 23:245–255.

[93] Butter, M., Rensma, A., Boxsel, J., Kalisingh, S., Schoone, M., Leis, M., Gelderblom, G., & Cremers, G. (2008, Oct). Robotics for healthcare: Final report. eHealth, Retrieved from http://www.scribd.com/doc/10269005/Rob optic's-for-Healthcare.

[94] https://www.roboticsbusinessreview.com/robo-dev/exoskeletons-uses-beyond-healthcare/ last accessed on 12 February 2022)

[95] Mekki, M., Delgado, A. D., Fry, A., Putrino, D., & Huang, V. (2018). Robotic rehabilitation and spinal cord injury: a narrative review. *Neurotherapeutics*, *15*(3), 604–617.

[96] Miller, L. E., Zimmermann, A. K., & Herbert, W. G. (2016). Clinical effectiveness and safety of powered exoskeleton-assisted walking in patients with spinal cord injury: systematic review with meta-analysis. *Medical devices (Auckland, NZ)*, *9*, 455.

[97] Gorgey, A. S., Sumrell, R., & Goetz, L. L. (2019). Exoskeletal assisted rehabilitation after spinal cord injury. *Atlas of Orthoses and Assistive Devices*, 440–447.

[98] Gorgey, A. S., Wade, R., Sumrell, R., Villadelgado, L., Khalil, R. E., & Lavis, T. (2017). Exoskeleton training may improve level of physical activity after spinal cord injury: a case series. *Topics in spinal cord injury rehabilitation*, *23*(3), 245–255.

[99] Gorgey, A. S., Wade, R., Sumrell, R., Villadelgado, L., Khalil, R. E., & Lavis, T. (2017). Exoskeleton training may improve level of physical activity after spinal cord injury: a case series. *Topics in spinal cord injury rehabilitation*, *23*(3), 245–255.

[100] Del-Ama, A. J., Gil-Agudo, Á., Pons, J. L., & Moreno, J. C. (2014). Hybrid FES-robot cooperative control of ambulatory gait rehabilitation exoskeleton. *Journal of neuroengineering and rehabilitation*, *11*(1), 1–15.

[101] Del-Ama, A. J., Koutsou, A. D., Moreno, J. C., De-Los-Reyes, A., Gil-Agudo, Á., & Pons, J. L. (2012). Review of hybrid exoskeletons to restore gait following spinal cord injury. *Journal of Rehabilitation Research & Development*, *49*(4).

[102] https://interestingengineering.com/15-medical-robots-that-are-changing-the-world(last accessed on 20 February 2022

[103] https://www.roboticsbusinessreview.com/opinion/what-robots-need-to-succeed-machine-learning-to-teach-effectively/ (last accessed on 20 February 2022).

8

Artificial Intelligence and Machine Learning Approach for Development and Discovery of Drug

Shweta Kumari[1], Akhalesh Kumar[2], Pawan Upadhyay[2], Ruchi Singh[1], Sudhanshu Mishra[3], Smriti Ojha*[3], and Neeraj Kumar[4]

[1]Narayan Institute of Pharmacy, Gopal Narayan Singh University, Jamuhar, Rohtas India
[2]Department of Pharmaceutical Sciences, Maharishi University of Information Technology, India
[3]Department of Pharmaceutical Science and Technology, Madan Mohan Malaviya University of Technology, Gorakhpur, India
[4]Faculty of Pharmacy, AIMST University, Malaysia
*Corresponding Author
Smriti Ojha
Email; smritiojha23@gmail.com

Abstract

Artificial intelligence (AI) is the combination of human brain power, machine, and computer techniques. In the healthcare and pharmaceutical industries, AI has been used in a variety of ways, with an emphasis on research, wearables, drug discovery, and virtual assistants. It also helps in maintaining patient data, regular monitoring of health issues, and diagnosis, all of which contribute to a better lifestyle and mental health. AI is a blessing for the pharmaceutical industry in terms of drug discovery, identification, validation, and improving the R&D efficiency in analyzing biomedical information. In the drug development process, it contributes to the initial steps of research by predicting 3D structural protein with effective receptor binding or targeted protein. Various applications are available for estimating possible drug interactions like Chem Tapper, SEA, etc. Similarly, instead of preliminary studies followed by a clinical trial, several web applications are used to identify the toxicity of new moieties while saving money and time. Furthermore, several AI tools are available to use, including DeepChem, DeepTox, organic PotentialNet, and Hit Dexter. This chapter focuses on AI-assisted platforms and tools that are changing the way drugs are discovered and bringing innovation to healthcare facilities in the future.

8.1 Introduction

In today's scenario, artificial intelligence (AI) is elevating day by day in various sectors, especially in the pharmaceutical Industry. Also, AI has shown an effective role in drug discovery and development. There are various tools and techniques in AI which makes it more effective in further aspects of the pharmaceutical industry [1]. It is very useful for drug repurposing, clinical trials, and improving pharmaceutical productivity. AI reduces the human workload and manpower by achieving targets in a short duration. In the pharmaceutical industry, digitalization is in great demand with acquiring and executing the knowledge for solving serious clinical problems. Interestingly, AI leads to automation as it can store large volumes of data [2]. AI is a technology-based system that can mimic human intelligence but cannot threaten to replace manpower completely. Recently, AI has been able to evaluate, interpret, and understand input data with the use of systems and software to conclude correct and independent verdicts. In the pharmaceutical sector, the applications of AI are gradually extending, and, in the future, it will completely modify the work culture [3]. With the help of deep learning and machine learning, it has made a significant contribution to the progress of drug and vaccine development [4].

AI is a game-changing tool for the development of new therapies. Additionally, deep learning generates a lot of buzz in the AI community. Many researchers suggest that machine learning will speed up drug discovery and development and the application of AI procedures; however, there is still a gap and a lack of understanding [5]. In the era of personalized medicine, data collection and management in pharmaceutical industries are becoming increasingly important. Thus, their information should be instantly and efficiently assembled, analyzed, and characterized. AI is a promising application for drug discovery, diagnosis, research, and clinical trials [6]. AI also helps in maintaining quality control and quality assurance, which maintains batch-to-batch consistency; thus, many pharmaceutical companies employ this technique in a combination of AI and human power. It is being regulated in line dosage forms fabricating process to attain the specific product standard. Besides, it also helps in marketing and designing pharmaceutical products. It creates a special identity for each product to attract customers for buying them. Indeed, many companies already employ this method, and it is projected that AI will promote smart work with massive innovative techniques in the coming days, acting as a fuel for industrial progress in any circumstance. Globally, AI is pivotal in all leading biopharmaceutical companies; for example, Sanofi signed a deal with a UK firm, Exscientia, to use its AI platform for advanced metabolic disease therapies, while Pfizer is using IBM Watson system that utilizes machine learning to improve immune-oncology drugs [6]. Likewise, GNS Healthcare in Cambridge, MA, USA, is providing an AI system for Roche subsidiary, Genetech, to hunt multinational company's deals for cancer treatments [7]. Currently, precision medicine helps researchers to co-relate and identify relationships between different datasets and medical lab reports through dynamic visualizations. AI utilizes symbolic programming for problem-solving with vast applications in healthcare, engineering, and business [8]. AI is integrated with software science that greatly evolved into a modern science. The main objective of AI is to deal with the problem solving, design, and application for learning, analyzing, and interpreting data. AI has distinct fields, including pattern recognition, statistics, similarity-based methods, clustering, and machine learning [9]. In the pharmaceutical sector, AI is a flourishing technology that utilizes automated algorithms to perform various tasks [9]. Traditionally, AI relies on human intelligence in multiple aspects of life and industry. Over the last three years, the pharmaceutical and biotechnology sectors have been using this powerful technology to reshape how scientists approach a disease, discover new and innovative approaches to generate new medications, and much more [10]. Figure 8.1 summarizes various steps for a drug development process that could be linked and assisted with AI to improve result efficacy and save time and money.

8.2 Tools of AI used to Emphasize Pharmacy

8.2.1 The robot pharmacy

The UCSF Medical Center utilized robotic technology for the preparation and tracking of medications to improve the safety of patients. Interestingly, the robotic technology has efficiently prepared 400,000 medication doses (oral and injectable) without any error [11]. As a result, UCSF pharmacists and nurses are free to use their knowledge by focusing on direct patient care and cooperating with physicians on their patients' health. This demonstrates that robots are superior to humans in terms of shape, size, and ability to deliver proper and accurate medications [11].

8.2.2 The MEDi robot

MEDi is an abbreviation for medicine and engineering designing intelligence. Tanya Beran, a professor of community health sciences at the University of

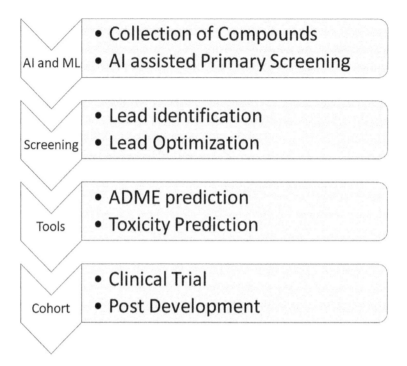

Figure 8.1 A schematic diagram of drug discovery.

Calgary, Calgary, AB, Canada, had a project to develop a pain management robot. After working in hospitals, she had the idea for children and infants who scream during treatment or medical procedures [12]. Although the robot is unable to think, determine, plan, or reason, the robot quietly establishes a rapport with the children and explains what to expect during therapy or medical procedures, indicating the existence of most advanced and effective AI tools [13].

8.2.3 The erica robot

Hiroshi Ishiguro, a professor at Osaka University, Suita, Japan, developed a new care robot namely, Erica. It has a blend of Asian and European facial features [14]. Besides, it has various other features, including the ability to speak Japanese, animated films look, a wish to visit south-east Asia, and a desire for a life partner to chat with. Erica has human-like facial expressions, and it can understand and answer questions but cannot walk independently [15].

8.2.4 The TUG robots

Aethon TUG robots can autonomously travel all over the hospital and deliver medications, meals, and all other necessities. Also, it can carry heavy things. It is designed in such a way that it is a very flexible and utilizable resource for loading various carts or different racks [16].

8.3 Applications

8.3.1 Modifying drug release

Controlled Drug Release: Hussain and his coworkers at Cincinnati University, Cincinnati, OH, USA, performed the modeling of pharmaceutical formulations in neural networks. Subsequently, in various *in vitro* studies, they used hydrophilic polymers for modeling the release characteristics of drugs and the range of drugs dispersed in matrices. With a single hidden layer in neural networks, all of these experiments were able to predict the range of drug release with reasonable accuracy [17]. Recently, researchers at pharmaceutical company Krka d.d., and the University of Ljubljana, Ljubljana, Slovenia formulated the Diclofenac sodium matrix tablet. They used neural networks and 2–3-dimensional response surface analysis to optimize and predict the rate of drug release [18].

Immediate Release Tablets: Almost 2–3 years ago, the University of Marmara and the University of Cincinnati employed statistics and neural networks to model hydrochlorothiazide tablet formulations [19].

8.3.2 Product development

The process of pharmaceutical drug development is a huge multivariate optimization problem as shown in Figure 8.2. The most beneficial aspect of artificial neural networks is their ability to generalize the system. These characteristics are ideal for overcoming and resolving obstructions in optimizing the formulations during drug development [20].

8.4 Benefits [21, 22]

New medicine development and discovery of active pharmaceutical ingredient.

- Analysis of data with speed, reproducibility, and accuracy.

- Effective utilization of incomplete datasets.

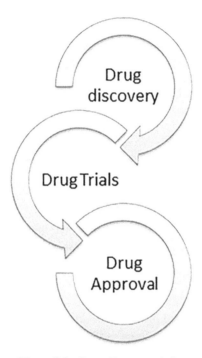

Figure 8.2　Drug discovery chain.

- Accommodate constraints and preferences.
- Generate understandable rules.
- Performance at low cost with enhancing drug quality.
- Improvement in confidence level.
- Improvement in customer response.
- The time duration is short for market.
- The very low error rate in comparison to humans.
- The accuracy, precision, and speed level are incredible.
- No effect on hostile or external environments.

8.5 AI-Integrated Medicine Development

The primary purpose of drug development research is to find medicinal agents that have beneficial effects on the body and can be used to diagnose, prevent, or treat various diseases [23]. Drugs are small chemically synthesized molecules or a molecule obtained from natural origin or a semisynthetic moiety, which can significantly attach to a target receptor (a proteaceous 3D structure) that are directly or indirectly involved in the pathogenesis of disorders [24].

To discover these active molecules, researchers and scientists primarily execute a large screening of available literature to identify the lead with a promising role in the field of life sciences. There are various steps to analyze the effectiveness of these active leads and to develop into a promising compound with patient safety and efficacy [25]. Nowadays, more precise structure-oriented drug design approaches are in common practice. As a result, the initial screening of new molecules will be easier as compared to the previous method. However, still, scientists and researchers need to design, synthesize, and test a variety of chemicals to find possible new drugs [26]. For example, even if a new molecule demonstrates its potential therapeutic impact in laboratory testing, clinical trials may still fail. Indeed, only about 10% of drug candidates make it to the market after Phase I studies [27]. It would be simple to predict the biological action of a lead molecule on 3D portentous structures using AI drug design. Accordingly, AI could predict the biological action of a lead molecule on 3D portentous structures, as well as side effects and toxicity levels, before their synthesis or production, by analyzing the chemical domain of proteinous 3D structures for the probable target of lead and its interaction with various biomolecules [28]. Moreover, AI has the potential to save billions for the development of medicinal agents and their suitable dosage forms by the next decade [29].

8.6 Role of Active Learning and Machine Learning in Drug Discovery

The implementation of machine learning can be explained by the introduction of many companies into this field to expand their business areas. The primary purpose of machine learning is to use high throughput techniques to screen lead compounds, thus reducing rigorous efforts of drug discovery [30]. Through *in silico* models and techniques, machine learning has reduced the use of *in vivo* animal testing. Besides, the data obtained from chemical structures, physiological pathways, and biological activities were used to fabricate and design accurate as well as advanced approaches [31]. They also provide a computational and comparative way to analyze and correlate the results based on the structure of the lead compound [32].

The input data normally contains many medicinal features, including cellular toxicity level, heterogeneity level of cellular structures, the efficacy of available animal model with its on-target activity, and pharmacokinetic parameters like elimination rate constant, elimination half-life, MRT, VRT, and cytochrome P450 metabolic rates [33]. These input data along with orthogonal data and its application within a probable domain with the integration of data for terminating programs are used to design algorithms. This previously designed algorithm predicts results as well as illustrates improvements required in current ongoing methodologies and procedures to get more efficient and reliable predictions [34]. These techniques are modified by repeating the same procedures until a final product is designed with desirable biological efficacy. The algorithm principles used in the active learning concept economically enable product screening [35]. Furthermore, active learning via machine learning is a useful approach for multi-dimensional optimization and drug discovery.

The active-learning is an area of computer science; currently, it is successfully used for drug discovery with its proven practical applicability in fabrication and drug development [36]. The most considered topics of active learning which are associated with drug discovery are cost-effective and cost-aware learning, re-labeling of existing molecules, and selection of various batches. The active learning algorithms of machine learning aid to reduce the high noise ratio and the probability of false negatives [37]. Semisupervised learning is a part of active learning algorithms which is successfully used for complex objectives.

Active learning algorithms are a representation of a promising concept for drug discovery with a wide range of practical applicability [38]. With future advancements in algorithmic technologies, this technique may enable lead discovery in an automated and rational decision-making way [39].

8.7 Explainable AI

Explainable AI is a branch of AI in which researchers and scientists can comprehend the outcomes of diverse issues. In the current state of AI-assisted and structure-based drug discovery, there is a high demand of such algorithms [40]. Explainable AI techniques are now emerging due to a lack of interpretation from certain machine learning tools as well as a need to enhance human reasoning and decision-making ability [41]. The area of explainable AI is developing, and it will prove its relevance in the upcoming years. The recent explainable AI research along with its advantages, limited data options, and prospects for drug discovery is explained here [42]. A summary of selected terminologies used in explainable AI is provided in Table 8.1.

Table 8.1 AI terminologies with their explanations used frequently in drug development.

S. No.	AI terminologies	Explanation
1	Active learning for drug design	Active learning is a subfield of machine learning in which algorithms for an underlying model with selected data are used to improve their models. Actively, the aim of active learning is to label new data and to understand the task more efficiently [43, 44].
2	Algorithm	An algorithm is a mathematical formula or mathematical expression that depicts the link between two or more variables in linear, exponential, or any other form of relationship. Algorithms can be aimed at a simple set of instructions with a finite end goal of producing a result [45].
3	Bots	A bot or chatbot is a program that runs on a website or app and interacts with users directly to assist them with simple tasks [46].
4	Cluster	Any collection of people or things who have something in common [46, 47].
5	*In silico* fragment-to-lead screening	To find a lead chemical, this computational strategy involves screening low molecular weight compounds against various macromolecular targets (typically high molecular weight proteins) of clinical importance [46–48].
6	Functional group and spectral deep learning	Molecule functional group that is responsible for characteristic chemical reactions. Screening of multi-level network of functional groups [49].
7	Image recognition	Computers can be programmed to grasp what is happening in an image, processed for image magnification with clarity, which is one of the most sophisticated machine learning techniques [43–45].
8	Natural language processing	This technology enables machines to decipher what humans are saying through text or voice [42, 45].
9	Lead optimization	The process of improving a compound's potency, selectivity, and pharmacokinetic characteristics [43, 46].
10	Neural networks	This AI model, which is designed to resemble the human brain, uses natural language processing and deep learning to recognize faces in photographs and analyze handwriting [43].
11	Databases	It combines a data warehouse for processing and storage [42–44].
12	Molecular mapping and representations	Graphical representation and mapping of the compound topology at the molecular and cellular level [47, 48].
13	Pharmacophore fingerprint: pharmacoprint	The combination of numerous chemical properties is required for a ligand's particular interaction with a biological receptor [47, 48].
14	3D pharmacophore modeling	3D assemblies with well-defined ligand interactions [47–49].
15	Toxalerts	Compounds' undesirable characteristics, toxicity, and undesired reactivity are connected to functional groups and other molecular substructures. These are collectively stored and predicted [50, 51].
16	Descriptor-free QSAR model	These are approaches for predicting lead's physicochemical and biological properties as a result of its molecular structure, bonds, and atomic bonding, applying descriptor selected algorithms [52, 53].

8.8 Computational Approaches for Explainable AI

8.8.1 Feature attribution

It helps to determine the local feature of the active ingredient which could be important for the prediction of its potency, safety, and efficacy [54–56]. Gradient-based methods, surrogate models, and perturbation-based methods are used to predict feature attribution of lead compounds [57]. These approaches have found their application to identify ligands, identification, and structure-based data for various correlated adverse effects in the prediction of protein–ligand interaction profiling [58].

8.8.2 Instance-based approach

This method is based on calculating a set of criteria that must be present or absent to anticipate the lead compound's activity. Anchors, counterfactual examples, and contrastive explanations are used in computation methods. However, its application in the field of drug discovery is not yet reported [59].

8.8.3 Graph convolution-based approach

It interprets various drug activities based on the message-passing framework models. Subgraph-based techniques and attention-related methods are used in this approach [48]. This method has been used in the studies of retrosynthesis, toxicophore and pharmacophore discovery, ADMET, and reactivity prediction [60].

8.8.4 Self-explaining

This approach develops models that are explained by design. Various methods used for this approach include prototype, self-explaining neurological networks, concept learning, and natural learning [61].

8.8.5 Uncertainty estimation

This approach is aimed to quantify the reliability of various approaches used for this purpose. It includes ensemble and probabilistic-based approaches. It has been used in reaction prediction, molecular-structure-based activity prediction, and active learning [62].

8.8.6 In silico molecular modeling

In silico molecular modeling is mainly based on 3D structures of proteins and receptors to predict the rate and time course of absorption, distribution, metabolism, elimination, and toxicity of the compound. Molecular modeling mainly works on ligand- and structure-based modeling [63]. It receives information from the electronic properties of molecules, their shape in lattice space, inhibitors, substrates, and metabolites' conformational structures [64]. These data are used for designing pharmacophore, density function, and shape focus models for the illustration of the spatial as well as chemical nature of the ligand. Structure-based models mainly use binding properties of atoms and conformational changes of receptors. Moreover, molecular modeling helps to discover a stable and reliable method to predict absorption, distribution, metabolism, elimination, and toxicities [65, 66].

8.9 AI Networks and Associated Tools

Life sciences have benefitted immensely from advances in AI. AI has a lot of potential to enhance and accelerate drug discovery. In 2020, a British start-up, Exscientia, and a Japanese pharmaceutical firm, Sumitomo Dainippon Pharma, used AI to develop a drug for obsessive–compulsive disorders [67]. The typical drug development process takes around five years to reach the trial stage; however, this drug took only one year. Additionally, cheminformatics has also grown by leaps and bounds in the last decade [68].

8.9.1 AlphaFold

Proteins are made up of amino acids and are the building blocks of life. The unique 3D structure of proteins largely defines their function. In the critical assessment of structure prediction, the AlphaFold has been recognized as a solution for various protein folding problems [69]. AlphaFold developed an attention-based neural network system to interpret the structure of a protein's spatial arrangements. It uses related amino acid sequences, multiple sequence alignment of its monomer units, and a representation of amino acid residue pairs, to refine the graph [70]. The AI system has developed strong predictions of the underlying 3D structure of the protein through iterating the physiological bioprocess [71]. AI can look into how protein structure predictions can help us learn more about diseases by identifying the proteins that fell into disrepair. Such insights could accelerate drug development efforts. Besides this, protein structure prediction is also helpful in pandemic response efforts [72].

8.9.2 DeepChem

DeepChem is a drug discovery framework based on open-source deep learning. The python-based framework includes a set of features for using deep learning in drug development. It creates deep learning neural networks with Google TensorFlow and scikit-learn. It also employs the RDKit Python framework for simple molecular data operations like converting SMILES strings to molecular graphs [73].

8.9.3 ODDT

The Open Drug Discovery Toolkit (ODDT) is a free open-source program for computer-assisted drug development (CADD). To create CADD pipelines, ODDT employs machine learning scoring functions (RF-Score and NN-Score) [74, 75] and is available as a Python library. ODDT can support a multitude of formats by boosting the use of Cinfony, a common API that unites molecular toolkits like RDKit and OpenBabel and makes interfacing with them more Python-like [73]. Numpy arrays are used to hold all-atom data collected from the underlying toolkits, providing speed and versatility. The ODDT is available under a three-clause BSD license that is suitable for both academic and industry use [76].

8.9.4 Cyclica

A biotech firm MatchMaker from Cyclica uses reams of biochemical and structural data to quickly compare candidate compounds against the full proteome. Pareto-optimal embedded modeling (POEM) is a parameter-free supervised learning method for creating property prediction models with less overfitting and higher interpretability. "If you're developing a chemical, it behooves you to examine the other 299 interactions that might have terrible impacts on humans," said CyclicA CEO, Naheed Kurji [77].

8.9.5 DeepTox

The clinical trials phase of drug development constitutes another bottleneck, taking a long time and money before a drug reaches the market. With the aid of machine learning and algorithm-based tools, clinical trial design becomes much easier and more economical. DeepTox is deep learning for the prediction of toxicities during clinical studies [74, 75]. DeepTox works by normalizing the chemical formula-based representation to compute a sufficient quantity of chemical descriptors. These descriptors are input data for machine learning [75].

8.9.6 Deep neural net QSAR

This is a correlation between deep neural networks and other ligand-based virtual screening. In deep neural net QSAR, the machine learning method uses active algorithms for the screening of the active lead and concurrent predictions from various QSAR models [76]. Chemoinformatics is employed in these models to present data and make predictions; nonetheless, they have significant drawbacks, such as the inability to be self-explainable and the prioritization of structural features with high activity [75, 76].

8.9.7 Organic

This tool helps digitalize the organic synthesis. Organic is an AI planning tool that uses a robotic platform to do planned and flow synthesis of organic structures that have the potential to act as lead [77].

8.9.8 PotentialNet

PotentialNet is designed to analyze protein–ligand binding affinity. It seeks input data as the distance between two adjacent atoms in an angstrom with a restriction against its chemical bond predictions. Neighbor type and noncovalent interactions are viewed for predictions. PotentialNet consists of three types of propagations, including covalent-bond-based propagation, noncovalent and covalent propagation, and third ligand-based propagations [78].

8.9.10 Hit dexter

Hit Dexter models are used for the prediction of large fractions of promising compounds among approved drugs. This technology works to find out and estimate the trigger effect of small molecules in biochemical processes and various biological assays [79].

8.10 Technical Obstacles and Prospects

The qualitative properties of input data used for designing algorithms are the main technical challenge for the application of AI at drug discovery platforms. The published results of various models lack reproducibility and are erroneous [80]. Subsequently, the complexity of the dataset for the prediction of 3D structures and their spatial arrangement is not easy to understand and predict. Furthermore, many published data are proprietary, and the inventors have highlighted legal concerns about the use of AI and machine learning technologies in this field [81]. On the other side, a lack of datasets in the field of drug discovery as well as a scarcity of skilled workers are two problems limiting the market's growth [82]. The global AI in the drug discovery market is expected to grow at a CAGR of 40.8% from USD

259 million in 2019 to USD 1434 million by 2024 [83, 84]. This growth is being driven by a growing number of cross-industry collaborations and partnerships, the increasing need to control drug discovery and development costs and reduce the overall time taken in this process, the rising adoption of cloud-based applications, and services, and the impending patent expiry of blockbuster drugs [85, 86].

8.11 Conclusion

A human being is the most sophisticated and complex machine that can ever be created in the whole universe. On the other hand, AI has changed and modified the pharmaceutical profession considerably. The need for advanced technology will increase continuously as the healthcare sector gets more sophisticated. Globally, in today's scenario, AI is known as the application of the algorithm and the best technique for the analysis and interpretation of data. The advancement of AI, along with its astonishing tools, is constantly aimed at reducing obstacles faced by pharmaceutical firms, affecting the medication development process as well as the total lifespan of the product, which may explain the rise in the number of start-ups in this area [87]. Using the latest AI-based technologies will not only reduce the time it takes for products to reach the market but also improve product quality and overall safety of the manufacturing process. Furthermore, this will provide better resource utilization and cost-effectiveness, highlighting the importance of automation [88]. Moreover, AI will develop further in the future, allowing it to reach its full potential and assist the pharmaceutical business.

8.12 Acknowledgment

I would like to thank my co-authors for contributing their knowledge and time and giving their support in compiling the work.

8.13 Funding

The authors received no specific grants from any funding agencies.

8.14 Conflicts of Interest

The authors declare no conflict of interest.

References

[1] Abràmoff, M. D., Lavin, P. T., Birch, M., Shah, N., & Folk, J. C. (2018). Pivotal trial of an autonomous AI-based diagnostic system for detection of diabetic retinopathy in primary care offices. *NPJ digital medicine, 1*(1), 1–8.

[2] Bhattacharyya, C., & Keerthi, S. S. (2001). Mean field methods for a special class of belief networks. *Journal of Artificial Intelligence Research, 15*, 91–114.

[3] Brafman, R. I. (2001). On reachability, relevance, and resolution in the planning as satisfiability approach. *Journal of Artificial Intelligence Research, 14*, 1–28.

[4] Lim, S., Lu, Y., Cho, C. Y., Sung, I., Kim, J., Kim, Y., ... & Kim, S. (2021). A review on compound-protein interaction prediction methods: data, format, representation and model. *Computational and Structural Biotechnology Journal, 19*, 1541–1556.

[5] Bod, R. (2002). A unified model of structural organization in language and music. *Journal of Artificial intelligence research, 17*, 289–308.

[6] Costello, J. C., Heiser, L. M., Georgii, E., Gönen, M., Menden, M. P., Wang, N. J., ... & Stolovitzky, G. (2014). A community effort to assess and improve drug sensitivity prediction algorithms. *Nature biotechnology, 32*(12), 1202–1212.

[7] Ghasemi, F., Mehridehnavi, A., Perez-Garrido, A., & Perez-Sanchez, H. (2018). Neural network and deep-learning algorithms used in QSAR studies: merits and drawbacks. *Drug discovery today, 23*(10), 1784–1790.

[8] Bui, H. H., Venkatesh, S., & West, G. (2002). Policy recognition in the abstract hidden markov model. *Journal of Artificial Intelligence Research, 17*, 451–499.

[9] Fox, M., & Long, D. (2003). PDDL2. 1: An extension to PDDL for expressing temporal planning domains. *Journal of artificial intelligence research, 20*, 61–124.

[10] Cheng, J., & Druzdzel, M. J. (2000). AIS-BN: An adaptive importance sampling algorithm for evidential reasoning in large Bayesian networks. *Journal of Artificial Intelligence Research, 13*, 155–188.

[11] Chen, X., Yan, C. C., Zhang, X., Zhang, X., Dai, F., Yin, J., & Zhang, Y. (2016). Drug–target interaction prediction: databases, web servers and computational models. *Briefings in bioinformatics, 17*(4), 696–712.

[12] Console, L., Picardi, C., & Duprè, D. T. (2003). Temporal decision trees: Model-based diagnosis of dynamic systems on-board. *Journal of artificial intelligence research, 19*, 469–512.

[13] Cristani, M. (1999). The complexity of reasoning about spatial congruence. *Journal of Artificial Intelligence Research*, *11*, 361–390.

[14] Dieterich, T. G. (2000). Hierarchical reinforcement learning with the MAXQ value function decomposition. *Journal of artificial intelligence research*, *13*, 227–303.

[15] Nebel, B. (2000). On the compilability and expressive power of propositional planning formalisms. *Journal of Artificial Intelligence Research*, *12*, 271–15.

[16] Drummond, C. (2002). Accelerating reinforcement learning by composing solutions of automatically identified subtasks. *Journal of Artificial Intelligence Research*, *16*, 59–104.

[17] Elomaa, T., & Kaariainen, M. (2001). An analysis of reduced error pruning. *Journal of Artificial Intelligence Research*, *15*, 163–187.

[18] Simon, H., & Frantz, R. (2003). Artificial intelligence as a framework for understanding intuition. *Journal of Economic Psychology*, *24*(2), 265–277.

[19] Kundu, M., Nasipuri, M., & Basu, D. K. (2000). Knowledge-based ECG interpretation: a critical review. *Pattern Recognition*, *33*(3), 351–373.

[20] Lang, J., Liberatore, P., & Marquis, P. (2003). Propositional independence-formula-variable independence and forgetting. *Journal of Artificial Intelligence Research*, *18*, 391–443.

[21] Masnikosa, V. P. (1998). The fundamental problem of an artificial intelligence realization. *Kybernetes.*

[22] Nau, D. S., Au, T. C., Ilghami, O., Kuter, U., Murdock, J. W., Wu, D., & Yaman, F. (2003). SHOP2: An HTN planning system. *Journal of artificial intelligence research*, *20*, 379–404.

[23] Palomar, M., & Martínez-Barco, P. (2001). Computational approach to anaphora resolution in Spanish dialogues. *Journal of Artificial Intelligence Research*, *15*, 263–287.

[24] Oke, S. A. (2008). A literature review on artificial intelligence. *International journal of information and management sciences*, *19*(4), 535–570.

[25] Wilkins, D. E., Lee, T. J., & Berry, P. (2003). Interactive execution monitoring of agent teams. *Journal of Artificial Intelligence Research*, *18*, 217–261.

[26] Wray, R. E., & Laird, J. E. (2003). An architectural approach to ensuring consistency in hierarchical execution. *Journal of Artificial Intelligence Research*, *19*, 355–398.

[27] Xu, X., He, H. G., & Hu, D. (2002). Efficient reinforcement learning using recursive least-squares methods. *Journal of Artificial Intelligence Research*, *16*, 259–292.

[28] Yu, K. H., Fitzpatrick, M. R., Pappas, L., Chan, W., Kung, J., & Snyder, M. (2018). Omics AnalySIs System for PRecision Oncology (OASISPRO): a web-based omics analysis tool for clinical phenotype prediction. *Bioinformatics*, *34*(2), 319–320.

[29] Zanuttini, B. (2003). New polynomial classes for logic-based abduction. *Journal of Artificial Intelligence Research*, *19*, 1–10.

[30] Zhang, N. L., & Zhang, W. (2001). Speeding up the convergence of value iteration in partially observable Markov decision processes. *Journal of Artificial Intelligence Research*, *14*, 29–51.

[31] Zucker, J. D. (2003). A grounded theory of abstraction in artificial intelligence. *Philosophical Transactions of the Royal Society of London. Series B: Biological Sciences*, *358*(1435), 1293–1309.

[32] Bellucci, M., Delestre, N., Malandain, N., & Zanni-Merk, C. (2021). Towards a terminology for a fully contextualized XAI. *Procedia Computer Science*, *192*, 241–250.

[33] Murdoch, W. J., Singh, C., Kumbier, K., Abbasi-Asl, R., & Yu, B. (2019). Definitions, methods, and applications in interpretable machine learning. *Proceedings of the National Academy of Sciences*, *116*(44), 22071–22080.

[34] Goebel, R., Chander, A., Holzinger, K., Lecue, F., Akata, Z., Stumpf, S., ... & Holzinger, A. (2018, August). Explainable ai: the new 42?. In *International cross-domain conference for machine learning and knowledge extraction* (pp. 295–303). Springer, Cham.

[35] Fiore, M., Sicurello, F., & Indorato, G. (1995). An integrated system to represent and manage medical knowledge. *Medinfo. MEDINFO*, *8*, 931–933.

[36] Reker, D., & Schneider, G. (2015). Active-learning strategies in computer-assisted drug discovery. *Drug discovery today*, *20*(4), 458–465.

[37] Doss, C. G. P., Chakraborty, C., Narayan, V., & Kumar, D. T. (2014). Computational approaches and resources in single amino acid substitutions analysis toward clinical research. *Advances in protein chemistry and structural biology*, *94*, 365–423.

[38] Yang, H., Lou, C., Sun, L., Li, J., Cai, Y., Wang, Z., ... & Tang, Y. (2019). admetSAR 2.0: web-service for prediction and optimization of chemical ADMET properties. *Bioinformatics*, *35*(6), 1067–1069.

[39] Yang, H., Lou, C., Sun, L., Li, J., Cai, Y., Wang, Z., ... & Tang, Y. (2019). admetSAR 2.0: web-service for prediction and optimization of chemical ADMET properties. *Bioinformatics*, *35*(6), 1067–1069.

[40] Yongye, A. B., & Medina-Franco, J. L. (2013). Systematic characterization of structure–activity relationships and ADMET compliance: a case study. *Drug Discovery Today*, *18*(15-16), 732–739.

[41] Paul, D., Sanap, G., Shenoy, S., Kalyane, D., Kalia, K., & Tekade, R. K. (2021). Artificial intelligence in drug discovery and development. *Drug Discovery Today*, *26*(1), 80.

[42] Jiménez-Luna, J., Grisoni, F., & Schneider, G. (2020). Drug discovery with explainable artificial intelligence. *Nature Machine Intelligence*, *2*(10), 573–584.

[43] Yu, J., Li, X., & Zheng, M. (2021). Current status of active learning for drug discovery. *Artificial Intelligence in the Life Sciences*, *1*, 100023.

[44] Arabi, A. A. (2021). Artificial intelligence in drug design: algorithms, applications, challenges and ethics. *Future Drug Discovery*, *3*(2), FDD59.

[45] Gilvary, C., Madhukar, N., Elkhader, J., & Elemento, O. (2019). The missing pieces of artificial intelligence in medicine. *Trends in pharmacological sciences*, *40*(8), 555–564.

[46] Sturm, N., Mayr, A., Le Van, T., Chupakhin, V., Ceulemans, H., Wegner, J., ... & Chen, H. (2020). Industry-scale application and evaluation of deep learning for drug target prediction. *Journal of Cheminformatics*, *12*(1), 1–13.

[47] Bote-Curiel, L., Munoz-Romero, S., Gerrero-Curieses, A., & Rojo-Álvarez, J. L. (2019). Deep learning and big data in healthcare: A double review for critical beginners. *Applied Sciences*, *9*(11), 2331.

[48] Lin, Y., Mehta, S., Küçük-McGinty, H., Turner, J. P., Vidovic, D., Forlin, M., ... & Schürer, S. C. (2017). Drug target ontology to classify and integrate drug discovery data. *Journal of biomedical semantics*, *8*(1), 1–16.

[49] Exner, T. E., Keil, M., & Brickmann, J. (2002). Pattern recognition strategies for molecular surfaces. II. Surface complementarity. *Journal of computational chemistry*, *23*(12), 1188–1197.

[50] Zhou, H., Dong, Z., & Tao, P. (2018). Recognition of protein allosteric states and residues: Machine learning approaches. *Journal of computational chemistry*, *39*(20), 1481–1490.

[51] Cao, X., Hu, X., Zhang, X., Gao, S., Ding, C., Feng, Y., & Bao, W. (2017). Identification of metal ion binding sites based on amino acid sequences. *PloS one*, *12*(8), e0183756.

[52] Schaller, D., Šribar, D., Noonan, T., Deng, L., Nguyen, T. N., Pach, S., ... & Wolber, G. (2020). Next generation 3D pharmacophore modeling. *Wiley Interdisciplinary Reviews: Computational Molecular Science*, *10*(4), e1468.

[53] Yang, H., Sun, L., Li, W., Liu, G., & Tang, Y. (2018). In silico prediction of chemical toxicity for drug design using machine learning methods and structural alerts. *Frontiers in chemistry*, *6*, 30.

[54] Lakkaraju, H., Kamar, E., Caruana, R., & Leskovec, J. (2017). Interpretable & explorable approximations of black box models. *arXiv preprint arXiv:1707.01154*.

[55] Deng, H. (2019). Interpreting tree ensembles with intrees. *International Journal of Data Science and Analytics*, 7(4), 277–287.

[56] Bastani, O., Kim, C., & Bastani, H. (2017). Interpreting blackbox models via model extraction. *arXiv preprint arXiv:1705.08504*.

[57] Maier, H. R., & Dandy, G. C. (1996). The use of artificial neural networks for the prediction of water quality parameters. *Water resources research*, 32(4), 1013–1022.

[58] Shukla, S. J., Huang, R., Austin, C. P., & Xia, M. (2010). The future of toxicity testing: a focus on in vitro methods using a quantitative high-throughput screening platform. *Drug discovery today*, 15(23–24), 997–1007.

[59] Socher, R., Bengio, Y., & Manning, C. D. (2012). Deep learning for NLP (without magic). In *Tutorial Abstracts of ACL 2012* (pp. 5-5).

[60] Srivastava, N., Hinton, G., Krizhevsky, A., Sutskever, I., & Salakhutdinov, R. (2014). Dropout: a simple way to prevent neural networks from overfitting. *The journal of machine learning research*, 15(1), 1929–1958.

[61] Settles, B. (2010). Active learning literature survey. University of Wisconsin. *Computer Science Department*.

[62] Warmuth, M. K., Rätsch, G., Mathieson, M., Liao, J., & Lemmen, C. (2001). Active learning in the drug discovery process. *Advances in Neural information processing systems*, 14.

[63] Bajorath, J. (2002). Integration of virtual and high-throughput screening. *Nature Reviews Drug Discovery*, 1(11), 882–894.

[64] Grave, K. D., Ramon, J., & Raedt, L. D. (2008, October). Active learning for high throughput screening. In *International Conference on Discovery Science* (pp. 185–196). Springer, Berlin, Heidelberg.

[65] Baram, Y., Yaniv, R. E., & Luz, K. (2004). Online choice of active learning algorithms. *Journal of Machine Learning Research*, 5(Mar), 255–291.

[66] Weber, L. (1998). Evolutionary combinatorial chemistry: application of genetic algorithms. *Drug Discovery Today*, 3(8), 379–385.

[67] Byvatov, E., Sasse, B. C., Stark, H., & Schneider, G. (2005). From virtual to real screening for D3 dopamine receptor ligands. *ChemBioChem*, 6(6), 997–999.

[68] Franke, L., Byvatov, E., Werz, O., Steinhilber, D., Schneider, P., & Schneider, G. (2005). Extraction and visualization of potential pharmacophore points using support vector machines: application to

ligand-based virtual screening for COX-2 inhibitors. *Journal of medicinal chemistry*, *48*(22), 6997–7004.

[69] Chapelle, O. et al. (2006) When can semi-supervised learning work? Semi-supervised Learning, *MIT Press,* pp. 2–3.

[70] Naik, A. W., Kangas, J. D., Langmead, C. J., & Murphy, R. F. (2013). Efficient modeling and active learning discovery of biological responses. *PLoS One*, *8*(12), e83996.

[71] Gaulton, A., Bellis, L. J., Bento, A. P., Chambers, J., Davies, M., Hersey, A., ... & Overington, J. P. (2012). ChEMBL: a large-scale bioactivity database for drug discovery. *Nucleic acids research*, *40*(D1), D1100–D1107.

[72] Schneider, P., & Schneider, G. (2003). Collection of bioactive reference compounds for focused library design. *QSAR & Combinatorial Science*, *22*(7), 713–718.

[73] Goh, G. B., Siegel, C., Vishnu, A., Hodas, N. O., & Baker, N. (2017). Chemception: a deep neural network with minimal chemistry knowledge matches the performance of expert-developed QSAR/QSPR models. *arXiv preprint arXiv:1706.06689.*

[74] Senior, A. W., Evans, R., Jumper, J., Kirkpatrick, J., Sifre, L., Green, T., ... & Hassabis, D. (2020). Improved protein structure prediction using potentials from deep learning. *Nature*, *577*(7792), 706–710.

[75] Riniker, S., & Landrum, G. A. (2013). Similarity maps-a visualization strategy for molecular fingerprints and machine-learning methods. *Journal of cheminformatics*, *5*(1), 1–7.

[76] Rudin, C. (2019). Stop explaining black box machine learning models for high stakes decisions and use interpretable models instead. *Nature Machine Intelligence*, *1*(5), 206–215.

[77] Schneider, P., Walters, W. P., Plowright, A. T., Sieroka, N., Listgarten, J., Goodnow, R. A., ... & Schneider, G. (2020). Rethinking drug design in the artificial intelligence era. *Nature Reviews Drug Discovery*, *19*(5), 353–364.

[78] Goodarzi, M., Dejaegher, B., & Heyden, Y. V. (2012). Feature selection methods in QSAR studies. *Journal of AOAC International*, *95*(3), 636–651.

[79] Patel, J. L., & Goyal, R. K. (2007). Applications of artificial neural networks in medical science. *Current clinical pharmacology*, *2*(3), 217–226.

[80] Yeh, J. Y., Coumar, M. S., Horng, J. T., Shiao, H. Y., Kuo, F. M., Lee, H. L., ... & Hsieh, H. P. (2010). Anti-influenza drug discovery:

structure– activity relationship and mechanistic insight into novel angelicin derivatives. *Journal of medicinal chemistry*, *53*(4), 1519–1533.

[81] Coley, C. W., Thomas III, D. A., Lummiss, J. A., Jaworski, J. N., Breen, C. P., Schultz, V., ... & Jensen, K. F. (2019). A robotic platform for flow synthesis of organic compounds informed by AI planning. *Science*, *365*(6453), eaax1566.

[82] Davies, I. W. (2019). The digitization of organic synthesis. *Nature*, *570*(7760), 175–181.

[83] Ponder, J. W., & Case, D. A. (2003). Force fields for protein simulations. *Advances in protein chemistry*, *66*, 27–85.

[84] Feinberg, E. N., Sur, D., Wu, Z., Husic, B. E., Mai, H., Li, Y., ... & Pande, V. S. (2018). PotentialNet for molecular property prediction. *ACS central science*, *4*(11), 1520–1530.

[85] Hamet, P., & Tremblay, J. (2017). Artificial intelligence in medicine. *Metabolism*, *69*, S36–S40.

[86] Vijayan, R. S. K., Kihlberg, J., Cross, J. B., & Poongavanam, V. (2021). Enhancing preclinical drug discovery with artificial intelligence. *Drug discovery today*.

[87] Chan, H. S., Shan, H., Dahoun, T., Vogel, H., & Yuan, S. (2019). Advancing drug discovery via artificial intelligence. *Trends in pharmacological sciences*, *40*(8), 592–604.

[88] Jämsä-Jounela, S. L. (2007). Future trends in process automation. *Annual Reviews in Control*, *31*(2), 211–220.

9

Artificial Intelligence in Boosting the Development of Drug

Deepika Bairagee[1]*, Poojashree Verma[1], Neelam Jain[1], Neetesh Kumar Jain[1], Sumeet Dwivedi[2], and Javed Ahamad[3]

[1]Pacific College of Pharmacy, Pacific University, Udaipur Rajasthan, India
[2]University Institute of Pharmacy, Oriental University, India
[3]Department of Pharmacognosy, Faculty of Pharmacy, Tishk International University, Iraq
*Corresponding Author
Deepika Bairagee
Email: bairagee.deepika@gmail.com

Abstract

The discovery and development of a drug is always influenced by a high degree of chance and serendipity. Drug development is the process of bringing a new drug molecule into clinical practice. Chemical entities with the potential to become therapeutic agents must be identified and rigorously evaluated during the drug development process, which is very complex and expensive. Several computational approaches have been established in recent years to minimize drug discovery timelines and costs, as well as to improve the quality and success rate of the development process. However, there is still more work to be done in terms of using innovative technology to simplify this process. Artificial intelligence (AI) has the potential to significantly improve the chances of finding novel drug candidates that can be marketed. AI can identify hit and lead compounds, allowing for rapid therapeutic target validation and structural design optimization. AI presents a huge technical advancement that might lead to a paradigm change in drug discovery and, eventually, clinical development. We think that developments that currently feel innovative will quickly become standard practice in terms of time of discovery, novelty, and commercial potential. This chapter focuses on AI methods that are used in the development of a drug. This chapter also discusses various tools and techniques used in AI and their challenges.

9.1 Artificial Intelligence

Artificial intelligence (AI) is increasingly being used in several aspects of society, including the pharmaceutical industry comprising drug investigation and development, drug repurposing, working on drug efficiency, and clinical preliminaries, among others; such use diminishes the human responsibility. We likewise examine crosstalk among the apparatuses and strategies used in artificial intelligence, continuous difficulties, and ways of defeating them, alongside the fate of AI in the drug business.

In recent years, the pharmaceutical business has seen a tremendous surge in information digitalization. Increasing digitalization, however, brings with it the issue of securing, analyzing, and applying that data to complex healthcare situations. This justifies the usage of artificial intelligence, which can cope with massive amounts of data, thanks to enhanced computerization. Computer primarily based intelligence is an innovation primarily based totally framework that includes specific improved gadgets and agencies that could emulate human knowledge. Simultaneously, it does not take steps to supplant human real presence totally [1, 2]. Artificial intelligence uses frameworks and programs that could interpret and take advantage of statistics for you to make impartial judgments for you to reap sure objectives. As this study shows, its uses are frequently sought in the pharmaceutical industry. According to the McKinsey Global Institute, fast breakthroughs in artificial-intelligence-directed robotization would most likely completely transform society's work culture [3].

Computer-based intelligence encompasses several technical areas, including thinking, information representation, and arrangement search, as well as a key artificial intelligence viewpoint (machine learning). Machine learning uses calculations that can comprehend styles in a formerly labeled series of data. Deep learning (DL) is a department of the device getting to know that entails the usage of artificial neural networks (ANNs). These are constructed of some of the interconnected state-of-the-art figuring components, such as "perceptions," which might be essentially equal to human herbal neurons and simulate electric motivation transmission inside the human mind [4]. ANNs are made up of many hubs, each of which receives various data and then switches them over to yield, either individually or in a multi-connected manner, using computations to solve problems. ANNs include different sorts, including multi-layer perceptron (MLP) organizations, repetitive neural networks (RNNs), and convolutional neural networks (CNNs), which use either administered or solo preparing methodology [5, 6].

The MLP network has applications including design acknowledgement, advancement helps, process recognizable proof, and controls, which are normally prepared by administered preparing methods working a solitary way, and can be utilized as general example classifiers. RNNs, like Boltzmann constants and Hopfield networks, are networks with a shut circle and the ability to retain and store data. CNNs are a collection of dynamic frameworks with close affiliations, as defined by geography that is used in image and video processing, organic framework demonstration of difficult cerebrum capacity management, design recognition, and current sign processing. Kohonen organizations, RBF organizations, LVQ groups, counter-proliferation organizations, and ADALINE networks are among the most complex structures. Figure 9.1 summarizes instances of artificial intelligence method spaces [5].

A few apparatuses have been created which are dependent on the organizations that structure the center engineering of artificial intelligence frameworks. The Watson supercomputer from International Business Machines (IBM) is an instance of synthetic intelligence-primarily-based equipment (USA). It changed into an advanced resource withinside the assessment of a patient's medicinal statistics and its interplay with a huge statistics collection, ensuing in remedy tips for malignant growth. This structure can also be implemented for the fast detection of diseases. This changed into testing with the aid of using its capacity to come across breast cancers in about 60 seconds [7].

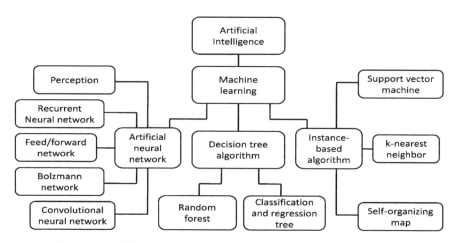

Figure 9.1 Different methods of artificial intelligence with their subfields.

9.2 Computer-Based Intelligence in the Lifecycle of Drug Items

Inclusion of artificial intelligence in the improvement of a drug item from the seat to the bedside can be envisioned, given that it can help judicious medication plan, aid dynamics, decide the right treatment for a patient, including customized meds, and deal with the clinical information created and using it for upcoming medication advancement. E-VAI is an insightful and dynamic AI stage created by Eularis, which utilizes ML calculations alongside a simple-to-utilize UI to make logical guides dependent on contenders, key partners, and at the presently held portion of the overall industry to foresee key drivers in deals of drugs, in this way helping advertising chiefs to assign assets for the greatest piece of the pie again, switching helpless deals and empowered them to expect where to make speculations. Various utilization of artificial intelligence in drug revelation and improvement are summed up in Figure 9.2.

9.3 Drug Development

In its broadest sense, drug improvement encompasses all levels of the process, from initial studies to finding an inexpensive subatomic goal to Phase III scientific trials that are useful resources inside the commercialization of

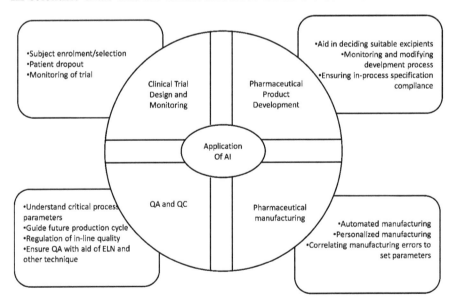

Figure 9.2   Application of artificial intelligence in diverse subfields.

the medicine to post-marketplace pharmacy surveillance and medicine repurposing considerations. Chemical components with the potential to become therapeutic specialists must be found and extensively examined during the medicine creation process, and the entire interaction must be lengthy and costly. Each new medicine that enters the market is projected to cost billions of dollars and take more than 10 years to create. As a result, methods for collaborating with and expediting the drug development process are in great demand.

The Food and Drug Administration (FDA) has currently made development in improving the usage of real-world data (RWD) in drug development. Data collected from assets other than usual exam settings, along with electronic health records (EHRs), authoritative cases, and billing facts, is cited as "RWD." These RWDs regularly include full-size affected person facts that are accompanied over time, together with contamination status, therapy, remedy adherence and results, comorbidities, and concomitant medicines. Data from RWD may be utilized to highlight beneficial development, outcomes research, patient consideration, security reconnaissance, and relative adequacy analysis. More crucially, RWD allows clinical scientists and administrative offices to reply to inquiries more rapidly, saving time and money while providing solutions that can be applied to a broader population. The adoption of EHR frameworks has increased in the United States during the previous decade. In the United States, technical developments and strategic adjustments have offered fertile ground for the usage of RWD in drug development. As a result, the FDA has released guidelines for using EHR data in scientific assessments, also including guidelines for combining RWD into administrative records.

Because of high-quality development in AI approaches, the location of AI, including machine learning and deep learning (ML/DL), has changed from hypothesis to real. Man-made intelligence has been broadly utilized in numerous ranges of the drug improvement procedure to discover novel targets, grow knowledge of sickness components, and increase new biomarkers, among different things. Many pharmaceutical agencies have begun to spend money on assets, advancements, and services, mainly inside the era and amassing of datasets to beautify AI and ML/DL research, and a lot of those datasets come from RWD foundations. There is a growing call for contemporary clinical development to observe a top-level view of AI and RWD convergence to outline the most up-to-date things, like discover current exploration gaps, and deliver bits of records [8].

9.4 The Drug Development Process

The drug development procedure in the United States is divided into five sections, each having several phases and stages (Figure 9.3). The five steps are as follows.

Step 1: Process of Discovery and Development
Drug discovery is the manner in which new pills are observed. All matters considered, pills have been commonly observed with the aid of using spotting dynamic fixings from commonplace medicines or sincerely with the aid of using a few coincidences. Traditional pharmacology was utilized a few years later to examine synthetic libraries including tiny atoms, natural items, and plant extracts to see which ones had therapeutic characteristics. Since the sequencing of human DNA, switch pharmacology has employed testing to uncover treatments for existing illnesses.

Throughout the cycle below, disease processes, atomic compound studies, current drugs with unanticipated side effects, and fresh advancements push drug discovery. To reduce probable medication side effects, today's medicine disclosure comprises screening hits, restorative science, and

Figure 9.3   Drug discovery and development process phases.

improvement of hits (expanding fondness and selectivity). At this level of the medication improvement process, viability or power, metabolic solidity (half-life), and oral bioavailability also are addressed [4].

a. Target Identification and Validation
Target ID identifies a protein (restorative specialist) of high quality that plays an important role in sickness. When restorative traits are identified, they are written down. Targets are safe, efficacious, and useable as drugs, and they fulfill clinical and commercial requirements. Some of the strategies that scientists use to favor objectives encompass *in vitro* genetic regulators, antibodies, and constituent genomics. Other techniques embody infection association, bioactive elements, cell-based absolute models, protein cooperation, weakening ways of research, and practical inspection of makings. The Duolink PLA and the Sanger Whole Genome CRISPER library are all right assets for drug development objectives [3].

b. Hit Discovery Process
Compound screening procedures are designed after objective approval.

c. Measure Development and Screening
Examinations take a look at frameworks that verify the brand new medication applicant's results at the cell, atomic, and natural stages.

d. High Throughput Screening
High throughput screening (HTS) makes use of superior mechanics, facts handling/managing programming, fluid looking after gadgets, and sensitive signs to quickly lead a large variety of pharmacological, substance, and hereditary tests, killing lengthy intervals of meticulous checking out via investigators. HTS distinguishes energetic mixtures, makings, or antibodies that impact social elements [4].

e. Hit to Lead
Minor particle hits from an HTS are analyzed and amplified in a controlled fashion into major complexes in the hit to lead (H2L) process. These combinations are then subjected to the principal streamlining technique.

f. Lead Optimization
The lead compounds discovered in the H2L interaction are integrated and adjusted in the number one lead optimization (LO) process to increase strength while reducing incidental effects. Lead streamlining

prepares the medication competitor by conducting exploratory testing with creature-viable models and ADMET devices.

g. Active Pharmaceutical Ingredients

Active pharmaceutical ingredients (APIs) are certainly dynamic fixes that motivate facts in a medicinal drug application. All medicines include the API or APIs, in addition to excipients. (Excipients are inert chemical materials that resource the drug's absorption into the human-oid body.) High potency active pharmaceutical ingredients (HP APIs) are debris with an extensively decreased detection restriction than tradi-tional APIs. They are utilized in multi-stage drug development and are classified by their potential for damage, pharmacological intensity, and occupational exposure limits (OELs) [4].

When one lead chemical for a pharmaceutical rival is discovered, the drug discovery process comes to an end, and the process of medication improvement starts.

Steps 2: Preclinical Research

Preclinical testing determines the medication's viability and safety when a lead chemical is discovered. Analysts choose the associated with regard to the medication:

- Retention, dispersion, application, and discharge data

- Expected benefits and systems of activity

- Best measurement and organization course

- Secondary effects/unfriendly events

- Consequences for sex, race, or identity gatherings

- Cooperation with different medicines

- Viability contrasted with comparable medications

Preclinical trials examine the novel medicine in nonhuman individuals for safety, toxicity, and pharmacokinetic (PK) data. Researchers focus on these preliminaries *in vitro* and *in vivo* using infinite measures [5].

a. Absorption, Distribution, Disposition, Metabolism, and Excretion

The pharmacokinetic characteristics of absorption, distribution, dispo-sition, metabolism, and excretion (ADME) are used to estimate how a new medicine affects the body. The individual impact is represented numerically in ADME.

b. Proof of Principle/Proof of Concept
Proof of principle (PoP) focuses on preclinical preliminaries and early well-being assessments that are effective. In drug disclosure and development initiatives, the terms proof of concept (PoC) and proof of principle (PoP) are used interchangeably. Effective PoP/PoC focuses on advancing the program through Phase II measurements research.

c. In Vivo, In Vitro, and Ex Vivo Assays
These three forms of research are focused on all living species or cells, including animals and people, as well as non-living substances and tissue removal. *In vivo* preclinical examination models are used to discover new medications employing *in vivo* preclinical examination animals such as mice, rats, and canines. *In vitro* research takes place in a laboratory setting. Individual cells or tissues from a non-living individual are used in *ex vivo* research [5]. *Ex vivo* studies have been conducted to select relevant malignant growth treatment specialists as well as to examine tissue attributes (physical, thermal, electrical, and optical) and establish the viability of innovative treatments. Because it provides an energetic, measured, and antiseptic atmosphere, a cell is constantly active as the basis for micro explant communities in an *ex vivo* investigation.

d. In Silico Assays
Test frameworks or organic exams carried out on a computer or using programming experience are referred to as *in silico* measures. With the continual advancements in computational power and conduct grasp of sub-atomic components and cell science, they are expected to become increasingly well-known.

e. Drug Delivery
Oral, cutaneous, film, intravenous, and internal breath are all-new medication delivery methods. Drug transport structures are used to manage the migration or appearance of novel medications. Biological barriers in animal or humanoid bodies may avoid medics from reporting to their assigned region or conveying when they are supposed to. The goal is to keep the drug from aiding in the mending of healthy tissues while it is still alive [6].

- **Oral:** Patients benefit from oral pharmaceutical administration since it is safe, cost-effective, and convenient. Oral medication transport may not be able to make precise calculations to the optimal location, but it is ideal for preventative immunizations and maintenance regimes. Patients should be aware of delayed movement, stomach complex destruction, maintenance discrepancies, or patients with gastrointestinal troubles or distress during association.

- **Skin:** The topical medicine movement includes balms, creams, lotions, and transdermal patches that provide a prescription to the body through maintenance. Patients like the non-interfering movement and their freedom to self-administer the medicine while using viable transportation.

- **Parenteral (IM, SC, or LP Membrane):** Medicine is given intramuscularly (IM), intraperitoneally (IP), or subcutaneously (SC). Because it avoids epithelial blockages, which make drug distribution difficult, it is frequently utilized in the treatment of unconscious patients.

- **Parenteral (Intravenous):** One of the most effective methods of consuming drugs is intravenous imbuement. In comparison to IM, SC, or LP film techniques, IV imbuement ensures that all prescription segments enter the circulatory system.

- **Parenteral (Inhalation):** The medicine quickly enters the mucosal lungs, nasal passages, throat, and mouth when inhaled. Internal breath transport issues include low mucosal surface areas and patient load, which make it difficult to transmit optimum estimates. In pneumonic internal breath medication conveyance, fine medicine powders or macromolecular prescription game plans are utilized. Small particles can be contained and transmitted into the flow framework because lung fluids mimic blood [2].

f. Detailing Optimization and Improving Bioavailability
    Throughout the preclinical and clinical stages, the plan is constantly evolving. It guarantees that pharmaceuticals are delivered to the appropriate location at the correct time and in the proper amount. Upgrades might help you combat dissolvability.

Stage 3: Clinical Development

Following preclinical research, scientists go on to experimental drug development, which entails preliminary experimental trials and volunteer trials to fine-tune the medication for humanoid usage.

a. The Complexity of Study Design, Associated Cost, and Implementation Issues

   Preliminary plans may be impacted at this time due to the intricacy of the clinical preliminary plan as well as related expenses and execution challenges. Preliminaries must be harmless and practicable, and they must be finished inside the drug development low-priced, with a mechanism in place to verify that the therapy performs as well as it can. This complete cycle must be well set up and attract a big number of helpers to be appealing [3].

b. Clinical Trials – Dose Acceleration, Single Ascending, and Multiple Dose Studies

   Prescription viability is determined by legitimate medicating, and clinical preliminary research examines portion acceleration, single increasing, and numerous portion studies to regulate the greatest silent measurement.

   - **Stage I** – Healthy volunteer study: Under 100 participants will aid analysts in analyzing the wellness and pharmacokinetics, retention, metabolic, and disposal repercussions for the body as well as slightly incidental impacts for harmless measurement limits.

   - **Stage II and Phase III** – The second phase of the trial examines the medicine's care and effectiveness in an additional 100–500 persons who may have been given a phone therapy or a normal prescription that has recently been utilized as treatment. While hostile occurrences and dangers are noted, an analysis of optimal part strength results in plans. Stage III involves the selection of 1000–5000 patients as well as the empowerment of prescription labeling and guidelines for proper medicine usage. In preparation for full-scale manufacturing and prescription approval, Stage III requires a large-scale collaborative effort as well as coordination and guidance from an Independent Ethics Committee (IEC) or an Institutional Review Board (IRB).

c. Organic Samples Collection, Storage, and Shipment

   Normal models are constructed, cared for, and sent from testing objections according to general guidelines and regulations during clinical primers. Dry ice packs or other temperature-settling processes can be used by natural model carriers to consolidate their models. Natural models are subject to a variety of criteria [3].

d. Pharmacodynamics (PD) Biomarkers
    PD biomarkers are nucleotides that link pharmacological standards with natural reactions in the real-world human environment. This data may be used to help choose objective subject matter expert combinations and improve prescription regimens and programs. Sensitivity and hypothesis testing power is improved by using PD endpoints in human starters.

e. Pharmacokinetic Analysis
    A preliminary evaluation of how medicine interacts with the human body is known as a pharmacokinetic assessment. Using compartmental presentation, the volume of movement, opportunity, and terminal half-life are all displayed.

f. Bioanalytical Method Development and Validation
    Bioanalytical procedures identify analyzers and metabolites in regular or human guides, such as medicine or biomarkers, to determine drug practicality and security. Test grouping, clean up, examination, and disclosure are all part of the full bioanalytical test.

g. Medicine (Analyte) and Metabolite Stability in Biological Samples
    The importance of adequacy in determining human prescription appropriateness necessitates the use of natural models. Prescriptions and pharmaceutical metabolites are prone to dilapidation, which might lead to drug center closure due to the prescription's existence.

h. Blood, Plasma, Urine, and Faces Sample Analysis for Drug and Metabolites
    To pick and look at changing qualities and effects of the medicine and its metabolites on persons, standard clinical starting models combine blood, plasma, urine, and faces.

i. Patient Protection – GCP, HIPAA, and Adverse Event Reporting
    To reliably get human patients, clinical essentials, as well as good clinical practices (GCP), the Health Insurance Portability and Accountability Act (HIPAA), and an opposing event explaining the IEC/IRB monitor and guarantee their care, should all be implemented [3].

Stage 4: FDA Review

It is a big step forward for the FDA to conduct a full survey after the new medicine has been tested for maximum viability and security and clinical preliminary results have been released. Currently, the FDA conducts an audit and either supports or rejects the pharmaceutical application submitted by the medication advancement group. The FDA approval process is mentioned in Table 9.1.

a. Administrative Approval Timeline

Depending on the wishes and needs of patients, the new medicine administrative endorsement course of events might be standard, rapid track, forward leap, and speed up endorsement or need an audit. If a standard or necessity audit is required, the approval process might take a lengthy time. Early endorsements, such as quick track, forward jump, or faster endorsements, might happen.

b. IND Application

IND presentations must be filed to the FDA before clinical studies may begin. Engineers can start clinical preliminaries if the preliminaries are ready to go and the FDA has not given the drug a negative response.

c. NDA/ANDA/BLA Applications

When clinical preliminaries indicate pharmacological security and viability, an NDA shortened new drug application (ANDA) or BLA is filed with the FDA. The FDA conducts surveys to gather information earlier determining whether or not to endorse a product. Formerly reaching a concluding determination, additional investigation, or a professional optional board may be mandatory.

d. Vagrant Drug

A vagabond medicine is intended to cure an illness that is so uncommon that financial backers are wary of promoting it under regular marketing conditions. It is possible that these medications will not be able to be supported promptly or at all.

e. Sped-up Approval

If there is compelling proof of a favorable outcome on an auxiliary endpoint rather than proof of result on the medication's honest therapeutic benefits, new medications may be given fast approval. The goal of endorsement is to show that the treatment may help treat significant or dangerous illnesses.

Table 9.1   The FDA's medication approval procedure in a nutshell.

Early Drug Discovery	Preclinical Studies	Clinical Development	FDA Review	Post-Market Monitoring
• Target identification & validation • Hit discovery • Assay development & screening • High throughput screening • Lead optimization	• *In vivo, in vitro, & ex vivo* assays • ADME • Proof of concept • Drug delivery • Formulation optimization, & bioavailability • Dose range finding • IND-enabling studies • IND application	• Phase I – Healthy volunteers study • Phase II and Phase III – Studies in patient population • Dose escalation, single ascending, & multiple dose studies • Safety & efficacy • Pharmacokinetic analysis • Bioanalytical method development and validation	• NDA/ANDA/BLA application • FDA approval • Drug registration	• FDA Adverse Event Reporting System (FAERS)

f. Purposes Behind Drug Failure

For a variety of reasons, new drug applications may fail, including toxicity, appropriateness, PH characteristics, bioavailability, or poor medicine execution.

Poisonousness: Due to security concerns about its use after assembly, the drug may be rejected if the danger of further medicine is too large in human or animal patients. If another medication's adequacy is not high enough, or if the proof is not clear, the FDA may reject it [3].

Medication can also fail an FDA audit due to PK characteristics or inadequate bioavailability as a result of low fluid solvency or excessive first-pass digestion. Insufficient activity duration and unanticipated human medication connections are two PK explanations for pharmacological disappointment.

Lack of Drug Performance: The FDA may discard a request for a plan that accomplishes improvement if the new treatment works as predicted but only to a limited extent.

Stage 5: Post-Market Monitoring

The FDA mandates drug corporations to analyze the care of their products utilizing the FDA Adverse Event Reporting System (FAERS) data collecting once they have been authorized and collated. FAERS is used by the FDA to assist in its post-advertising security reconnaissance program. This effort allows manufacturers, health experts, and customers to report concerns with supported medications.

Here is a quick rundown of the current FDA medication agreement procedure [9].

9.5 In Drug Discovery, Artificial Intelligence

The enormous substance space, which comprises more than 1060 particles, encourages the production of many medication atoms [10]. The pharmaceutical development process, on the other hand, is constrained by a lack of cutting-edge technologies, making it a time-consuming and costly process that AI can assist with [11]. Simulated intelligence can recognize hit and lead chemicals, enabling a more efficient medication structure plan and faster approval of the pharmaceutical target [12].

Notwithstanding its benefits, AI faces some critical information challenges, like the scale, development, variety, and vulnerability of the information. Traditional machine learning methods are unlikely to be able to manage the informative indexes accessible for drug discovery in pharmaceutical businesses, which might contain a large number of combinations. Using a computer model based on the quantitative design action connection, large volumes of mixes or basic physicochemical constraints, such as log P or log D, may be anticipated fast (QSAR). Regardless, these models are far from being able to forecast mind-boggling biological features such as combination viability and bad consequences. In addition, QSAR-based models handle issues such as a shortage of preparation sets, trial information problems in preparation sets, and trial approvals. Recently established AI procedures, like DL and substantial displaying exams, can be applied to undertake security and viability assessments of pharmaceutical particles based on huge data demonstrating and study to solve these challenges. Merck was the winner of a QSAR ML rivalry in 2012, which recognized the welfare of DL in the pharmaceutical industry's drug disclosure approach. DL models beat outdated ML methods for 15 absorptions, distribution, metabolism, excretion, and toxicity (ADMET) informative indices of pharmacological competitors [13, 14].

The virtual synthetic space is enormous, and it presents a particle topographical guide by expressing atom appropriations and attributes. Synthetic space modeling's purpose is to assemble positional data for bioactive mixes, confidentially and virtual screening (VS) aids in the assortment of appropriate atoms used for additional testing. PubChem, ChemBank, DrugBank, and ChemDB are examples of open access synthetic environments.

Using a combination of *in silico* techniques to computer-generated screen compounds from simulated synthetic spaces, as well as construction and ligand-based methodologies, researchers were able to obtain a better profile examination, a faster end of nonlead mixtures, and a better choice of medication particles, all while consuming less. To pick a lead ingredient, medication plan computations such as coulomb grids and sub-atomic finger impression acknowledgement reflect the physical, substance, and toxicological outlines [12, 15].

Different restrictions, such as prediction models, particle similarity, the atomizing method, and the usage of *in silico* methods, can be utilized to forecast the best synthetic assembly of a molecule [16]. When 95,000 distractions were tried against these receptors, Pereira *et al.* introduced DeepVS, a framework for docking 40 receptors and 2950 ligands that displayed good presentation. A multiobjective computerized substitution calculation was used to increase the strength profile by evaluating the form comparability, pharmacological activity, and physicochemical properties of a cyclin-subordinate kinase-2 inhibitor.

Straight discriminant analysis (LDA), support vector machines (SVMs), arbitrary timberland (RF), and decision trees are examples of AI-based QSAR procedures that have grown into QSAR exhibiting apparatuses that may be employed to speed up QSAR examination. When Lord *et al.* compared the capability of six AI computations to rank unknown mixes with natural action to that of conventional approaches, they found that the difference was statistically insignificant [17].

9.6 Artificial Intelligence in Drug Screening

The most common way of finding and fostering a medication can assume control for longer than 10 years and costs US$2.8 billion by and large. That being said, the vast majority of restorative atom bomb Phase II clinical preliminaries and administrative endorsement [1, 18] nearest-neighbor classifiers (RF), support vector machine machines (SVMs), and profound neural organizations (DNNs) are used for VS and can predict *in vivo* movement and risk based on amalgamation attainability [18, 20]. Bayer, Roche, and Pfizer are among the biopharmaceutical companies that have partnered with IT companies to create a platform for medication disclosure in parts like immuno-oncology and cardiovascular disease. The AI-enabled elements of VS are detailed later down [12].

9.6.1 The expectation of the physicochemical properties

Dissolvability, segment coefficient (logP), degree of ionization, and inborn porousness are physicochemical qualities of medicines that impact its pharmacokinetics properties and objective receptor family in a roundabout way and should be addressed while developing another drug. To predict physicochemical features, unique AI-based devices can be used. For example, ML makes use of massive informative indexes that are given throughout the compound augmentation process that has previously been completed to prepare the software [21]. In drug configuration calculations, subatomic parameters like SMILES strings, potential energy estimations, electron thickness everywhere the particle, and iota orientations in 3D are utilized to construct likely atoms using DNN and forecast their attributes [22].

Zang *et al.* established the estimation program interface (EPI) suite [21], which is a workshop approach for determining the six physicochemical characteristics of ecological synthetic chemicals offered by the

Environmental Protection Agency (EPA) using a measurable construction property relationship (QSPR). The lipophilicity and dissolvability of various combinations were predicted using neural networks based on the ADMET indication and the ALGOPS algorithm. To predict the dissolvability of atoms, DL techniques such as aimless chart recursive neural organizations and diagram-based convolutional neural organizations (CVNN) have been utilized [23].

ANN-based models, diagram sections, and part-edge-based models have all been used to predict the corrosive separation steady of mixes in a few cases [21]. Cell lines like Madin–Darby canine kidney cells and human colon adenocarcinoma (Caco-2) cells have also been utilized to produce cell penetrability data for a certain particle class, which is then analyzed by AI-assisted indicators.

Kumar *et al.* used 745 mixtures to create six extrapolative models [SVMs, ANNs, LDAs, probabilistic neural organization calculations, and fractional least square (PLS)], which were then applied to 497 mixtures to forecast their digestive absorptivity based on boundaries like sub-atomic surface region, sub-atomic mass, all-out hydrogen count, sub-atomic refractivity, sub-atomic volume, logP, and absolute value. In a similar line, human gastrointestinal digestion of a range of synthetic combinations was predicted using *in silico* models based on RF and DNN. As a result, AI plays an important role in drug research, anticipating not just a medication's ideal physicochemical properties but also its optimum bioactivity [25].

9.6.2 Forecast of bioactivity

The viability of medication atoms relies upon their partiality or the objective protein or receptor. Medication particles that do not show any communication or partiality toward the designated protein cannot convey the restorative reaction. In certain occurrences, it may likewise be conceivable that created drug atoms collaborate with accidental proteins or receptors, prompting poisonousness. Subsequently, drug target restricting liking (DTBA) is indispensable to anticipating drug–target associations. Artificial-intelligence-based techniques can gauge the limiting proclivity of medication by thinking about either the elements or likenesses of the medication and its objective. Element-based connections perceive the substance moieties of the medication, and the objective is to decide the component vectors. On the other hand, in similitude-based cooperation, the comparability among medication and target is thought of, and it is expected that comparative medications will associate with similar targets [26].

For predicting drug–target communications, web programs like as ChemMapper and the comparability gathering method (SEA) are available. KronRLS, SimBoost, DeepDTA, and PADME are just a few of the ML and DL methods that have been utilized to conclude DTBA. ML-based methods like Kronecker-regularized least squares (KronRLS) evaluate the resemblance among medications and protein atoms to predict DTBA. SimBoost, on the other hand, forecasts DTBA using relapse trees, taking into account both element-based and likeness-based cooperations. Any combination of SMILES medication features, ligand most extreme normal foundation (LMCS), expanded availability unique mark, or any combination of these can be evaluated [26].

DL approaches, as opposed to ML, have demonstrated improved execution since they use network-based strategies that do not rely on the accessibility of the 3D protein structure. Some of the DL approaches used to measure DTBA include DeepDTA, PADME, WideDTA, and DeepAffinity. DeepDTA takes drug data in the form of SMILES, which include an amino acid order for protein input as well as a one-dimensional illustration of the pharmacological structure [27]. WideDTA is a CVNN DL approach that considers ligand SMILES (LS), amino corrosive configurations, LMCS, and protein sections and themes for selecting the limiting fondness [28].

The methodologies addressed before [29] are DeepAffinity and protein and drug molecule association prediction (PADME). DeepAffinity is a customizable deep learning model that can be used with RNN and CNN, as well as unlabeled and labeled data. In terms of fundamental and physical qualities, it examines the molecule in the SMILES configuration and protein successions [30]. PADME is a DL-based stage that uses feed-forward neural networks to predict drug–target interactions (DTIs). It uses a mix of the drug's components and aim at protein's information to assess the degree of collaboration between the treatment and the target protein. To calculate the prescription and aim, the SMILES illustration and the protein arrangement composition (PSC) are used independently [29]. Medicines and mark proteins of recognized and unidentified compounds, as well as medication repurposing and atomic system comprehension, may all be studied using solo ML techniques like MANTRA and PREDICT. MANTRA bundles chemicals with comparable quality articulation profiles and groups those combinations that are believed to have a common component of activity and a regular chemical pathway using a CMap informative index. A medicine's bioactivity also contains information on ADME. XenoSite, FAME, and SMARTCyp are artificial-intelligence-based devices that are tasked with determining where medicine is digested. Furthermore, the programming tools CypRules,

MetaSite, MetaPred, SMARTCyp, and WhichCyp were utilized to find exact CYP450 isoforms that interfere with a process. The leeway route of 141 supported drugs was accurately completed using SVM-based metrics [31].

9.6.3 Expectation of poisonousness

To avoid unwanted repercussions, every medicine particle's poisonousness must be predicted. To determine a medicine's poisonousness, cell-based *in vitro* tests are frequently performed as initial research and monitored by animal studies, increasing the cost of pharmacological disclosure. LimTox, pkCSM, admetSAR, and Toxtree are some of the online cost-cutting tools accessible. Advanced AI-based systems seek for commonalities amongst blends or use input highlights to determine a substance's toxicity. The National Institutes of Health, the Environmental Protection Agency (EPA), and the US Food and Drug Administration (FDA) organized the Tox21 Data Challenge to test a few computer methods for assessing the toxicity of 12,707 natural combinations and medications. DeepTox, a machine learning system, beat all other approaches when it came to recognizing static and dynamic components inside synthetic atom descriptors such as molecular weight (MW) and Van der Waals volume. Table 9.2 highlights the different AI techniques that have been employed in drug research.

SEA was used to compare the health goal expectations of 656 featured medications to 73 unintentional targets that might have antagonistic effects. eToxPred was developed using a machine-learning-based technology and used to assess the harmfulness and union plausibility of small natural atoms, with an accuracy of 72% [31]. In essence, open-source devices like TargeTox and PrOCTOR are employed in the prediction of poisonousness [32]. TargeTox is an aim-based medical hazard estimating technique based on natural organizations that use the culpability by-affiliation standard, which asserts that elements in natural organizations with comparable practical properties share likenesses [33]. It may provide protein network data and utilize a machine learning classifier to predict drug toxicity by combining pharmacological and useful properties. The delegate was created using an RF model that comprised medicine likeliness characteristics, sub-atomic elements, target-based components, and attributes of protein which focuses to provide a "Delegate score," which indicated if a drug would fail clinical preliminaries because of its noxiousness. It also discovered FDA-approved drugs with potentially hazardous drug interactions. Using a sophisticated CVNN method, the Tox (R) CNN was used to assess the cytotoxicity of drugs given to DAPI-stained cells [34].

Table 9.2 Examples of artificial intelligence (AI) techniques in drug development.

Tools	Details
DeepChem	The MLP model was utilized to locate a feasible candidate for drug development using a python-based AI system.
DeepTox	A processer program that calculates the noxiousness of over 12,000 medicines.
DeepNeuralNetQSAR	A Python-based system that practices computational techniques to help in the identification of chemical molecular movement.
ORGANIC	A molecular plan tool that aids in the creation of molecules with certain characteristics.
PotentialNet	To forecast ligand binding affinity, NNs are employed.
Hit Dexter	A machine learning method is being utilized to forecast which molecules will reply to biological investigations.
Delta Vina	A machine learning technology is being utilized to anticipate which molecules will answer biological tests.
Neural graph fingerprint	Using a scoring system, re-scoring drug-ligand binding affinity. It can be used to anticipate novel molecular properties.
AlphaFold	Predicts protein 3D structures
Chemputer	Aids in the reporting of chemical synthesis procedures in a uniform style.

9.7 Artificial Intelligence in Planning Drug Particles

9.7.1 The expectation of the objective protein structure

When building a medication atom, it is vital to choose the right target for effective therapy. Various proteins have a role in the infection's growth, and they can be overexpressed at times. Consequently, for a specific focus of infection, foresee the construction of the objective protein to plan the medication atom. Artificial intelligence can help with structure-based medication disclosure by anticipating the 3D protein structure if the plan is in harmony with the synthetic climate of the aim protein site, assisting with foreseeing the impact of a complex on the aim as well as security thoughts before it is blended or created. AlphaFold, a DNN-based AI tool, was utilized to analyze the distance among neighboring amino acids and the comparison sites of the peptide securities to predict the 3D objective protein assembly, and it did a fantastic job, appropriately foreseeing 25 out of 43 structures.

AlQurashi utilized RNN to estimate the protein structure in a concentrate. After three steps (calculation, math, and assessment), the inventor identified an intermittent mathematical structure (RGN). The essential protein arrangement was encoded here, and the torsional plots for a certain buildup and, to a degree, finished spine obtained from the mathematical unit upstream were then deemed information and given another spine as yield. As a result of the last unit, the 3D construction was supplied as a yield. The predicted and test structures were compared using the distance-based root mean square deviation (dRMSD) measure. To keep the dRMSD between the test and projected designs low, RGN's bounds were enhanced [35]. AlQurashi anticipated that his AI approach will be faster than AlphaFold's when it comes to predicting protein structure. In any case, AlphaFold is expected to be extra precise in forecasting protein assemblies with comparable layouts to the reference assemblies [36].

A review was done to anticipate the 2D assembly of a protein using MATLAB and a nonlinear three-layered NN tool compartment based on feed-forward directed learning and backpropagation blunder computing. The NNs learned calculations and performed execution assessments while using MATLAB to prepare data and generate relevant indexes. The 2D structure has a prediction accuracy of 62.72%.

9.7.2 Foreseeing drug-protein communications

Collaborations amongst pharmaceutical proteins are important to a treatment's effectiveness. Forecasting a drug's communication with a receptor or protein is serious for determining its viability and efficacy and medication repurposing and evading polypharmacology. Numerous AI approaches have demonstrated their effectiveness in correctly forecasting ligand–protein interactions and delivering suitable repair. Wang *et al.* used the SVM technique to uncover nine unique mixtures and their communication with four critical targets based on 15,000 protein–ligand cooperations generated based on major protein groupings and fundamental features of small atoms [37].

Yu *et al.* used two RF models to forecast likely medicine–protein connections, comparing them to stages with high affectability and explicitness, such as SVM, and employing pharmacological and substance data. Furthermore, these modes may be used to forecast drug–target affiliations, which might be expanded to include target–sickness and target–target affiliations, allowing for a faster medication disclosure procedure [38]. Xiao *et al.* employed the synthetic minority over-sampling technique and the neighborhood cleaning rule to improve the data for the construction of drug targets. The iDrug-GPCR, iDrug-Chl, iDrug-Enz, and iDrug-NR sub-predictors may be used to distinguish between GPCRs, particle channels, chemicals, and atomic receptors (NR). Target-pocketknife testing was used to compare this indication to other indicators, and the former outperformed the latter in terms of forecast precision and consistency [39].

AI's capability to forecast medication–target communications has also been utilized to aid in drug repurposing and polypharmacy evasion. When a remedy is repurposed, it becomes eligible for Phase II clinical trials right away. Because it costs $8.5 million to relaunch an old drug vs. $41.2 million to introduce a new pharmacological component, utilization is limited. The "culpability by affiliation" method may be used to estimate the creative association between a treatment and an infection, which can be determined either by information or by calculation. The ML methodology, which utilizes algorithms like SVM, NN, computed relapse, and DL, is commonly used in a computationally determined organization. When repurposing a pharmaceutical, different relapse stages like PREDICT, SPACE, and other ML draws close comparison between drug–drug, ailment infection similitude, target atom similarity, compound construction, and quality articulation profiles [40].

DeepDTnet, a deep learning technology based on cell networks, was used to anticipate the therapeutic usage of topotecan, a topoisomerase inhibitor now in use. By blocking the humanoid retinoic acid corrosive receptor-related vagrant receptor-gamma t, it can also be used to treat multiple sclerosis (ROR-t). A provisional US patent presently protects this level. Self-assembling manuals (SOMs) are ML's sole classification and are employed in the repurposing of medications. They adopt a ligand-based strategy to deal with scanning innovative off-focuses for a set of pharmaceutical particles, preparing the framework on a predetermined number of mixes with perceived natural activities and then using it to study other combinations. According to current research, DNN was utilized to repurpose present medications with the confirmed movement against SARS-CoV, HIV, flu sickness, and 3C-like protease inhibitors. Extended network finger impressions (ECFPs), useful class fingerprints (FCFPs), and octanol-water segment coefficient were used to create the AI stage (ALogP count). Based on their cytotoxicity and viral limitation, of the examined medications may be progressed further, according to the findings [41].

Polypharmacology, defined as a medication atom's predisposition to interact with numerous receptors, resulting in asymmetric antagonistic effects, may also be predicted by drug–protein interactions. Artificial intelligence can aid in the era of more secure pharmaceutical particles by planning another atom based on polypharmacology reasoning. Computer-based intelligence stages, such as SOM, may be used to link a variety of mixes to various objectives and off-targets, thanks to the vast knowledge bases available. To create connections between the pharmacological properties of drugs and their prospective aims, Bayesian classifiers and SEA computations may be utilized.

Li *et al.* verified the application of KinomeX, a web-based artificial intelligence platform that uses deep neural networks (DNNs) to discover kinase polypharmacology based on chemical structures. This step makes use of DNN, which has over 300 kinases and 14,000 bioactivity data. By concentrating on a medication's overall selectivity for the kinase family and specific subfamilies of kinases, this method can help in the development of innovative synthetic modifiers. NVP-BHG712 was used as a model chemical in this study to precisely predict its primary targets and off-targets [42]. Ligand Express, a cloud-based proteome-screening AI tool developed by Cyclica, is used to identify receptors that can communicate with a certain small atom (represented by a SMILE string) and to build on-target and off-target connections. This makes thoughtful the medication's potential negative effects much easier.

Computer-based intelligence is one of the best medication plan over a couple of years, the new drug configuration approach has been generally used to configure drug particles. The conventional strategy for a new drug configuration is being supplanted by developing DL techniques, the previous having weaknesses of convoluted amalgamation courses and troublesome forecast of the bioactivity of the original particle. PC-supported union arranging can likewise propose a great many constructions that can be combined and predict a few diverse union courses for them.

The Chematica program was created by Grzybowski *et al.*, and it may encrypt a set of criteria into the machine and provide various combination courses for eight therapeutically significant targets. This program has demonstrated its effectiveness in terms of refining yield and cutting costs. It can also provide alternate integrating ways to protect components and is intended to aid in the mixing of yet-to-be-combined mixtures. Similarly, DNN is founded on natural science and retrosynthesis principles, which, when combined with Monte Carlo tree look and emblematic AI, allow for considerably quicker response prediction and medication disclosure and planning than older approaches [43, 44].

Coley *et al.* fostered a structure where an unbending forward response format was applied to a gathering of reactants to blend synthetically achievable items with a huge pace of response. Based on a score provided by the NNs, ML was used to determine the winning item. Putin *et al.* studied a DNN dubbed with the built-up ill-disposed neural PC (RANC) in the context of RL for the reorganization of small natural atoms. Particles addressed as SMILES strings were used to prepare this step. It then built particles with predefined synthetic properties at the time, such as MW, logP, and topological polar surface area (TPSA). RANC was associated with ORGANIC, a stage that outdone RANC in terms of producing interesting structures with little design length loss [45].

To recognize particles taken from the ChEMBL dataset and processed into SMILES strings, RNN also employed the long momentary memory (LSTM). This was used to create a VS particle library with a diverse collection of particles. This approach was used to secure 5-HT2A receptor focuses, *Staphylococcus aureus* focuses, and *Plasmodium falciparum* focuses [46].

For a novel pharmacological mix, Popova *et al.* designed the reinforcement learning for structural evolution technique, which employs generative and prophetic DNNs to encourage new combinations. The generative model uses a stack memory to produce more unique particles in the form of SMILE strings, but prophetic models are employed to forecast the properties of the freshly formed complex [47]. Merk *et al.* employed a generative AI model to create retinoid X and PPAR agonist atoms with the intended restorative belongings lacking the need for composite criteria. The scientists created five atoms, four of which showed considerable modulatory movement in cell testing, demonstrating the efficacy of generative AI in particle union. Because of its numerous advantages, AI's contribution to particle back planning can be valuable to the pharmaceutical industry, including providing internet learning and synchronous advancement of all-around educated information, as well as proposing possible combination courses for intensifying prompting quick lead planning and improvement [48].

9.8 Artificial Intelligence in Propelling Drug Item Advancement

The revelation of an original medication atom requires its ensuing consolidation in an appropriate dose structure with wanted conveyance qualities. Around here, AI can supplant the more established experimentation approach. Different computational instruments can resolve issues experienced in the plan region, for example, strength issues, disintegration, porosity, etc., with the assistance of QSPR. Rule-based frameworks are used to choose the kind, nature, and amount of excipients conditional on the medicine's physicochemical features. They work through an input component to screen and adjust the complete interaction periodically [48].

Guo *et al.* developed a crossover framework for the creation of piroxicam direct-filling hard gelatin cases based on the drug's disintegration profile using expert systems (ES) and artificial neural networks (ANN). The model expert system (MES) selects and offers plan progression based on the information limits. Backpropagation is used by ANN to find out how to connect plan limitations to the best reaction, all while being constrained by the control module to maintain smooth definition development [49].

The influence of the dust stream property on the pass on filling and cycle of pill pressure was investigated using a variety of numerical approaches, including computational liquid elements (CFD), discrete component displaying (DEM), and the finite element method. CFD may also be utilized to look at the impact of tablet computation on the disintegration profile of the tablet. The employment of these numerical models in combination with AI might help speed up the development of medications [50].

9.9 Artificial Intelligence in Drug Fabricating

As the complexity of assembling operations rises, along with the demand for more effectiveness and improved item quality, current assembling frameworks strive to deliver human input to machines, constantly adjusting the assembling practice. The pharmaceutical business may profit from AI addition in processing. Devices, such as CFD, employ Reynolds-averaged Navier–Stokes solvers technology to focus on fomentation and anxious emotions in various hardware (e.g., blending tanks) by automating many medicine jobs. Comparative frameworks, like direct mathematical reproductions and enormous swirl recreations, include progressed ways to deal with tackling confounded stream issues in assembling.

By connecting diverse substance codes and functioning using a prearranging language known as Chemical Assembly, the smart Chemputer stage permits computerized atom combination and assembly. It has been used to successfully mix and assemble sildenafil, diphenhydramine hydrochloride, and rufinamide, with profit and quality comparable to hand union [51]. AI breakthroughs should be able to efficiently finish granulation in granulators with sizes from 20 to 500. Basic components were connected to their reactions via neuro-fluffy thinking and innovation. In both mathematically comparable and distinct granulators, they inferred a polynomial condition for forecasting the extent of the granulation liquid to be added, needed speed, and impeller width.

DEM has been broadly used for the pharmaceutical industry to investigate the period consumed by tablets in the shower zone, the belongings of fluctuating cutting edge speediness and outline, anticipating the likely path of the tablets in the covering system, and focusing on the isolation of powders in a twofold combination, for example. Along with fluffy models, ANNs focused on the relationship between machine parameters and tablet covering in the assembly process to reduce tablet covering.

Meta-classifiers and tablet-classifiers are artificial intelligence (AI) devices that help check the final product's quality, highlighting a probable mistake in the tablet's manufacturing. A patent has been filed based on patient data that provides a system for selecting the best drug and measuring routine for each patient, as well as creating the best transdermal fix in the same manner.

Modern fabricating frameworks are attempting to supply human information to machines, constantly modifying the assembling practice, as assembling operations become more complex and there is a greater demand for efficiency and improved item quality. The application of AI industries might be a boon to the pharmaceutical industry. Apparatuses, like CFD, employ Reynolds-averaged Navier–Stokes solvers technology to focus on the influence of turbulence and anxiety in varied hardware (e.g., mixed tanks), making use of the robotization of many pharmacological activities. Comparative frameworks, like direct mathematical reproductions and huge whirlpool recreations, include progressed ways to deal with tackling confounded stream issues in assembling [52].

9.10 Artificial Intelligence in Quality Assurance and Control

Getting the most out of raw resources necessitates a delicate balance of numerous constraints. Manual impedance is required for quality control testing on the goods as well as the upkeep of bunch to group uniformity. While this is not always the ideal solution, it does show the present need for AI execution. The FDA changed the current good manufacturing practices (cGMP) by introducing a "quality by design" method to deal with the core activity as well as specified standards that oversee the last nature of the medicinal item.

Gams *et al.* utilized a grouping of humanoid and AI efforts to examine initial data from creation bunches and construct decision trees. These were also turned into regulations, which the administrators assessed to manage the formation cycle in the future [53]. With an inaccuracy of 8%, Goh *et al.* used ANN to predict the disintegration of the tested materials [54]. They focused on the disintegration profile, which has effectively foreseen the disintegration of the tested material with a mistake of 8% with an indication of the bunch to group uniformity of theophylline pellets.

Simulated intelligence may also be used to control in-line production processes so that the item meets the specified standard. An ANN that uses a mix of self-versatile advancement, proximity pursuit, and backpropagation calculations to watch the freeze-drying procedure is utilized. This may be utilized to forecast the temperature and parched cake width at a later time point ($t + t$) for a specific set of operating parameters, which can aid in final product quality control.

A mechanized information section stage, like an Electronic Lab Notebook, combined with current, astute methods may ensure the item's quality assurance. Similarly, under the total quality management master framework, information mining and other information revelation processes may be used as significant approaches in generating difficult decisions and developing innovations for astute quality control [55].

9.11 Artificial Intelligence in a Clinical Trial Plan

Clinical trials take 6–7 years to complete and need a substantial financial commitment. They are used to measure the security and feasibility of a drug in people with a specific infectious disease. Despite this, just one particle out of every ten that enters these early phases reaches effective freedom, which is a severe setback for the business. Unseemly understanding determination, a lack of specific requirements, and a powerless framework can all lead to disappointment. However, with the vast amount of sophisticated clinical data available, these disappointments can be reduced by implementing AI [56].

Patient enrollment accounts for 33% of the clinical preliminary course of events. The recruitment of relevant patients ensures the attainment of a clinical preliminary, which otherwise results in 86% of disappointment instances [57]. By utilizing patient-explicit genome–exposome outline investigation, which can aid in the initial forecast of the accessible medication focuses in the patients chosen, artificial intelligence can aid in the selection of only an exact ailing populace for enrollment in Phase II and III clinical preliminaries. Preclinical atom disclosure, also foreseeing lead compounds earlier at the beginning of clinical preliminaries using various parts of AI, such as prescient ML and additional thinking procedures [56], aids early forecasting of lead particles that would permit clinical preliminaries with the deliberation of the chosen persistent populace.

Existing patients from clinical preliminary tests resulted in a 30% failure rate, necessitating extra screening procedures to complete the preliminary, wasting time and money. This may be prevented by keeping a careful eye on the patients and assisting them in following the clinical preliminary's recommended protocol. In a Phase II trial, AiCure created portable software to track normal medication usage by schizophrenia patients, which enhanced patient adherence by 25%, assuring the study's success [57].

9.12 Artificial Intelligence in Drug Item Execution

9.12.1 Artificial intelligence in market situating

Market positioning is the most common approach for generating a personality for a product to convince customers to acquire it, making it a crucial component in nearly all business systems for enterprises to build their distinct brand. This method was employed in the promotion of the pioneering brand Viagra, which was intended not just for the conduct of erectile dysfunction in males but also for the treatment of other issues related to personal enjoyment.

It has been simpler for firms to acquire a distinct awareness of their image in the public eye, thanks to innovation and the internet as a platform. Organizations employ online search tools as one of the basic steps to achieve a prominent place in web-based advertising and aid in the location of the item on the lookout, according to the Internet Advertising Bureau. Businesses are always seeking to rank their websites more advanced than those of competitors to get rapid awareness for their brand.

Researchers were able to get a greater understanding of markets by combining factual inquiry methods and molecular swarm augmentation calculations (invented by Eberhart and Kennedy in 1995) with NNs. They could be able to assist in determining the best marketing plan for the creation based on the unique client request expectation [55].

9.12.2 Artificial intelligence in market expectation and investigation

The accomplishment of an organization lies in the persistent turn of events and the development of its occupational. Indeed, smooth with admittance to considerable assets, R&D yield in the drug business is dropping as a result of the disappointment of organizations to take on novel showcasing innovation. The advances in computerized advances alluded to as the "fourth modern transformation" are serving creative digitalized showcasing using a multicriteria dynamic methodology, which gathers and dissects measurable and numerical information and executes human surmising to settle on AI-based dynamic models to investigate new advertising systems [58].

Artificial intelligence additionally helped in an exhaustive examination of the principal prerequisites of an item according to the client's perspective just as understanding the need of the market, which helps in dynamic utilizing forecast instruments. It can likewise conjecture deals and examine the market. Computer-based intelligence-based programming connects with buyers and makes mindfulness among doctors by showing notices guiding them to the item site by a tick. What is more, these techniques utilize regular language-handling devices to examine watchwords entered by clients and relate them to the likelihood of buying the item [59, 60].

A few B2B organizations have unveiled self-administration innovations that permit for allowed glancing of health goods, which can be found by providing identity, placing orders, and tracking their delivery. Pharmaceutical companies are also promoting their web-based apps, such as 1 mg, Medline, Netmeds, and Ask Apollo, to meet patients' unmet requirements. Market expectations are also vital for various medication distribution organizations that may use artificial intelligence in the sector, such as "business keen smart sales prediction analysis," which mixes time series gauging and continuous application. This enables pharma businesses to anticipate the sale of items ahead of time, avoiding expenditures connected with surplus stock or customer loss due to deficiencies [58].

9.13 Artificial Intelligence in the Item Cost

The organization determines the final cost of the medical item based on market research and costs associated with its creation. The primary concept behind using AI to compute this cost is to take advantage of its capability to impress the rationale of a humanoid master to investigate the issues that impact estimation after a product has been manufactured. Patent expiry, cost of the reference item, and value fixative approaches control the cost of marked and nonexclusive medications. Variables such as consumption during innovative work of the medication, severe cost administrative plans in the worried nation, a distance of the selectiveness period, and a portion of the overall industry of the enhanced medication following a year prior are patent finish, and cost of the orientation item, as well as value fixing approaches.

Huge amounts of measurable data, such as item improvement cost, item interest on the lookout, stock expenditure, fabricating cost, and rivals' item cost, are broken down by product in ML, resulting in calculations for predicting item cost. Man-made intelligence stages, for example, competitor, dispatched by Intelligence Node (founded in 2012), is a completed retail aggressive insight stage that dissects the competitor value information and assists retailers and brands in screening the opposition. Savvy Athena and Navetti PricePoint allow customers to select how much their item is worth, and they urge that medicine companies do the same to aid with item pricing [60].

9.14 Conclusion

Drug research and development may be made more efficient and accurate with AI and machine learning. These technologies not only improve process efficiency but also reduce or eliminate the need for clinical trials in some cases by replacing them with simulations. They also allow researchers to study molecules more thoroughly without the use of trials, lowering costs and raising ethical concerns. Integrating AI and machine learning into drug research is expected to revolutionize drug development in the future, but there are still a lot of challenges to overcome, such as cleaning of unstructured and heterogeneous datasets, and occasional computing hardware incompetency, to name a few. Once these hurdles are removed, AI and machine learning developments may be more broadly deployed and enhanced, ushering in a new era for the pharmaceutical industry.

9.15 Acknowledgment

The authors are grateful to the administration of Oriental University, Indore, for their assistance.

9.16 Funding

There is no funding issued.

9.17 Conflict of Interest

There is no potential for a conflict of interest.

References

[1] Yang, Y., & Siau, K. L. (2018). A qualitative research on marketing and sales in the artificial intelligence age. *MWAIS 2018 Proceedings, 41*.

[2] Wirtz, B. W., Weyerer, J. C., & Geyer, C. (2019). Artificial intelligence and the public sector—applications and challenges. *Int. J. Public Adm. 42*, 596–615.

[3] Lamberti, M. J., Wilkinson, M., Donzanti, B. A., Wohlhieter, G. E., Parikh, S., Wilkins, R. G., & Getz, K. (2019). A study on the application and use of artificial intelligence to support drug development. *Clin. Ther. 41*, 1414–1426.

[4] Beneke, F., & Mackenrodt, M. O. (2019). Artificial intelligence and collusion. *IIC-Int. Rev. Intellect. Prop. Comput. Law. 50*, 109–134.

[5] Bielecki, A. (2019). *Models of neurons and perceptrons: selected problems and challenges.* Springer International Publishing.

[6] Kalyane, D., Sanap, G., Paul, D., Shenoy, S., Anup, N., Polaka, S., ... & Tekade, R. K. (2020). Artificial intelligence in the pharmaceutical sector: current scene and future prospect. In *The Future of Pharmaceutical Product Development and Research* (pp. 73–107). Academic Press.

[7] Mishra, V. (2018). Artificial intelligence: the beginning of a new era in pharmacy profession. *Asian Journal of Pharmaceutics (AJP): Free full-text articles from Asian J Pharm, 12.*

[8] Chen, Z., Liu, X., Hogan, W., Shenkman, E., & Bian, J. (2021). Applications of artificial intelligence in drug development using real-world data. *Drug Discov. Today, 26,* 1256–1264.

[9] Pandey, A. (2020). Drug Discovery and Development Process. *Learning Center, June.*

[10] Mak, K. K., & Pichika, M. R. (2019). Artificial intelligence in drug development: present status and future prospects. *Drug Discov. Today, 24,* 773–780.

[11] Duch, W., Swaminathan, K., & Meller, J. (2007). Artificial intelligence approaches for rational drug design and discovery. *Curr. Pharm. Des. 13,* 1497–1508.

[12] Baronzio, G., Parmar, G., & Baronzio, M. (2015). Overview of methods for overcoming hindrance to drug delivery to tumors, with special attention to tumor interstitial fluid. *Front. Oncol. 5,* 165.

[13] Ciallella, H. L., & Zhu, H. (2019). Advancing computational toxicology in the big data era by artificial intelligence: data-driven and mechanism-driven modeling for chemical toxicity. *Chem. Res. Toxicol. 32,* 536–547.

[14] Chan, H. S., Shan, H., Dahoun, T., Vogel, H., & Yuan, S. (2019). Advancing drug discovery via artificial intelligence. *Trends Pharmacol. Sci. 40,* 592–604.

[15] Sellwood, M. A., Ahmed, M., Segler, M. H., & Brown, N. (2018). Artificial intelligence in drug discovery. *Future Med. Chem. 10,* 2025–2028.

[16] Zhang, L., Tan, J., Han, D., & Zhu, H. (2017). From machine learning to deep learning: progress in machine intelligence for rational drug discovery. *Drug Discov. Today. 22,* 1680–1685.

[17] Álvarez-Machancoses, Ó., & Fernández-Martínez, J. L. (2019). Using artificial intelligence methods to speed up drug discovery. *Expert Opin. Drug Discov. 14,* 769–777.

[18] Fleming, N. (2018). How artificial intelligence is changing drug discovery. *Nature, 557,* S55–S55.

[19] Dana, D., Gadhiya, S. V., St Surin, L. G., Li, D., Naaz, F., Ali, Q., ... & Narayan, P. (2018). Deep learning in drug discovery and medicine; scratching the surface. *Molecules*, *23*, 2384.

[20] Yang, X., Wang, Y., Byrne, R., Schneider, G., & Yang, S. (2019). Concepts of artificial intelligence for computer-assisted drug discovery. *Chem. Rev. 119*, 10520–10594.

[21] Hessler, G., & Baringhaus, K. H. (2018). Artificial intelligence in drug design. *Molecules*, *23*(10), 2520.

[22] Kumar, R., Sharma, A., Siddiqui, M. H., & Tiwari, R. K. (2017). Prediction of human intestinal absorption of compounds using artificial intelligence techniques. *Curr. Drug Discov. Technol. 14*, 244–254.

[23] Chai, S., Liu, Q., Liang, X., Guo, Y., Zhang, S., Xu, C., ... & Gani, R. (2020). A grand product design model for crystallization solvent design. *Comput. Chem. Eng. 135*, 106764.

[24] Thafar, M., Raies, A. B., Albaradei, S., Essack, M., & Bajic, V. B. (2019). Comparison study of computational prediction tools for drug-target binding affinities. *Front. Chem.*, 782.

[25] Öztürk, H., Özgür, A., & Ozkirimli, E. (2018). DeepDTA: deep drug-target binding affinity prediction. *Bioinform. 34*, i821–i829.

[26] Mahmud, S. H., Chen, W., Jahan, H., Liu, Y., Sujan, N. I., & Ahmed, S. (2019). iDTi-CSsmoteB: identification of drug–target interaction based on drug chemical structure and protein sequence using XGBoost with over-sampling technique SMOTE. *IEEE Access*, *7*, 48699–48714.

[27] Lang, J. (2018). Proceedings of the Twenty-Seventh International Joint Conference on Artificial Intelligence (IJCAI 2018).

[28] Feng, Q., Dueva, E., Cherkasov, A., & Ester, M. (2018). Padme: A deep learning-based framework for drug-target interaction prediction. *arXiv preprint arXiv:1807.09741*.

[29] Karimi, M., Wu, D., Wang, Z., & Shen, Y. (2019). DeepAffinity: interpretable deep learning of compound–protein affinity through unified recurrent and convolutional neural networks. *Bioinform. 35*, 3329–3338.

[30] Pu, L., Naderi, M., Liu, T., Wu, H. C., Mukhopadhyay, S., & Brylinski, M. (2019). eToxPred: a machine learning-based approach to estimate the toxicity of drug candidates. *BMC Pharmacol. Toxicol. 20*, 1–15.

[31] Basile, A. O., Yahi, A., & Tatonetti, N. P. (2019). Artificial intelligence for drug toxicity and safety. *Trends Pharmacol. Sci. 40*, 624–635.

[32] Lysenko, A., Sharma, A., Boroevich, K. A., & Tsunoda, T. (2018). An integrative machine learning approach for prediction of toxicity-related drug safety. *Life Sci. Alliance. 1*.

[33] Jimenez-Carretero,D.,Abrishami,V.,Fernandez-de-Manuel,L.,Palacios,I., Quilez-Alvarez, A., Diez-Sanchez, A., ... & Montoya, M. C. (2018). Tox_ (R) CNN: Deep learning-based nuclei profiling tool for drug toxicity screening. *PLoS Comput. Biol. 14*, e1006238.

[34] AlQuraishi, M. (2019). End-to-end differentiable learning of protein structure. *Cell systems. 8*, 292–301.

[35] Hutson, M. (2019). AI protein-folding algorithms solve structures faster than ever. *Nature.*

[36] Wang, F., Liu, D., Wang, H., Luo, C., Zheng, M., Liu, H., ... & Jiang, H. (2011). Computational screening for active compounds targeting protein sequences: methodology and experimental validation. *J. Chem. Inf. Model. 51*, 2821–2828.

[37] Yu, H., Chen, J., Xu, X., Li, Y., Zhao, H., Fang, Y., ... & Wang, Y. (2012). A systematic prediction of multiple drug-target interactions from chemical, genomic, and pharmacological data. *PloS one, 7*, e37608.

[38] Xiao, X., Min, J. L., Lin, W. Z., Liu, Z., Cheng, X., & Chou, K. C. (2015). iDrug-Target: predicting the interactions between drug compounds and target proteins in cellular networking via benchmark dataset optimization approach. *J. Biomol. Struct. Dyn. 33*, 2221–2233.

[39] Park, K. (2019). A review of computational drug repurposing. *Transl. Clin. Pharmacol. 27*, 59–63.

[40] Ke, Y. Y., Peng, T. T., Yeh, T. K., Huang, W. Z., Chang, S. E., Wu, S. H., ... & Chen, C. T. (2020). Artificial intelligence approach fighting COVID-19 with repurposing drugs. *Biomed. J. 43*, 355–362.

[41] Li, Z., Li, X., Liu, X., Fu, Z., Xiong, Z., Wu, X., ... & Zheng, M. (2019). KinomeX: a web application for predicting kinome-wide polypharmacology effect of small molecules. *Bioinform. 35*, 5354–5356.

[42] Grzybowski, B. A., Szymkuć, S., Gajewska, E. P., Molga, K., Dittwald, P., Wołos, A., & Klucznik, T. (2018). Chematica: a story of computer code that started to think like a chemist. *Chem, 4*, 390–398.

[43] Klucznik, T., Mikulak-Klucznik, B., McCormack, M. P., Lima, H., Szymkuć, S., Bhowmick, M., ... & Grzybowski, B. A. (2018). Efficient syntheses of diverse, medicinally relevant targets planned by computer and executed in the laboratory. *Chem, 4*, 522–532.

[44] Putin, E., Asadulaev, A., Ivanenkov, Y., Aladinskiy, V., Sanchez-Lengeling, B., Aspuru-Guzik, A., & Zhavoronkov, A. (2018). Reinforced adversarial neural computer for de novo molecular design. *J. Chem. Inf. Model. 58*, 1194–1204.

[45] Segler, M. H., Kogej, T., Tyrchan, C., & Waller, M. P. (2018). Generating focused molecule libraries for drug discovery with recurrent neural networks. *ACS Cent. Sci. 4*, 120–131.

[46] Popova, M., Isayev, O., & Tropsha, A. (2018). Deep reinforcement learning for de novo drug design. *Sci. Adv. 4*, eaap7885.

[47] Merk, D., Friedrich, L., Grisoni, F., & Schneider, G. (2018). De novo design of bioactive small molecules by artificial intelligence. *Molecular informatics*, *37*(1–2), 1700153.

[48] Guo, M., (2002). A prototype intelligent hybrid system for hard gelatin capsule formulation development. *Pharm. Technol.* 6, 44–52.

[49] Chen, W., Desai, D., Good, D., Crison, J., Timmins, P., Paruchuri, S., ... & Ha, K. (2016). Mathematical model-based accelerated development of extended-release metformin hydrochloride tablet formulation. *AAPS PharmSciTech*, *17*, 1007–1013.

[50] Steiner, S., Wolf, J., Glatzel, S., Andreou, A., Granda, J. M., Keenan, G., ... & Cronin, L. (2019). Organic synthesis in a modular robotic system driven by a chemical programming language. *Sci.*, *363*, eaav2211.

[51] P. J. Das, P. J., Preuss, C., Mazumder, B., Mandlik, V., Bejugam, P. R., and Singh, S. (2016). Artificial Neural Network for Drug Design, Delivery and Disposition.

[52] Gams, M., Horvat, M., Ožek, M., Luštrek, M., & Gradišek, A. (2014). Integrating artificial and human intelligence into tablet production process. *Aaps Pharmscitech*, *15*, 1447–1453.

[53] Goh, W. Y., Lim, C. P., Peh, K. K., & Subari, K. (2002). Application of a recurrent neural network to prediction of drug dissolution profiles. *Neural Comput. Appl. 10*, 311–317.

[54] Paul, D., Sanap, G., Shenoy, S., Kalyane, D., Kalia, K., & Tekade, R. K. (2021). Artificial intelligence in drug discovery and development. *Drug Discov. Today*, *26*, 80.

[55] Harrer, S., Shah, P., Antony, B., & Hu, J. (2019). Artificial intelligence for clinical trial design. *Trends Pharmacol.l Sci. 40*, 577–591.

[56] Fogel, D. B. (2018). Factors associated with clinical trials that fail and opportunities for improving the likelihood of success: a review. *Contem. Clin. Trials Commun. 11*, 156–164.

[57] Singh, J., Flaherty, K., Sohi, R. S., Deeter-Schmelz, D., Habel, J., Le Meunier-FitzHugh, K., ... & Onyemah, V. (2019). Sales profession and professionals in the age of digitization and artificial intelligence technologies: concepts, priorities, and questions. *J. Pers. Sell. Sales Manag. 39*, 2–22.

[58] Davenport, T., Guha, A., Grewal, D., & Bressgott, T. (2020). How artificial intelligence will change the future of marketing. *J. Acad. Mark. Sci. 48*, 24–42.

[59] Syam, N., & Sharma, A. (2018). Waiting for a sales renaissance in the fourth industrial revolution: Machine learning and artificial intelligence in sales research and practice. *Ind. Mark. Manag. 69*, 135–146.

[60] De Jesus, A. (2019). AI for Pricing–Comparing 5 Current Applications. *Emerj Artif. Intell. Res. 2.*

10

Artificial Intelligence in Medical Image Processing

Mohamed Yousuff*, Rajasekhara Babu, and Thota Ramathulasi

School of Computer Science and Engineering, Vellore Institute of
Technology, India
yousuffrashid@gmail.com
*Corresponding Author
Mohamed Yousuff
Research Scholar, School of Computer Science and Engineering, Vellore
Institute of Technology (VIT University), India
Email: yousuffrashid@gmail.com

Abstract

Image processing and medical science domains have seen an increase in
using the term artificial intelligence (AI) over the past decade. Although AI
is a relatively new concept, it has been formalized since the 1940s. In its
simplest form, AI refers to computer algorithms that can mimic human intelligence's problem solving and learning abilities. In particular, AI applications
based on machine learning (ML) algorithms have experienced remarkable
innovation in the field of computer vision during the past decade. As a result
of these extraordinary developments, the medical community has created
new steps in medical care or assisted with treatment planning. AI and ML
research works have shown promising results in a wide range of medical
applications. Diagnosis, image segmentation, outcome prediction, and many
more tasks are transformed by the advent of AI. Research and clinical teams
and companies have been working together to develop clinical AI solutions
since ML tools have matured enough to meet clinical requirements in recent
years. Today, we are closer than ever to seeing AI used in clinical settings;
so learning the basics of this technology is a "need" for any medical professional. This chapter provides an overview of AI, paying particular attention

to the methods utilized based on medical imaging (MI) analysis. It also discussed how current ML and deep learning (DL) methods could be used to automate and improve different steps of clinical practice. Function approximation, like regression and classification, is a typical supervised task. For example, pathology can be classified as present or absent in an image. In case of regression, we can also enhance images pixel by pixel or map images to each other (e.g., mapping an input computed tomography (CT) image to its corresponding dose distribution output).

10.1 Introduction

AI models are gaining traction in biomedical research and therapy, demonstrating their utility in various applications, including risk determination and assessment, personalized monitoring, diagnostics (such as categorization of molecular illness subclasses), prognostication reaction to the treatment, and progression. Incorporating much information flows from disparate repositories may have therapeutic potential for these innovatory breakthroughs. These data contributors encompass medical images, which account for the majority of patient information (particularly in oncology) but also risk factors in infections, multi-omics data, therapeutic practices, and timely information. Integrating these sources effectively into models results in superior medical services and promotes human intellect and AI synchronization [1].

All of these areas of research have the potential to significantly advance the present situation approaching precision medicine, culminating in much more credible and personalized methodologies that have a significant impact on diagnosis and treatment trajectories. This entails a turning point away from statistical and population-oriented forecasting toward individual prognostication, permitting even more efficacious prophylactic and curative initiatives. Although numerous regulations on the construction and use of AI models are presented, approved, and published, prospective AI techniques are multiple and diverse. Some difficulties and areas need to be explained regarding "how to build AI models" for clinical decision-making. Implementation of AI on MI covers many topics such as computing sample size, data curation, data augmentation strategies to be followed while handling comparatively less and unbalanced datasets, data harmonization, labeling, and annotation of radiomics data [2].

The Coronavirus Disease 2019 (COVID-19) pandemic has necessitated the search for rapid, widespread, precise, and minimal cost assessments, and lung imaging is a critical adjunctive resource for COVID-19 diagnostics and monitoring. According to the American College of Radiology and Fleischner

Society Consensus Statements, imaging of COVID-19 is suggested in the event of aggravating breathing problems and, in resource-limited scenarios, for prioritizing patients with reasonable to severe pathological symptoms and a strong likelihood of malady. This entails two primary responsibilities. The first is diagnostics, which incorporates casual assessment and offers supporting corroboration in medical scenarios involving a suspicious false-negative reverse transcription–polymerase chain reaction (RT-PCR) test. The second objective is to contribute to evaluating treatment consequences, progression of the disease, and expected diagnosis. In the perspective of COVID-19, the domain of AI in MI is advancing, and expectations are elevated that AI could indeed assist health care professionals and radiologists with all these functions [3].

Echocardiography (ECCG) has developed into a critical imaging tool for anesthetists in monitoring and diagnosing central thoracic pathology in cardiovascular surgery patients in the course of the perioperative spell. Moreover, it is a expert reliant procedure that includes specialized skills in order to construe the information adequately. The contribution of AI methodologies is widening in enabling anesthetists to handle complex and confusing electrocardiography data and assuring more precise and reliable perspectives in a short time span, with the intent of enhancing the consequences of inmates with circulatory system illness at the time of surgery [4]. With the accumulation of huge collective ultrasound (US) conventional datasets and the immersion of AI approaches and their missing patterns, it is presumed that AI-oriented US will progressively develop into an essential learning resource, describing its diagnosing premise, assisting the US diagnosis process, and, inevitably, favoring both patients and clinicians [5].

This chapter is organized to cover a plethora of applications and implementations of AI in the field of MI. Section 10.2 addresses the utilization of AI strategies to deal with MRI data. Section 10.3 elaborates the significance of various ML and DL models in COVID-19 chest computed tomography (CCT) data and COVID-19 chest X-ray (CXR) data. The usage of AI approaches in the field of ECCG and US image analysis is discussed in Sections 10.4 and 10.5, respectively. The conclusion of the chapter is presented in Section 10.6.

10.2 Magnetic Resonance Imaging (MRI)

Magnetic resonance imaging (MRI) is a type of biomedical approach used to provide visuals of the interior of impenetrable body parts in living creatures and determine the quantity of adsorbed water in geological formations.

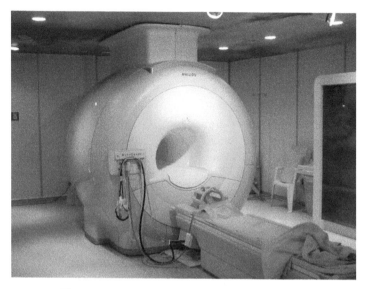

Figure 10.1 MRI machine with its complete setup.

It is mostly used to visualize diseased or other biological changes in live tissues. It is currently a widely utilized technique for MI. MRI instruments use extremely strong magnets to polarize and energize hydrogen nuclei present in the molecules of water found in living organisms, generating a constructive interference that is regionally recorded and culminating in visuals of the physique. MRI produces a two-dimensional view of a lean carve of the body, whereas, nowadays, it is possible to generate three-dimensional frames. There are most certain health hazards linked with cellular inflaming due to radio frequency exposure, especially in cases of an embedded device (for example, pacemaker) in the body. These dangers are rigorously handled as part of the instrument's design and scanning methods [6]. This section of the chapter elaborates on the role of AI in MRI image processing for various diseases. Figure 10.1 depicts the MRI machine, whereas Figure 10.2 shows an MRI image of a human head in the side view.

10.2.1 Alzheimer's disease (AD)

AD is a chronic, incurable brain condition that gradually deteriorates memory and cognitive abilities. It is among the most prevalent neurodegenerative disorders in those over the age of 66 globally. Several AI-oriented computer-aided diagnostic (CAD) techniques based on brain imaging data have been developed to achieve precise and appropriate diagnostic and to diagnose AD

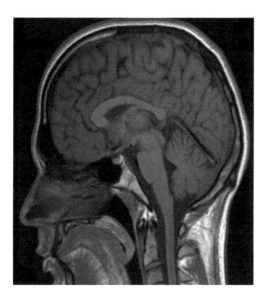

Figure 10.2   MRI visual of human head side view.

in its initial phases [7]. The scientific community is actively pursuing the goal of developing automated methods for achieving an earlier and definite diagnostic. A contemporary multinational contest for AD predictors is conducted under the title "A Machine Learning Neuroimaging Challenge for Automated Diagnosis of Mild Cognitive Impairment" (MLNeCh). The event is on the basis of preprocessed collections of T1-biased MRI image repository that have been categorized into four classes: persistent AD, patients with mild cognitive impairment (MCI), patients with MCI turned to AD, and healthy individuals.

A strategy is presented for timely detection of AD, which is assessed using the MLNeCh data repository. Due to the fact that instantaneous categorization of AD is relying on the input (feature vectors) data of high-dimensional nature, various methodologies of choosing the features and reducing the dimensions are evaluated by comparing to evade the "curse-of-dimensionality" challenge [8]. The classification technique is then accomplished as a conjunction of support vector machine (SVM) models, trained on various subsets of the actual collection of data meant for training. The proposed tetra-classifier strategy excels all self-contained methods assessed in the evaluations. The ultimate ensemble is constructed using a collection of classifiers, each of which has been trained on a separate subset of the complete training dataset. The presented ensemble model holds the significant benefit of executing effectively when only a subset of the dataset is utilized [8].

Figures 10.1 and 10.2 represent the setup of MRI machine and the visual of head side view, respectively.

Structural MRI (sMRI) and resting-state functional MRI (rs-fMRI) have already shown potential outcomes in diagnosing AD, but the usefulness of combining sMRI and rs-fMRI is still not substantially investigated. The performances of rs-fMRI and sMRI in mono- and multi-mode methods are assessed for categorizing individuals with MCI whose health condition transitions into likely AD-MCI as converter (MCI-C) and those having MCI without chances of getting AD as non-converters (MCI-NC). The approach incorporates cortical and subcortical proportions as input features to the model. The rs-fMRI and sMRI characteristics are fed to train and evaluate a SVM model to differentiate MCI-C from MCI-NC. The proposed model for categorizing MCI-C and MCI-NC made use of a limited set of optimum features and attained sMRI Ă = 90%, rs-fMRI accuracy (Ă) = 94%, and sMRI and rs-fMRI combined Ă = 98%. This study examines the integration of rs-fMRI and sMRI to detect initial level of AD [9].

AD is caused due to the death of neurons in the brain because of beta-amyloid accumulation and the quick dissemination of tau proteins in the portions of the brain. Contemporary diagnosis techniques are either too expensive or incapable of detecting AD's histopathology characteristics. As a result, computing information model (CIM) is proposed for AD diagnostics. The brain's MRIs are preprocessed with an adaptable histogram and fragmented into four inherent modal operations (IMOs) using "bidirectional empirical mode decomposition." For each IMO, indigenous duple patterns (IDPs) are generated, and the histograms are joined together. The dataset balancing is achieved using adaptable simulated sampling, and a correlated pair t-test is used to choose the most important characteristics for every fold of 10-fold cross-validation. SVM-Poly 1 and random forest (RF) are implemented to classify the input observations, with each achieving the maximum Ă = 94%. The approach suggests that the proposed CIM is effective in hospitals for automatically classifying AD versus healthy MRI images [10].

The new approach is proposed to determine the utility of rule mining in assessing AD by utilizing decision trees (DT) and RF methods and incorporating the obtained rules into an argumentation-oriented rationale scheme to facilitate the interpretation and explanation of the results. On the brain MRIs collected from healthy subjects (HSs) and AD individuals, the DT and RF techniques are implemented. The DT is computed using the KNIME analytical framework, while the RF is computed using the R package. The argumentation strategy executed in the Gorgias environment attained an accuracy of 92%, outperforming the DT and RF algorithms. Ultimately, the effectiveness of all algorithms throughout this analysis is consistent with previous research. Additionally, the interpretation provided by the suggested technique for the many feasible forecasts results in a better, accurate, and comprehensive evaluation of the patient's situation. The approach revealed the utility of rule extraction in assessing AD using MRI data and the benefits of using the argumentation-oriented symbolical rationale to compose and evaluate ML results [11].

Currently, DL-oriented techniques for classifying neuroimaging datasets associated with AD have been developed and substantial growth is achieved. The endwise learning that maximizes the effects of DL has received little consideration because of the congenital difficulty of neuroimaging driven by data paucity. Thus, a method is proposed for endwise learning of a volumetrical CNN model for a two-class determination such as AD versus HS, gradual MCI (gMCI) versus HS, persistent MCI (pMCI) versus HS, and gMCI versus pMCI versus HS on MRI data repository. The suggested methodology employs a convolutional autoencoder (CAE) prominent unsupervised learning to resolve the AD versus HS categorization job and regulated TL approach to tackle the gMCI versus pMCI determination task. To identify the maximum significant biomarkers associated with AD and gMCI, a gradient-oriented visualization technique is utilized to represent the geographical relevance of the convolution neural network (CNN) model's conclusion. To evaluate the experimental contributions, the experiments are carried out on the ADNI data repository. The findings indicate that the suggested method outperformed previous network models with $Ă = 87\%$ and $Ă = 74\%$ for the AD and gMCI discrimination tasks, respectively [12].

10.2.2 Skeletal issues

Considering the popularly accepted perception that structural changes are associated with ail in knee joint [osteoarthritis (OA)], a causal correlation is not clearly demonstrated. Modern research indicates the presence of various subcategories of OA pain patterns in the knees. Although few individuals experience increasing aggravation of their pain, others endure considerable anguish stabilization. Formulating a direct link with image-oriented indicators and pain escalation is beneficial for prognosis. The proposed objective is divided into two sections: (i) to denote various pain pathways in OA sufferers; and (ii) to examine the relationship between MRI bioindicators obtained with the help of three-dimensional (3D) CNN and the denoted pain itineraries [13].

The OA schemes gathered recurrent scales of "knee injury and OA outcome score" for a pair of knee joints over 10 years' duration from 4797 patients. The imaging biomarker finding process utilized 3D "double echo steady state" photos of the knee joint from standard patients. Specific pain curves are leveled with the help of the regression approach with quadratic polynomials of the unit and dual degree to minimize the innate distortion in the collected observations and to manage missing features. Clustering analysis is performed using Bayesian Gaussian mixture method which takes standardized computed variables as input. Silhouette score is utilized to select the optimum model which can obtain better unique pain patterns. The regression attributes are altered toward the "Gaussian mixture model's" means in order to determine the posterior distribution for every anguish curve describing group affiliation [13].

The built model is a 3D version of the DenseNet-121 architecture. The data augmentation procedures are implemented randomly on the training data repository image files in order to avoid class imbalance problems. Kaiming He weight initialization procedure is followed. The layers are trained to understand the posterior likelihood of pain patterns. Mean square error (MSE) is considered as a cost function for regression. The expectation–maximization approach was used to optimize a total of 26 possible models with a varied number of items and various forms of correlation (complete, linked, diagonal, and sphere). The Gaussian combination (with three factors) with linked correlation is picked as an ideal clustering model because it rendered the greatest average silhouette score. Clustering evaluation revealed unique pain pathways: persistent, deteriorating, and gradually deteriorating. The MSE values are found to be 0.0149 for training, 0.1557 for testing, and 0.1550 for validation. The accuracy performance metric is used as a measure to ensure that the model precisely estimates the appropriate and likely clusters with Ă = 99% for the training phase while 81% and 78% for validation and testing phases [13].

OA is predominantly diagnosed through variations in hyaline cartilage on MI, technological impediments such as distortion, aberrations, and modalities create a significant barrier to the most accurate, reliable, and effective earlier identification of OA. Due to contemporary improvements, deep neural networks (DNNs) are demonstrating exceptional performance in this specific application space. A DeepKneeExplainer, a unique explicable approach, is proposed for diagnosing knee joint OA primarily using radiography and MRI images. The methodology starts from a deeply built stacked transformation approach to thoroughly preprocess radiography images and MRIs against distortion and aberrations. Next, using a U-Net structural model implementing residual neural network (ResNet) framework, the feature points are extracted [14].

DenseNet and visual geometry group (VGG) models are trained using the portion of interest to distinguish the associations. Ultimately, a gradient-directed category activating maps and layered architecture is used to emphasize category features, accompanied by presenting domain expert interpretable justifications for the estimates. The technique outperforms equivalent cutting-edge methods by 92% in extensive trials using the multi-modal OA study collaborators. It is anticipated that the findings would inspire medicinal practitioners and researchers to adopt easily interpretable methodologies and DNN-oriented analytical workflows, resulting in a greater endorsement and approval of AI-aided solutions in medical care for enhanced knee joint OA diagnosis [14]. Due to the intrinsic slowness of MRI recording, two distinct speeding approaches have been developed: a continuous collection of numerous correlating observations (parallel imaging) and the acquirement of less than needed for typical signal computation approaches (compressed sensing). Both technologies complement one another in terms of expediting MRI procurement.

A new technique for integrating a classic parallel imaging approach with DNN is proposed which generates good feature reconstructions also at massive accelerating factor. The suggested technique, dubbed GrappaNet, accomplishes gradual restoration just by modeling the reconstruction issue to an easier thing that can be rectified using only a conventional parallel imaging technique utilizing a neural network (NN), then applying the parallel imaging approach, and eventually refining the outcome using another NN. End-wise training is feasible for the complete network. The evaluation metrics are reported on the newly published fast MRI collection, demonstrating GrappaNet that can produce nice, enhanced, and good reconstructions over alternative approaches [15].

10.2.3 Brain illness diagnosis

Cancer of the brain or central nervous system is one of the top ten risk factors of mortality. Globally, brain tumors are not the main cause of death, but 40% of other forms of cancer result in brain tumors because of metastasis dissemination. Despite the fact that biopsy is regarded as the ideal marker for cancer diagnostics, it has a number of limitations, including low sensitivity/specificity values, health threats associated with the biopsy process, and comparatively more time duration is spent to view the biopsy results. Due to the growing number of individuals with brain cancer, there exists an undeniable need for non-invasive and intelligent CID method capable of precisely diagnosing and grading a tumor within a no-time [16].

Clinically related six-class dataset is collected and utilized for transfer learning (TL) based AI approach using a CNN which resulted in a higher standard of performance metrics in brain cancer rating and discrimination using MRI image data repositories. The CNN-based DL model surpasses all the considered ML models in this multi-class classification task. AlexNet architecture-based model reported amazing performance metrics for K3 ($\check{A}$ = 100%), K7 ($\check{A}$ = 96), and K10 ($\check{A}$ = 97) cross-validation. The average area under the curve (AUC) of DL and ML are registered to be 0.998 and 0.887, respectively, for $p < 0.0001$, and the proposed model demonstrated a 13% betterment compared to ML. The best model is evaluated statistically using a tumor isolation factor and on a simulated dataset containing seven categories [16].

To enhance compressive sensing MRI (CS-MRI) techniques in the context of finer feature attrition at large accelerating factors, a recursive feature refining (RFR) model endowed with predefined transformations for the purpose of restoring significant features and minutiae. Nonetheless, the suggested RFR-CS does have a few drawbacks, including the requirement for hyper-parameter choosing, a prolonged rebuilding duration, and a constant sparsifying transformation. To address these concerns, the RFR processes are unfurled in RFR-CS into a supervisory model-oriented network termed RFR-Net. When both the regularization variable and the maximum attribute improvement function in RFR-CS are supplied with training datasets, they acquire learning. Furthermore, in order to extrapolate the sparseness-imposing operation, CNN-oriented inverting modules are investigated in the sparseness-enforcing denoising component. Rigorous testing on both generated and brain invitro magnetic resonance repository have demonstrated that the suggested model is capable of capturing picture features and preserving spatial features while performing quick reconstruction [17].

A stable technique for 3D picture segmentation retains the benefits of both types of learning techniques, but it also overcomes its drawbacks by efficaciously incorporating supervised and unsupervised learning models. The suggested strategy is used for the segmentation of brain MRI images in a wide range of studies using numerous publicly available 3D brain MRI data repositories. The practical results demonstrate that the suggested model surpasses previous contemporary segmentation techniques under both types of learning when implemented on unique MRI observations or on forge observations excluding the requirement to retrain the model with the help of data repository annotations metadata [18].

InfiNet is a unique, parameter-efficacious, and pragmatic complete CNN model for meaningful segmentation of newborn brain MRI data at the iso-acute level. InfiNet may readily be adapted for various multi-modality segmentation applications. The T1 and T2 data of the scans are considered as input meant for dual encoder branches of InfiNet. The articulate-decoder branch gets terminated in the classification layer. The decoder portion of the InfiNet acts inventively by upsampling the less-resolution input features received through many encoder branches. To accomplish nonlinear interpolation, the aggregated values produced in the max-pooling levels of every encoder unit are connected to the associated decoder unit. To generate densely connected attribute maps, the sparse markers are merged with intermediary encoder approximations (jump links) and convolved using learnable channels. InfiNet is trained from start to finish in order to optimize the generic dice losses, which is appropriate in situations involving higher class-label imbalances. InfiNet accomplishes the entire segmentation in less than 60 seconds and outperforms a variety of modernistic DNN structures and its multimodal versions [19].

10.2.4 Cancer and other disease analysis

Brain imaging methods are critical in specific diagnostics because they deliver unique perspectives on the brain structure, allowing for a better understanding of the brain's functioning and activity. In the field of medical science, image processing is utilized to aid in earlier identification and medication of life-threatening sickness. The purpose of the approach is to offer a method for cancer identification using brain MRI samples by combining CNN and shallow stacking autoencoders. This pairing is proven to have a substantial impact on the categorization process's accuracy and efficacy. The technique is implemented in MATLAB and validated using a sample of 125 MRI images. The observed findings demonstrated that the suggested classifier is highly remarkable at distinguishing and rating malignancy MRI images [20].

Cancer is the second leading cause of death, after cardiovascular illnesses. Particularly, brain tumor has the lowest survivability among all types of cancer. The classification of tumors is determined by a variety of variables, including textures, form, and position. Healthcare professionals had already recommended much more suitable therapies depending on a precise diagnosis of the malignancy. Due to the varying form, position, scale, and textures of brain tumors, the procedure of subdividing the MRI is highly complicated throughout their examination. Surgeons and radiologists can quickly identify and classify cancers if a platform integrating CAD and AI is available. A new approach is proposed for automatic segmentation that facilitates the extraction of tumors from MRI images and also improves the effectiveness of categorization and segmentation. The preliminary tasks involve data preprocessing and fragmentation techniques for subdividing normal and malignant tumors or cells via data expansion and grouping [21].

A current learning-oriented strategy was used to process automatic segmentation in composite MRI samples in order to detect tumors; thus, the clustering technique termed "Bat Algorithm with Fuzzy C-Ordered Means (BAFCOM)" was used to endorse subdividing of tumors. The Bat procedure computes the first centroids and distances between a pixel in the BAFCOM grouping method, which further obtains the tumor by calculating the distance between the carcinoma region of interest (RoI) and the non-carcinoma RoI. Following that, the MRI samples were evaluated using the "enhanced capsule networks (ECN)" approach to determine whether it is healthy or a tumor. Finally, the ECN algorithm is used to evaluate the effectiveness through discriminating between two types of tumors in MRI data. Additionally, a genetic algorithm (GA) is used to perform the automated cancer grade categorization, increasing classification results [21].

Characterization of the left and right ventricular chambers and cardiac muscle using cardiovascular MRI is a typical diagnostic procedure. Thus, automating the similar duties has been a focus of considerable research for past 10 years. The "Automatic Cardiac Diagnosis Challenge" data is the largest freely accessible and completely labeled data repository for cardiac MRI (CMRI) analysis. The collection includes information of 155 CMRI observations using multiple instruments as well as benchmark values and categorization by two health professionals. The primary goal is to determine the extent to which cutting-edge DL approaches can segment the cardiac muscles and the ventricles and categorize diseases in CMRI. The findings suggest that the appropriate systems precisely duplicate expert analysis, with an average correlation level of 0.98 for automated medical parameters retrieval and $\check{A} = 96\%$ for automated diagnostic. These findings pave the way for amazingly precise and automated CMRI assessment. Additionally, the circumstances in which DL algorithms continue to fail is also identified [22].

10.3 Radiography-Based COVID-19 Diagnosis

Radiography is an imaging approach that uses radiation to observe an entity's internal structure. Medical radiography is utilized for diagnostics and treatment. In the context of this chapter, CCT scan and CXR image datasets are used. The CCT scan and CXR image of the patient is shown in Figures 10.3 and 10.4.

Given the critical nature of early patient identification in terms of treating patients and isolating infected people to avoid virus spread, numerous

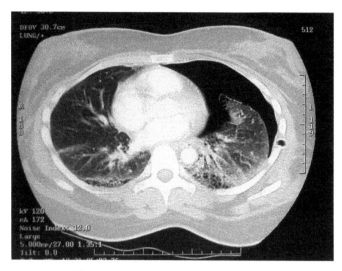

Figure 10.3 CCT scan image of a patient

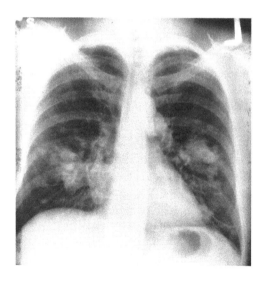

Figure 10.4 CXR image of a patient.

studies have focused on creating methods for more rapid and cost-effective patient identification. Researchers are trying to find better substitutes for screening since RT-PCR consumes more time and is limited in quantity. Radiography images (CCT and CXR) of the COVID-19 infected individuals contain crucial information. Pneumonia caused by virus pathogens appears uniquely in radiography images. So, AI is expected to deliver better analysis and performance while handling these images. COVID-19 patient's diagnostics using CCT or CXR datasets is a multi-step classification task. To begin with, the lungs images are preprocessed. The features are then retrieved using CNN or another approach. Finally, the excerpted features are employed to make a diagnosis via a classifier system. This section addresses the applications of AI to CCT and CXR image processing for COVID-19 diagnosis.

10.3.1 ML-based approach

The main goal is to obtain early detection of COVID-19 by combining five image filters with a "composed hybrid feature selection model" using a standard CCT image collection. This provides the benefits of three different ways to limit feature extraction and improves the genetic pseudo-code by optimizing the first parent or sample generation and genetic operators. Thus, the outcomes of filter strategy are utilized as preceding knowledge and J48 DT model as an evaluation function to accelerate convergence in order to pick the most delicate features. A stack hybrid classification approach is employed on the short-listed features to improve prediction and performance metrics. The model outperforms conventional classification techniques in terms of optimal feature selection and discrimination process betterment and efficiently minimized the false-negative observations with good accuracy of 96% using a Naive–Bayes classification model. The model's output demonstrates a high degree of accuracy in classifying COVID-19 CCT scans. These outcomes demonstrate the feasibility of employing AI to extract radiological parameters for the accurate and prompt diagnosis of COVID-19 [23].

The deep studying-based methodology on CXR is typically suggested for the identification of COVID-19 infected individuals. Deep features, along with the assistance of vector gadget COVID-19 CXR images, are easily distinguished from other illnesses. The approach is beneficial for practicing physicians since it allows the early diagnosis of COVID-19 infection in patients. The proposed technique of multi-level thresholding combined with SVM demonstrated good accuracy in classifying COVID-19-infected lungs. All images are identical in size and saved in the JPEG image format with 512×512 pixels resolution. The lung classification performance metrics such as sensitivity ($\hat{S}$), specificity ($\c{S}$), and $\breve{A}$ are found to be 96%, 99%, and 97%, respectively [24].

A unique "joint classification and segmentation (JCS)" approach is proposed for pragmatic and explicable COVID-19 CCT diagnostic. To prepare the JCS model, a vast COVID-19 dataset for classification and segmentation is produced, consisting of 144,168 CCT images from 410 COVID-19 individuals and 355 healthy subjects. 3860 CCT scans from 205 patients are labeled using pulverized pixel-level annotations for opacities, which are enhanced fading of mesenchymal tissue of the lung. Additionally, lesion counts, opacification areas, and positions are marked, which aids in many facets of diagnostics. Numerous trials reveal that the JCS diagnostic is quite effective at classifying and segmenting COVID-19 [25].

10.3.2 DNN algorithms for diagnosis

DL has been a burgeoning area of research in AI in recent years. These techniques have been hailed as a highly effective means of automatically detecting disease using CCT scan and CXR images. Many of these methodologies begin by training CNN on a considerable collection of CXR image repositories; then, fine-tuning procedures are carried out with COVID-19 observations at a smaller scale.

10.3.2.1 DNN and chest CT scan

A group of viral pneumonia cases occurring within a short period of time may signal a flare-up. Expeditious and reliable identification of viral pneumonitis will surely aid in the prevention of epidemics. Due to virus evolution and the emergence of new mutations, datasets shift, limiting the performance of classifiers; the challenge of distinguishing viral from non-viral pneumonia is articulated as a one-class anomaly detection issue. A methodology is proposed for confidently detecting anomalies comprising a feature exercise component, an anomaly monitoring component, and a confidence-predicting component. Finally, the categorization task is carried out using DL algorithms [26].

A DL model is designed for automatically detecting anomalies in CCT scans of COVID-19 patients and evaluating its quantitative performance to that of radiological physicians. The DL method for lesion recognition, segmentation, and localization was developed and validated in 14,435 patients who had CCT images and a confirmed pathogen diagnosis. The approach was evaluated on a non-overlapping observation of 96 definite COVID-19 patients who were admitted to three hospitals throughout China during the pandemic. The quantitative identification performance of the model is evaluated with three radiological clinicians with two proficient radiologists' and performance is measured. Out of 96 individuals, 88 exhibited pneumonia lesions on CCT imaging, while 8 had no abnormalities. The model demonstrated an impressive $\hat{S}$ of 1.00 on a per-patient basis. With an average computation time of 20 seconds per case, the algorithm outperformed radiology physicians in evaluating CCT images. Additionally, the DL technology can help radiologists in making faster and more correct diagnoses [27].

CCT scans can be used to identify lung infections. However, the problems associated with these traits, such as image quality and infection features, restrict their usefulness. By utilizing AI technologies and computer vision algorithms, detection accuracy can increase, resolving these concerns. A multi-task deep-learning-based technique is proposed for segmenting lung infections using CCT scans. The procedure begins by segmenting the infected lung areas. Following that, segment the infections within these locations. Additionally, the suggested model is trained to utilize the two-stream inputs to accomplish multi-class segmentation. Multi-task learning enables us to address the scarcity of labeled data. Additionally, the multi-input stream enables the model to learn on a variety of features, which can enhance the outcomes. Numerous characteristics have been utilized to assess the approach. The results demonstrate that the proposed system can segment lung infections with a fair degree of precision even when data and tagged images are scarce [28].

The CCT scan plays a vital role in diagnostics. It is not feasible to determine the surface area and location of lesions reliably and laboriously. DL-based software is developed to assist in the identification, emplacement, and determination of COVID-19 pneumonitis. Between February 12 and March 17, 2020, a total of 2461 severe acute respiratory syndrome Coronavirus 2 (SARS-CoV-2) positive individuals are noticed retroactively at Huoshenshan Hospital in Wuhan and the fundamental medical peculiarities are examined. The uAI intelligent assistant analysis system is applied and evaluated on CCT dataset resulting in precise evaluation of pneumonia in COVID-19 patients [29].

10.3.2.2 DNN and chest X-ray

Recent research from CXR imaging indicates that these images possess pertinent information concerning the COVID-19 virus. Advanced AI techniques combined with CXR imaging aid in the efficient espial of disease and, thus, help in vanquishing the issue of deficit health experts in rural areas. A model is developed for autonomous COVID-19 diagnosis utilizing raw CXR. The suggested approach is designed to offer a precise diagnosis for binary (COVID-19 versus nil-outcomes) and categorization (COVID-19 versus nil-outcomes versus pneumonitis). The DarkNet model was utilized as a classifier in a newfangled object identification system called You Only Look Once (YOLO). The model is implemented with 17 convolutional layers with distinct filtering upon every layer. The model achieved Ă = 98% for binary class and Ǎ = 87% for multi-class situations. Further, this model can be used to support radiologists in verifying their preliminary screening, as well as to screen patients instantly via cloud [30].

A DL computer-assisted diagnostics (CAD) system uses a YOLO algorithm for contemporaneously finding and analyzing COVID-19 as well as the other eight respiratory ailments such as collapsed lung, pulmonary infiltration, pulmonary air leaks, mass, pulmonary nodule, pleural effusion, inflated heart, and pneumonitis. The CAD system is evaluated using two distinct CXR datasets, namely COVID-19 and ChestX-ray8. 50,495 labeled CXR images are used to train the CAD model. The CAD system instantaneously identifies and classifies COVID-19 dubious spots across the complete set of CXR images, reaching ultimate identification and classification Ǎ of 97% and 98%, respectively. The challenging scans of COVID-19 and certain diverse lung illnesses are predicted exactly, with a ground truth value of more than 90%. By incorporating DL regularizers for data stabilization and augmentation, diagnosis efficiency is enhanced [31].

Furthermore, the CAD system demonstrates its capability to diagnose each CXR image in less than 9 milliseconds. Thus, the described CAD system is capable of predicting 108 frames per second during deployment. The DL CAD system demonstrates its competence and dependability in diagnosing COVID-19 and other considered lung disorders. Thus, the model appears to be dependable for assisting medical systems, patients, and clinicians invalidating their practices [31].

10.3.2.3 New DNN models on chest CT scan

Accurate COVID-19 screening remains a significant issue because of the spatial intricacy of three-dimensional (3D) volumes, the challenge of marking sites of infection, and the modest variation in CCT between COVID-19 and other viral respiratory disorders. A few pioneering efforts have achieved substantial strides; they usually require manual labeling of infectious areas and are incapable of being interpreted. It is plausible to get remarkably precise and intelligible COVID-19 detection through CCT with weak marking. "An attention-based deep 3D multiple instance learning (AD3D-MIL)" method is proposed in which a patient-level tag is given to a 3D CCT that is regarded as a bundle of entities. AD3D-MIL is capable of semantically inducing deep 3D entities in the vicinity of a potential disease location [32].

Additionally, AD3D-MIL employs a care-oriented combined sharing approach on 3D entities to render perceptiveness into the participation of each entity to the bundle label. Finally, AD3D-MIL understands the categorical distributions of the bag-level tags to facilitate the learning process. 462 CCT images are used, including 232 images from 80 individuals with COVID-19 infection, 103 images from 103 individuals with usual pneumonia, and 128 images from 128 persons without pneumonia. The model achieves the best Ǎ of 98%, an AUC of 99%, and a Cohen kappa value of 96% after a sequence of empirical investigations. These benefits qualify the method as an effective assisting tool for COVID-19 screening [32].

Recent research indicates that individuals parasitized with SARS-CoV-2 have specific imaging structures noticed on their CCT scans. A publicly available SARS-CoV-2 CCT scan dataset has been assembled, consisting of 1253 CCT scans confirmed for SARS-CoV-2 illness and 1231 CT scans of persons who were not affected with SARS-CoV-2, for a total of 2484 CCT scans. This image dataset is gathered from actual patients hospitalized in the hospitals of Sao Paulo, Brazil. This repository of data holds a strong purpose such as exploring and innovating AI systems capable of determining whether a person is affected with SARS-CoV-2 by analyzing their corresponding CCT scans. As a benchmark result for this image dataset, a new DL model named the explainable deep neural network (xDNN) technique is implemented, and the model attained an F1-score of 97%, which is somewhat encouraging [33].

Rapid and precise segmentation of COVID-19 from CCT is critical for early diagnosis and patient monitoring purposes. A novel U-Net-based segmentation structure that makes use of an attention mechanism is proposed because not all encoder components are suitable for segmentation. The new approach is intended to integrate an attention mechanism, which consists of a spatial and channel attention subsystem, into a U-Net framework to reweight the feature delineation spatially and channel-wise in order to apprehend affluent context-specific associations for enhanced feature depiction [34].

Additionally, the pivotal Tversky loss is inducted to handle the segmentation of tiny lesions. The experimental findings, which are assessed against a COVID-19 CCT segmentation image dataset containing 474 CCT slices, demonstrate that the novel method is capable of producing an exact and timely segmentation outcome on COVID-19. Segmenting a single CCT slice ends up taking 0.3 seconds only. Dice score and Hausdorff distance computed are found to be 83% and 19, respectively [34].

10.3.2.4 New DNN models on chest X-ray

Many proficient physicians believe, based on research, that it is intricate to detect COVID-19 at its genesis using CXR because the infection's traces are detectable only after the sickness has transitioned to the symptomatic or moderate, or acute stage. A compact trending classifier, the convolutional support estimator network (CSEN) model, is created due to its adaptability for classification tasks using fewer data. Finally, this approach offers a new benchmark observation termed Early-QaTa-COV19, which contains 175 initial-stage COVID-19 pneumonitis data (with very few or no infectious indications) annotated by medical experts, as well as 1579 normal stage observations. A comprehensive set of trials demonstrates that the CSEN reports the highest $\hat{S} = 99\%$ and $Ş \geq 96\%$ [35].

One of the most frequently used and efficacious procedures employed by researchers is the analysis of CXR of the respiratory system for COVID-19. However, personally inspecting each record requires multiple radiology professionals and time, which is one of the cumbersome duties during an outbreak. A DL-based method termed nCOVnet is offered as a possible quick screening option for identifying COVID-19 by evaluating patients' CXR and seeking visual markers from the CXR of infected individuals. The model is adequate to detect a COVID-19 positive individual in less than 5 seconds. With such a small quantity of data, the model obtains a true positive rate of 98% [36].

Additionally, nCOVnet overcomes the problem of lack of RT-PCR kits, as it requires only a CXR apparatus, which is usually available in the majority of hospitals worldwide. As a result, nations and states will no longer be required to await massive supplies of RT-PCR kits. Rapid diagnosis of COVID-19 enables isolation of COVID-19 patients and reduction of communal dissemination. The COVID-19 outbreak seems to be in Stage 2 in numerous places throughout the world, and they will be unable to purchase the excessively priced kits, and the COVID-19 epidemic will not abate fairly soon until all countries establish effective testing methods. Numerous preliminary works that claimed accuracy of up to 98%–99% did not account for the likelihood of data leaking, which is tackled during the training phase of nCOVnet, ensuring that the results are fair. This model may aid hospital or clinical administration and medical specialists in taking the appropriate procedures to handle COVID-19 patients following their rapid detection [36].

10.3.3 Transfer learning (TL) approach

TL is a technique in which the cognition acquired while resolving a previous challenge is applied to another that is related but not alike. This strategy is especially appealing when there is insufficient data available to train the DL model. VGG16 or VGG19, ResNet50, and GoogLeNet (InceptionV3) are some of the famous TL models. Figure 10.5 depicts the TL approach.

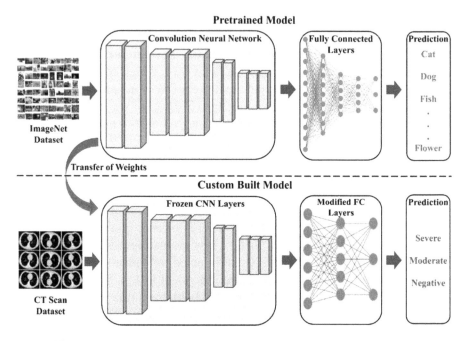

Figure 10.5   TL approach implemented for the classification of CCT scans.

10.3.3.1 Implementing TL approach on chest CT scan

Based on the radiographic alterations associated with COVID-19 in CCT images, DL algorithms can grasp COVID-19's unique, vivid properties and facilitate diagnostics before the pathogenic trials, thereby reducing significant duration for infection prevention. 1065 CCT scans of COVID-19 cases with infection confirmation (325 images) and those diagnosed recently with normal viral pneumonia (740 images) are gathered for analysis. The inception TL model is altered to create the intended model, which is then validated internally and externally. Internal validation resulted in a Ă of 89.5%, while external testing obtained 79.3%. These findings establish a shred of credible evidence for AI algorithms implementation to extract radiological characteristics for the expeditious and correct diagnostics of COVID-19 [37].

COVID-19 has a detrimental effect on the lungs, and the severity of the illness can be determined utilizing a selected imaging technique. TL is used to identify COVID-19 from CCT scans that have been disintegrated to three levels using a type of transform known as a stationary wavelet. To increase the recognition rate, a three-phase identification approach is presented. Phase 1 involves stationary wavelets in action to perform data augmentation, Phase 2 involves COVID-19 identification using a previously trained

CNN architecture, and Phase 3 involves peculiarity positioning in CCT scan images. For the experimental analysis, this approach used prominent pre-trained CNN models such as SqueezeNet architecture and ResNet architecture with 18, 50, and 101 layers, respectively. 70% of the dataset is meant to train and update the weights of the network, while 30% is used to evaluate the model. The performance analysis reveals that the ResNet18 TL-based model transcended with the highest classification Ă of 99% during training and testing, while it reports Ă of 97% during validation [38].

A simple two-dimensional DL framework termed the "fast-track COVID-19 classification network (FCONet)" is designed to identify COVID-19 pneumonitis from a solitary CCT scan. FCONet is constructed using TL by selecting one model from the four [Visual Geometry Group (VGG16), ResNet-50, Google Inception Version3, and Extreme Inception (Xception)] contemporary pre-trained DL models as a core. 3994 CCT scans of people with COVID-19 pneumonitis, another pneumonitis, and non-pneumonitis disorders are accumulated for training and testing FCONet architecture. These CCT scans are divided into an 8:2 ratio for training and testing. The testing data is used to examine the diagnosis efficacy of four pre-trained FCONet models for COVID-19 pneumonitis. ResNet-50 surpassed the competing pre-trained models with Ŝ = 99%, Ş = 100%, and Ă = 99% during the testing phase.

10.3.3.2 Implementing TL approach on chest X-ray

A classification model for CXR using dense convolutional networks and TL with three categories, namely COVID-19, pneumonitis, and normal, is developed. The neural networks are refined on the ImageNet model and then transferred twice with the NIH ChestX-ray14 dataset. The unique approach of output neuron retention is used, which modifies the twice TL technique. To illustrate the models' method of operation, heatmaps are generated using layer-wise relevance propagation (LRP). An accuracy of 100% on a test dataset is exceptional. Twice, TL and output neuron retention demonstrated encouraging results in terms of performance, particularly at the early stages of the training phase [39].

The LRP demonstrated that words on CXR could impact network estimations; this effect had a negligible influence on the accuracy metric. Although additional laboratory tests and more observation are required to assure adequate generalization, the cutting-edge results demonstrate that, when combined with AI, CXR can turn into a low-cost and efficient supplementary approach for COVID-19 diagnosis. Heatmaps produced by LRP enhance the interpretability of DNN and point the way to future diagnostic research. Twice, TL with output neuron maintained performance improvements [39].

Leveraging the detection rate of CXR becomes crucial. Patient prioritization is essential, and DL can help accelerate CXR image diagnosis by recognizing COVID-19 instances. COVIDPEN, a TL strategy based on an amended EfficientNet-oriented framework for the identification of COVID-19 situations, is developed. The *post-hoc* analysis is used to extrapolate the suggested model to determine the interpretability of the predictions. The suggested model's efficacy is proved using a systematic observation of CXR. Experimental results involving many baseline assessments demonstrate that the procedure is comparable and provides clinically describable situations for healthcare practitioners [40].

10.3.3.3 Smartphone apps

Comfortable diagnosis of COVID-19 using any technical device, such as a smartphone, can be pretty beneficial. It is an intriguing technique to use AI to identify COVID-19 from CXR on an android mobile device. In MATLAB, a CNN model is constructed and subsequently transformed to a TensorFlow Lite (TFLite) model for deployment on a mobile device operating on the android operating system. The developed android application utilizes the TFLite model to distinguish COVID-19 in CXR. The detection phase achieved a precision of 99% and an $F1$-score of 99% as a result of five-fold cross-validation [41].

COVID-MobileXpert is a lightweight DNN-based mobile application that is used to screen for COVID-19 cases and prediction. The design and construction of a new knowledge transmit and distilment system includes a previously trained serving physician network, retrieving CXR attributes and properties from a vast dataset of lung illnesses. A calibrated NN that understands the vital CXR attributes is needed to distinguish COVID-19 from pneumonitis or healthy subjects with less quantity of observations or data. To address the issue of enormously alike predominant foreground and background in clinical images, innovative cost functions and training techniques for the layers to understand reliable properties are inducted. COVID-MobileXpert has demonstrated immense capability for faster deployment through rigorous testing with a variety of design and tuning parameter configurations [42].

10.4 Echocardiogram Analysis and Classification using AI

The heart is visualized through the use of sound waves in an ECCG. Heart rate and blood flow can be observed by performing this prevalent test. An ECCG is a diagnostic tool the surgeon uses to look for signs of heart disorders

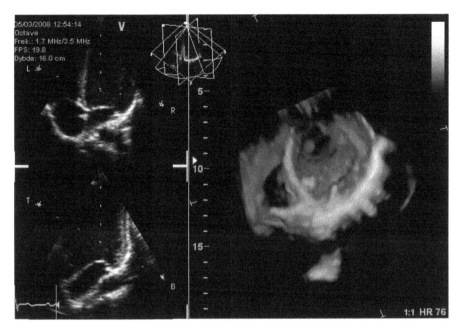

Figure 10.6 ECCG of a heart perceived from the apex.

as shown in Figure 10.6. For example, a physician generally recommends an ECCG to look for heart valve or chamber problems or to determine whether symptoms like respiratory distress or pain in the chest are due to a heart condition, or to screen for congenital abnormalities prior to actual childbirth (fetal ECCG) [43]. AI is becoming a primary priority in medical science that can be applied to ECCG in order to overcome the troubles of discrepancy and intra- and inter-spectator fluctuation while obtaining and interpreting the images. ECCG is prone to inter-observer variations and a heavy reliance on expertise level contradicting other methodologies, like CCT, CXR, and MRI. ECCG, a type of cardiovascular imaging, is becoming increasingly popular and intricate. The indispensable requirements are to enhance efficacy, reduce acquirement time, and decrease echocardiographic variations. Patients, sonography technicians, and heart specialists will all benefit from AI in this regard. AI will never substitute sonographers; instead, it will become more productive and precise due to its use [4].

Heart disease can be detected using myocardial contrast ECCG (MCECCG), an imaging modality that evaluates left ventricular operation and myocardial permeability to identify atherosclerotic and coronary issues. Obtaining reliable myocardial segmentation (MS) from blurry and time-varying image data is difficult when performing automatic MCECCG reperfusion quantitation on the myocardium. There have been numerous flourishing applications of RF to clinical image segmentation challenges. On the other hand, the pixel-wise RF discriminator does not take into account contextual correlations among labeled responses of individual pixels. In addition, RF that merely makes use of localized appearance aspects is vulnerable to input that has experienced significant intensity changes. To address the shortcomings of classic RF, a unique model is proposed that includes the presentation of a wholly automated segmentation pathway for MS in comprehensive 2D MCECCG data [44].

Furthermore, a statistical contextual structure model is employed in order to offer contextual structure prior knowledge, which is used to assist the RF segmentation in two different directions. As a first step, an innovative contextual structure feature is introduced into the RF architecture, allowing for a more precise RF likelihood map. Second, the contextual structure model is suited to the RF likelihood map in order to refine and confine the ultimate segmentation to myocardial forms that are likely to occur. Additionally, the inclusion of a locality preserving detection technique as a preprocessing phase in the segmentation process results in another compelling performance establishment. This technique on a 2D visual is even further generalized to 2D + *t* series, which assures that the ultimate sequence segmentations are persistent in terms of temporal consistency. On medical MCECCG datasets, the provided method significantly enhances segmentation performance metrics and excels other conventional technologies, such as the traditional RF and its versions [44].

The apical four-chamber (A4C) vision obtained using ECCG during prenatal development is a critical step in assessment and prompt treatment of congenital heart anomalies, and it should be used whenever possible. A novel approach aims to automate the segmentation of cardiovascular formations in ultrasonic A4C stances, specifically the thorax, aorta, ventricles, atrium, and epicardium, in order to aid physicians during an early fetal inspection. Moreover, when it comes to segmentation tasks, the following difficulties are frequently encountered. There are three problems with this image: (i) pixelated or sparse distribution of pixels resulting in low-resolution imaging; (ii) ambiguous tissue demarcation; and (iii) reasonably low crispness and contrast. In order to overcome these problems, a cascaded U-net (CU-net) model is proposed that uses "structural similarity index measure (SSIM)" loss [45].

First and foremost, the CU-net with two separate formative assessments aids in the establishment of unambiguous tissue boundaries as well as the assuagement of the gradient disappearing situation faced by the expansion of network depth. Second, the refinement of segmentation results is assured since the prior information gets transmitted from initial layers to deeper layers of CU-net architecture. Third, the approach makes use of SSIM loss in an attempt to uphold delicate structural features while also obtaining clear boundary definitions. On a data source of 1713 A4C stances, experiments conducted illustrate that the presented approach accomplishes 0.93 as pixel Ã, 3.4 as Hausdorff distance, and 0.86 as dice coefficient, demonstrating its performance and possibilities as a therapeutic technique [45].

ECCG allows doctors to see the apparent movement of the heart's chambers and valves and to diagnose cardiovascular disease. DL has played a significant role in the development of many medical computer-abetted diagnostics systems. DL techniques must be used to improve the performance of this kind of system, which is a demanding necessity. The proposed DL-based system is helpful in the task of classifying multiple ECCG views and identifying the physiological spot. To begin with, every structure in the echo-motion is used to fetch the spatial CNN characteristics. Second, the neutrosophic temporal motion properties are retrieved. A ResNet model is employed to retrieve the CNN features. Then, using the features composition method, both spatial and neutrosophic temporal CNN characteristics are combined. Eventually, the merged CNN characteristics are sent into a long short-term memory (LSTM) network in order to categorize and locate echo-cardio views. The experiment is conducted on freely accessible ECCG data that included 433 recordings for nine cardio-views. Deploying the ResNet architecture resulted in the highest overall $\tilde{A} = 97\%$ and $\hat{S} = 96\%$ across many pre-trained models [46].

Accurate recognition of end-systolic (ES) and end-diastolic (ED) frames in an ECCG cine sequence can be a challenging but crucial preprocessing phase in forming autonomous cardiovascular parameter measurement systems. Given the disparity in cardiac architecture and pulse rate that are commonly linked with pathological situations, the identification process is challenging. This task is described as a regression problem and offers multiple DL-based strategies for localizing the ED and ES frames by minimizing a unique universal extrema structural loss function. The suggested architecture combines a model for visual feature excerption based on CNN and a model for periodic interdependence between frames in a series using RNN. The performance of two CNN models, namely DenseNet and ResNet, and four RNN models such as LSTM, bi-directional LSTM, Gated Recurrent Unit (GRU), and bi-GRU are compared. The ideal DL model is made up of a DenseNet and a GRU which has been trained using the specified loss function. Correspondingly, the frame disparities of 0.21 and 1.44 for the ED and ES frames are present in the observed inter-observer variation for manual identification of these frames [47].

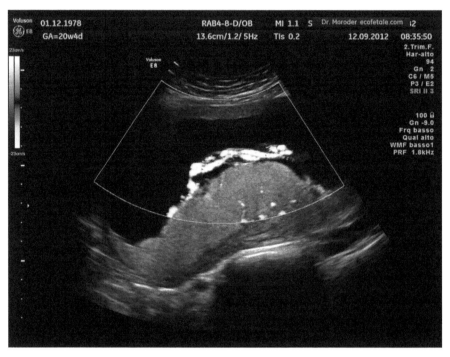

Figure 10.7   US image of birth cord and placenta in human.

10.5 AI-Based Ultrasound Imaging Analysis

US, a versatile green imaging method, is gaining international recognition as a principal imaging modality in a range of diagnostic disciplines (as shown in Figure 10.7) as a result of the ongoing advent of sophisticated ultrasonic systems and the implanted US-oriented digital healthcare services. Indeed, experienced clinicians should personally gather and physically examine images to diagnose, identify, and track diseases in the US. Diagnosis effectiveness is necessarily compromised due to the innate attribute of significant operator dependency in the US. In comparison, AI succeeds at automatically detecting complicated patterns and quantifying imaging datasets, indicating a significant possibility for assisting clinicians in obtaining more precise and reliable outcomes. ML and DL are extensively implemented in the domain of US on the respective vascular structures, namely throat, chest, childbirth and fetal cardiology, abdomen, and pelvis, muscular and skeletal system, by covering visual quality control, physiology characterization, object recognition, lesion edge detection, and morphological segmentation [5].

Contrast-enhanced ultrasound (CEUS) is frequently used to diagnose liver injuries (LI). Unconventional liver cancer (LC) is hard to determine from focal nodular hyperplasia (FNH) on CEUS footage even among LI. As a result, a new approach termed feature fusion is suggested and analyzed as a solution for resolving this issue. From the CEUS motion clips data, the suggested methodology excerpts a collection of hand-crafted characteristics and deep features. The two distinct sorts of characteristics are then combined. Eventually, a classifier is used to determine if the patient has LC or FNH. Numerous classification algorithms have demonstrated superior efficiency, emphasizing the advantage of fused features. Additionally, compared to conventional CNN, the suggested fused features are more interpretable [48]. A ResNet is utilized to excerpt features from carotid ultrasound images (CUSIs) and to classify plaques within the given images. The experimental results demonstrated that when applied to 1829 CUIS data, the model based on SVM achieved a significantly lesser plaque recognition accuracy than the ResNet model. SVM and ResNet showed 75% and 81% accuracies in classifying CUSI, respectively [49].

Thyroid nodules are extremely common, and only a tiny proportion of them are malignant. Numerous non-invasive strategies have been implemented with the assistance of the Internet of Medical Things (IoMT) to accelerate the amount of malignant tumor detection. These methodologies can be broadly classified into two categories such as radiomics-based and DL-based methods. In essence, DL models based on CNN have demonstrated impressive performance metrics in a wide range of medical image processing, analysis, and classification tasks. First, radiomics-based approach is implemented to extract 303-dimensional statistical properties from preprocessed image dataset at maximum throughput. The classification step is executed after reducing and extracting more appropriate features using dimensionality reduction techniques. Later, a technique based on DL is designed and evaluated by refining and tweaking the pre-trained VGG16 model. US dataset of 3122 nodules (1842 benign and 1394 malignant nodules) are retroactively gathered from 1042 case scenarios. The 80-20 rule of splitting the data for training and testing is adapted. The DL-based approach showed comparatively better performance metrics ($Ă = 75\%$) than the counterpart performance metrics ($Ă = 67\%$) [50].

Cirrhosis is an acute liver disorder that compromises patients' lives and general well-being. Presently, the CAD platform frequently uses US imaging to detect cirrhosis. DL methods for cirrhosis diagnostic tests using US image data repository have appeared as a consequence of the rapid advent of AI. Moreover, owing to the difficulty and fluctuation of US images, this input is frequently required to be physically annotated. A study presents a top DL approach termed LiverTL for instantaneous cirrhosis US classification tasks to address these shortcomings. LiverTL contains an immediate

RoI detection module that facilitates the retrieval of RoIs from multiple US images. Concurrently, the categorization module makes use of RoI regions and the TL infrastructure to acquire the cirrhosis diagnostics results. On the test dataset, LiverTL reaches the highest level of classification accuracy. The cirrhosis dataset experiments showed that a well-designed TL model is critical for obtaining precise and reliable classification results. These observations may lead to future advancements in the detection and treatment of cirrhosis [51].

10.6 Conclusion

A wide range of ML and DL strategies are being used to evaluate AD based on observable brain radical reforms on MRI with hyper-good performance. However, since these frameworks are black boxes without explicit and specific unambiguous knowledge representation, generating rudimentary expository imaging structures is challenging. A DL framework associates automatically learned MRI imaging biological markers with ankle pain datasets. It is possible to provide the estimated spot and the inherent dubiety in the challenge using the architecture. The DL model should be enhanced to include additional variables such as demographic data that may assist pain management.

Conversely, because of their layered nonlinear and complicated structures, DNN is largely murky and is regarded as black-box methodologies, raising a slew of professional and practical considerations. Additionally, these strategies cannot rationalize specific diagnostic choices in the way that subject experts or clinicians do, posing an increased danger in the healthcare context. With the fast development of ML in the current times, image segmentation approaches based on unsupervised and supervised learning have shown significant strides and successfully applied to various fields, including MI. Both approaches, moreover, will get their own set of strengths and weaknesses.

COVID-19 testing is typically performed using the RT-PCR procedure, which is not quite reliable. Other indicators of respiratory illness include urine and blood tests, clinical signs such as flu, muscle soreness, absence of odor, and CCT and CXR imaging. The innovation of ensemble models that consider all types of discriminatory inputs from multiple tests to produce quite precise results is necessary. Including almost every area that is examined, the data sources are growing exponentially. Each day, new evidence and information outcome emerge that, in some instances, contradict prior data. In this context, incremental learning models are a beneficial strategy.

This is especially true for data relating to epidemics and CCT/CXR images. As a result, AI frameworks should be developed to adjust to changing data inputs.

It is improbable that computers will substitute human specialists in the inference, perception, and interpretation of echocardiographic data. Whereas, AI, ML, and DL models have revealed the private details of the machine's correlations to pathological conditions. Human experts must remain an integral part of clinical care and the diagnostic process to protect patients from harm to safeguard against machine errors. Considering its drawbacks, AI remains a crucial concept for the long term of echocardiography, and extra added research is required. Anesthetists must be familiar with the benefits and disadvantages of AI innovation in echocardiography. AI approaches, fueled by the high requirement for US image analysis in medical care and sophisticated technology, will undeniably have a bright future in US. Regrettably, a complete and precise comprehension of this new topic is necessary. Because AI-based US prognosis lacks the ability of a human mind to achieve top affiliations and the individualistic care and logical arguments endorsed in the treatment process, radiologists cannot be completely substituted by AI.

10.7 Acknowledgment

We are thankful to the coauthor for their contribution to this chapter.

10.8 Funding

None.

10.9 Conflict of Interest

There is no conflict of interest.

References

[1] I. Castiglioni *et al.*, "AI applications to medical images: From machine learning to deep learning," *Phys. Medica*, vol. 83, pp. 9–24, 2021, doi: 10.1016/j.ejmp.2021.02.006.

[2] A. Barragán-Montero *et al.*, "Artificial intelligence and machine learning for medical imaging: A technology review," *Phys. Medica*, vol. 83, pp. 242–256, 2021, doi: 10.1016/j.ejmp.2021.04.016.

[3] J. Born *et al.*, "On the role of artificial intelligence in medical imaging of COVID-19," *Patterns*, vol. 2, no. 6, p. 100269, 2021, doi: 10.1016/j. patter.2021.100269.

[4] X. Chen *et al.*, "Artificial Intelligence in Echocardiography for Anesthesiologists," *J. Cardiothorac. Vasc. Anesth.*, vol. 35, no. 1, pp. 251–261, 2021, doi: 10.1053/j.jvca.2020.08.048.

[5] Y.-T. Shen, L. Chen, W.-W. Yue, and H.-X. Xu, "Artificial intelligence in ultrasound," *Eur. J. Radiol.*, vol. 139, p. 109717, 2021, doi: 10.1016/j. ejrad.2021.109717.

[6] R. W. Brown, Y.-C. N. Cheng, E. M. Haacke, M. R. Thompson, and R. Venkatesan, *Magnetic resonance imaging: physical principles and sequence design*. John Wiley & Sons, 2014.

[7] X. Zhao, C. K. E. Ang, U. R. Acharya, and K. H. Cheong, "Application of Artificial Intelligence techniques for the detection of Alzheimer's disease using structural MRI images," *Biocybern. Biomed. Eng.*, vol. 41, no. 2, pp. 456–473, 2021, doi: 10.1016/j.bbe.2021.02.006.

[8] L. Nanni, A. Lumini, and N. Zaffonato, "Ensemble based on static classifier selection for automated diagnosis of Mild Cognitive Impairment," *J. Neurosci. Methods*, vol. 302, pp. 42–46, 2018, doi: 10.1016/j.jneumeth.2017.11.002.

[9] S. H. Hojjati, A. Ebrahimzadeh, A. Khazaee, and A. Babajani-Feremi, "Predicting conversion from MCI to AD by integrating rs-fMRI and structural MRI," *Comput. Biol. Med.*, vol. 102, pp. 30–39, 2018, doi: 10.1016/j.compbiomed.2018.09.004.

[10] J. E. W. Koh *et al.*, "Automated detection of Alzheimer's disease using bi-directional empirical model decomposition," *Pattern Recognit. Lett.*, vol. 135, pp. 106–113, 2020, doi: 10.1016/j.patrec.2020.03.014.

[11] K. G. Achilleos, S. Leandrou, N. Prentzas, P. A. Kyriacou, A. C. Kakas, and C. S. Pattichis, "Extracting Explainable Assessments of Alzheimer's disease via Machine Learning on brain MRI imaging data," in *2020 IEEE 20th International Conference on Bioinformatics and Bioengineering (BIBE)*, 2020, pp. 1036–1041, doi: 10.1109/BIBE50027.2020.00175.

[12] K. Oh, Y.-C. Chung, K. W. Kim, W.-S. Kim, and I.-S. Oh, "Classification and Visualization of Alzheimer's Disease using Volumetric Convolutional Neural Network and Transfer Learning," *Sci. Rep.*, vol. 9, no. 1, p. 18150, 2019, doi: 10.1038/s41598–019-54548–6.

[13] J. J. Lee, F. Liu, S. Majumdar, and V. Pedoia, "Can AI predict pain progression in knee osteoarthritis subjects from structural MRI," *Osteoarthr. Cartil.*, vol. 27, p. S24, 2019, doi: 10.1016/j.joca.2019.02.036.

[14] M. R. Karim *et al.*, "DeepKneeExplainer: Explainable Knee Osteoarthritis Diagnosis From Radiographs and Magnetic Resonance

Imaging," *IEEE Access*, vol. 9, pp. 39757–39780, 2021, doi: 10.1109/ACCESS.2021.3062493.

[15] A. Sriram, J. Zbontar, T. Murrell, C. L. Zitnick, A. Defazio, and D. K. Sodickson, "GrappaNet: Combining Parallel Imaging With Deep Learning for Multi-Coil MRI Reconstruction," in *2020 IEEE/CVF Conference on Computer Vision and Pattern Recognition (CVPR)*, 2020, pp. 14303–14310, doi: 10.1109/CVPR42600.2020.01432.

[16] G. S. Tandel, A. Balestrieri, T. Jujaray, N. N. Khanna, L. Saba, and J. S. Suri, "Multiclass magnetic resonance imaging brain tumor classification using artificial intelligence paradigm," *Comput. Biol. Med.*, vol. 122, p. 103804, 2020, doi: 10.1016/j.compbiomed.2020.103804.

[17] Y. Liu, Q. Liu, M. Zhang, Q. Yang, S. Wang, and D. Liang, "IFR-Net: Iterative Feature Refinement Network for Compressed Sensing MRI," *IEEE Trans. Comput. Imaging*, vol. 6, pp. 434–446, 2020, doi: 10.1109/TCI.2019.2956877.

[18] X. Wang, C. Guo, and X. Zhou, "Robust Segmentation of 3D Brain MRI Images in Cross Datasets by Integrating Supervised and Unsupervised Learning," in *2020 10th International Conference on Information Science and Technology (ICIST)*, 2020, pp. 194–201, doi: 10.1109/ICIST49303.2020.9202117.

[19] S. Kumar, S. Conjeti, A. G. Roy, C. Wachinger, and N. Navab, "InfiNet: Fully convolutional networks for infant brain MRI segmentation," in *2018 IEEE 15th International Symposium on Biomedical Imaging (ISBI 2018)*, 2018, pp. 145–148, doi: 10.1109/ISBI.2018.8363542.

[20] M. Thachayani and S. Kurian, "AI Based Classification Framework For Cancer Detection Using Brain MRI Images," in *2021 International Conference on System, Computation, Automation and Networking (ICSCAN)*, 2021, pp. 1–4, doi: 10.1109/ICSCAN53069.2021.9526456.

[21] A. M. Alhassan and W. M. N. W. Zainon, "BAT Algorithm With fuzzy C-Ordered Means (BAFCOM) Clustering Segmentation and Enhanced Capsule Networks (ECN) for Brain Cancer MRI Images Classification," *IEEE Access*, vol. 8, pp. 201741–201751, 2020, doi: 10.1109/ACCESS.2020.3035803.

[22] O. Bernard *et al.*, "Deep Learning Techniques for Automatic MRI Cardiac Multi-Structures Segmentation and Diagnosis: Is the Problem Solved?," *IEEE Trans. Med. Imaging*, vol. 37, no. 11, pp. 2514–2525, 2018, doi: 10.1109/TMI.2018.2837502.

[23] A. A. Farid, G. I. Selim, and H. A. A. Khater, "A novel approach of CT images feature analysis and prediction to screen for corona virus disease

(COVID-19)," *Int. J. Sci. Eng. Res.*, vol. 11, no. 3, 2020, doi: 10.14299/ijser.2020.03.02.

[24] L. N. Mahdy, K. A. Ezzat, H. H. Elmousalami, H. A. Ella, and A. E. Hassanien, "Automatic X-ray COVID-19 Lung Image Classification System based on Multi-Level Thresholding and Support Vector Machine," *medRxiv*, 2020, doi: 10.1101/2020.03.30.20047787.

[25] Y.-H. Wu *et al.*, "JCS: An Explainable COVID-19 Diagnosis System by Joint Classification and Segmentation," *IEEE Trans. Image Process.*, vol. 30, pp. 3113–3126, 2021, doi: 10.1109/TIP.2021.3058783.

[26] J. Zhang *et al.*, "Viral Pneumonia Screening on Chest X-ray Images Using Confidence-Aware Anomaly Detection," 2020, doi: arXiv:2003.12338.

[27] Q. Ni *et al.*, "A deep learning approach to characterize 2019 coronavirus disease (COVID-19) pneumonia in chest CT images," *Eur. Radiol.*, vol. 30, no. 12, pp. 6517–6527, 2020, doi: 10.1007/s00330–020-07044–9.

[28] O. Elharrouss, N. Subramanian, and S. Al-Maadeed, "An encoder-decoder-based method for COVID-19 lung infection segmentation," 2020, doi: arXiv:2007.00861.

[29] H.-T. Zhang *et al.*, "Automated detection and quantification of COVID-19 pneumonia: CT imaging analysis by a deep learning-based software," *Eur. J. Nucl. Med. Mol. Imaging*, vol. 47, no. 11, pp. 2525–2532, 2020, doi: 10.1007/s00259–020-04953–1.

[30] T. Ozturk, M. Talo, E. A. Yildirim, U. B. Baloglu, O. Yildirim, and U. Rajendra Acharya, "Automated detection of COVID-19 cases using deep neural networks with X-ray images," *Comput. Biol. Med.*, vol. 121, p. 103792, 2020, doi: 10.1016/j.compbiomed.2020.103792.

[31] M. A. Al-antari, C.-H. Hua, and S. Lee, "Fast Deep Learning Computer-Aided Diagnosis against the Novel COVID-19 pandemic from Digital Chest X-ray Images," *Res. Sq.*, 2021, doi: 10.21203/rs.3.rs-36353/v1.

[32] Z. Han *et al.*, "Accurate Screening of COVID-19 Using Attention-Based Deep 3D Multiple Instance Learning," *IEEE Trans. Med. Imaging*, vol. 39, no. 8, pp. 2584–2594, 2020, doi: 10.1109/TMI.2020.2996256.

[33] E. Soares, P. Angelov, S. Biaso, M. H. Froes, and D. K. Abe, "SARS-CoV-2 CT-scan dataset: A large dataset of real patients CT scans for SARS-CoV-2 identification," *medRxiv*, 2020, doi: 10.1101/2020.04.24.20078584.

[34] T. Zhou, S. Canu, and S. Ruan, "Automatic COVID-19 CT segmentation using U-Net integrated spatial and channel attention mechanism," *Int. J. Imaging Syst. Technol.*, vol. 31, no. 1, pp. 16–27, 2021, doi: 10.1002/ima.22527.

[35] M. Ahishali *et al.*, "Advance Warning Methodologies for COVID-19 using Chest X-Ray Images," 2021, doi: arXiv:2006.05332.

[36] H. Panwar, P. K. Gupta, M. K. Siddiqui, R. Morales-Menendez, and V. Singh, "Application of deep learning for fast detection of COVID-19 in X-Rays using nCOVnet," *Chaos, Solitons & Fractals*, vol. 138, p. 109944, 2020, doi: 10.1016/j.chaos.2020.109944.

[37] S. Wang *et al.*, "A deep learning algorithm using CT images to screen for Corona Virus Disease (COVID-19)," *medRxiv*, 2020, doi: 10.1101/2020.02.14.20023028.

[38] S. Ahuja, B. K. Panigrahi, N. Dey, V. Rajinikanth, and T. K. Gandhi, "Deep transfer learning-based automated detection of COVID-19 from lung CT scan slices," *Appl. Intell.*, pp. 1–15, 2020, doi: 10.1007/s10489–020-01826-w.

[39] P. R. A. S. Bassi and R. Attux, "A deep convolutional neural network for COVID-19 detection using chest X-rays," *Res. Biomed. Eng.*, 2021, doi: 10.1007/s42600–021-00132–9.

[40] A. K. Jaiswal, P. Tiwari, V. K. Rathi, J. Qian, H. M. Pandey, and V. H. C. Albuquerque, "COVIDPEN: A Novel COVID-19 Detection Model using Chest X-Rays and CT Scans," *medRxiv*, 2020, doi: 10.1101/2020.07.08.20149161.

[41] K. F. Bushra, M. A. Ahamed, and M. Ahmad, "Automated detection of COVID-19 from X-ray images using CNN and Android mobile," *Res. Biomed. Eng.*, vol. 37, no. 3, pp. 545–552, 2021, doi: 10.1007/s42600–021-00163–2.

[42] X. Li, C. Li, and D. Zhu, "COVID-MobileXpert: On-Device COVID-19 Patient Triage and Follow-up using Chest X-rays," 2020, doi: arXiv:2004.03042.

[43] A. Alam and A. S. Rahyab, "Chapter 25 - Echocardiography," in *A Medication Guide to Internal Medicine Tests and Procedures*, G. Hughes, Ed. Philadelphia: Elsevier, 2022, pp. 113–116.

[44] Y. Li, C. P. Ho, M. Toulemonde, N. Chahal, R. Senior, and M.-X. Tang, "Fully Automatic Myocardial Segmentation of Contrast Echocardiography Sequence Using Random Forests Guided by Shape Model," *IEEE Trans. Med. Imaging*, vol. 37, no. 5, pp. 1081–1091, 2018, doi: 10.1109/TMI.2017.2747081.

[45] L. Xu, M. Liu, J. Zhang, and Y. He, "Convolutional-Neural-Network-Based Approach for Segmentation of Apical Four-Chamber View from Fetal Echocardiography," *IEEE Access*, vol. 8, pp. 80437–80446, 2020, doi: 10.1109/ACCESS.2020.2984630.

[46] A. I. Shahin and S. Almotairi, "An Accurate and Fast Cardio-Views Classification System Based on Fused Deep Features and LSTM," *IEEE Access*, vol. 8, pp. 135184–135194, 2020, doi: 10.1109/ ACCESS.2020.3010326.

[47] F. Taheri Dezaki *et al.*, "Cardiac Phase Detection in Echocardiograms With Densely Gated Recurrent Neural Networks and Global Extrema Loss," *IEEE Trans. Med. Imaging*, vol. 38, no. 8, pp. 1821–1832, 2019, doi: 10.1109/TMI.2018.2888807.

[48] J. Zhou, F. Pan, W. Li, H. Hu, W. Wang, and Q. Huang, "Feature Fusion for Diagnosis of Atypical Hepatocellular Carcinoma in Contrast- Enhanced Ultrasound," *IEEE Trans. Ultrason. Ferroelectr. Freq. Control*, vol. 69, no. 1, pp. 114–123, 2022, doi: 10.1109/TUFFC.2021.3110590.

[49] W. Ma, R. Zhou, Y. Zhao, Y. Xia, A. Fenster, and M. Ding, "Plaque Recognition of Carotid Ultrasound Images Based on Deep Residual Network," in *2019 IEEE 8th Joint International Information Technology and Artificial Intelligence Conference (ITAIC)*, 2019, pp. 931–934, doi: 10.1109/ITAIC.2019.8785825.

[50] Y. Wang *et al.*, "Comparison Study of Radiomics and Deep Learning-Based Methods for Thyroid Nodules Classification Using Ultrasound Images," *IEEE Access*, vol. 8, pp. 52010–52017, 2020, doi: 10.1109/ ACCESS.2020.2980290.

[51] H. Yang, X. Sun, Y. Sun, L. Cui, and B. Li, "Ultrasound Image-Based Diagnosis of Cirrhosis with an End-to-End Deep Learning model," in *2020 IEEE International Conference on Bioinformatics and Biomedicine (BIBM)*, 2020, pp. 1193–1196, doi: 10.1109/BIBM49941.2020.9313579.

11

Advancement of AI in Cancer Management: Role of Big Data

Shilpa Rawat[1], Akanksha Pandey[1], Rishabha Malviya[1], Shivkanya Fuloria[2], Swati Verma[1], Sonali Sundram[1], and Bhuneshwar Dutta Tripathi[3]

[1]Department of Pharmacy, School of Medical and Allied Science, Galgotias University, India
[2]Faculty of Pharmacy, AIMST University, Malaysia
[3]Department of Pharmacy, Narayan Institute of Pharmacy, India
*Corresponding Author
Akanksha Pandey
Research Scholar, Department of Pharmacy, School of Medical and Allied Science, Galgotias University, India
Email: pandey.akanksha1611@gmail.com

Abstract

Cancer is one of the leading causes of death in both men and women. According to a new global analysis show that 1.93 million new cases and ten million deaths from melanoma. Tumor patients have a range of treatment choices available to them, depending on their stage of the disease. Big data analysis could be utilized to aid in the resolution of critical healthcare concerns. Due to the plethora of data sources, consistency in data collecting is crucial. Data standardization results in more consistent and complete datasets, which facilitates their connection to other data sources. The rate of adoption of state-tested and certified EHR applications within the care industry is nearing a halt. The availability of numerous government-certified EHR programs, on the other hand, has inhibited knowledge exchange and sharing. The initial objective is now to gain unjust insights from the massive amounts of data generated by EMRs.

11.1 Introduction

According to the most recent global estimates, cancer is one of the leading causes of death, affecting both men and women, with 193 hundred thousand new cases and 10 million melanoma fatalities (20% of all fatalities) per year. Cancer is a broad group of disorders (approximately one hundred well-known tumors) characterized by aberrant cell proliferation with a non-heritable propensity to spread uncontrollably and invade surrounding tissues, interrupting vital activities and ultimately resulting in death. Cancer cells' genetic features are driven by six basic talents: continual regenerative output, development inhibitor avoidance, tolerance against killing, replication of eternity, initiation of developmental stages, and triggering the malignant transformation. Tumor patients can choose from a variety of treatment choices depending on the severity of their illness. Patients with localized and early-stage tumors typically undergo surgery, benefit from careful waiting or vigorous police work, as in glandular cancer, or rely on medicine to sustain rather than cure, as in the case of such abnormal blood malignancy. Patients with severe stages of sickness, on the other hand, would be treated with chemotherapy, molecularly focused medical help, or a combination of such treatments. NASA scientists invented the term "big data" in 1997 to characterize the difficulty of storing huge volumes of data generated by a new data-intensive form of computer activity. A study published in 2008 titled "Achieving Fundamental Advancements across Business, Research, as well as Community with BD Technology," credited with popularizing the term, emphasized the ease with which content solutions may be integrated into a variety of scenarios, ranging from Wal-computer Mart's memory units (4000 trillion bytes) understanding storage facility to the 15 petabytes of data proposed to be produced annually by the huge essential particle accelerator project 2.

The term "big data" is used in oncology to describe the quick collection as well as the collection of vast volumes of information that is typically derived via public tumor entries, EHR, and sometimes huge genotyping research. Multidisciplinary collaboration as well as dataset management to combine a variety of data sources and give trustworthy statistics to capture essential information are some of the challenges in adopting BD in cancer research. Big data approaches may be employed in tumor studies as well as in transforming knowledge into new ways to make better cancer treatment and delivery decisions. In this chapter, we discuss the connection between cancer and big data. Especially, what role does big data play in oncology, and what new treatments are emerging as a result? [4].

11.1.1 Big data

To date, the term "big data" has a hazy definition, but it generally refers to datasets that are sufficiently big to be analyzed by humans. Datasets are included within the range in size from 1012 to 1018 bytes. Large amounts of data are inherently heterogeneous. Consider the electronic medical record (EMR). A single patient's EMR contains test results, examination results, radiography feature-based images (per pixel), and patient records complete with transcription faults and spelling errors. In general, formatted information (test results or CPT codes) and unstructured information (material from a doctor's belief about the concept) can be found in big datasets. This implies that an investigation program must first identify a question before building an investigation to respond to it. On either approach, big data investigation may be methodology-driven, with approaches performed to the data first to identify causal links. It could contribute to a list of relationships with different degrees of association that can then be thoroughly examined. As a result, big data analysis could be utilized to jumpstart the methodological approach before the key questions have been defined. Methodologies for BD analytics including data mining and machine learning differ from earlier CER procedures in that they can be utilized to harness massive data assets [5–11].

"Big data" refers to a large amount of patient-level data obtained for another reason, including patient history or compensation claims, in the framework of treating cancer. To acquire fresh insights, the data elements are integrated or processed. "Real-world proof" can be found in big data derived from electronic health records [12]. Clinical evidence obtained from de-identified real-world datasets gathered as part of routine care rather than a planned randomized clinical trial is known as "real-world evidence." Processing such a massive volume of data needs a high level of knowledge due to its complexity and diversity. To turn enormous amounts of information into useful evidence to make healthy choices and to treat patients, proper study designs and competent analysis approaches are essential. Because big data often contains real-world data, it could help bridge the gap between clinical and translational research. This is especially significant in cancer research because pharmaceutically funded studies usually include poor control arms, forcing patients to choose between a new treatment and the best current standard [13]. Furthermore, it is well known that randomized controlled trials underrepresent minorities and other underserved groups, with just around 4% of all cancer patients taking part on average. Dr. Norman E. Shapeless, Director of the NCI, highlighted big data as one of the four areas of remarkable possibility for researchers, during the 2018 ASCO Annual Meet. As a result, scientists will have to shift their focus from passive data transmission to data aggregation. This will improve our understanding of cancer therapy and present real-world evidence [12–17].

One of the ten major revolutions projected to occur in the next ten years is the usage of large amounts of data, and the technological innovation required to evaluate it [18]. Big data is a phenomenon that influences almost every sort of company. Businesses that rely on information technology, including IBM, Google, Facebook, and Amazon, were among the first to make significant use of it. Several of the world's biggest information technology companies have created algorithms that use neural networks and machine learning algorithms to predict people's behavior and then use that information for targeted marketing. Big data has caught the interest of health insurance firms, governments, and being used in the field of life sciences [18–21].

As a result, information acquisition, procedures, and technology are all focused in order to ensure acceptable quality. Whereas, recently, the emphasis has been on elevated information such as monitoring of genomics, getting equally important to have high-quality medical statistical results, if not more crucial [22]. To preserve reliability, a continuous learning health system must handle and assure data quality. Some of these applications use information from registries, EMRs, organ repositories, genetic decoding, care plans, and optical elements, among other sources, to reduce treatment disparities, ensure quality and safety for specific patients, and make adjustments to drugs based on the outcomes of patients who have had the same disease [23]. The objective is to develop a medical system that is receptive to learning augmenting and improving the treatment that professionals can give to their patients. In radiation oncology, there are three categories of data, including diagnostic, therapy, and symptom management. The objective is to impact treatment decisions quantitatively based on diagnostic and prognostic criteria to improve future patient outcomes. With each new patient, the system learns more [24]. The system's data can assist us in understanding treatment risks and results but not biological processes. As a result, every big data project should be evaluated with known and unknown biological processes. More scientific research is needed to properly understand the biological mechanisms that trigger them [22–26].

Cancer therapeutic development today faces a massive and growing demand for proof. Clinical trials are essential in the development of new medications, but they are not without downsides. They are, for example, both expensive and time-consuming. Clinical trials enroll a minority of sufferers, and those who are not generally representative of the general population. Cancers are now divided into subtypes based on molecular characteristics that are too tiny to analyze in a randomized trial. Furthermore, faster FDA approvals of innovative medications are associated with bigger post-approval obligations. The traditional evidence-generation strategy, which is primarily based on prospective clinical trials, will be unable to close the significant evidence gap. Big data analysis can help to solve several problems in cancer treatment and medication research. Using data collected during ordinary treatment and real-world instances, it is feasible to learn from each patient. This strategy demands an interpretation mechanism in the EHR for both aggregation and data analysis. Almost every cancer patient's care is recorded electronically. However, these electronic health records were not designed with research in mind, and the majority of data necessary for research is in unstructured formats such as physician narratives, radiology reports, and biomarker test results, which are difficult to acquire and analyze. To make sense of the data in electronic health records, both structured data (such as height, weight, and chemotherapy regimens) and unstructured data must be used. Previously, researchers had to read and interpret these records, which impeded their expansion [27].

Big data might be valuable in the design and modification of disease preventive measures, particularly in the medical industry. The combination of large genetic and environmental data will help determine if individuals or groups are at risk of specific chronic diseases such as cancer. This might lead to tailored interventions aimed at altering environmental and behavioral factors that contribute to health concerns in certain groups. Big data may also be used to examine current preventative measures and reveal fresh insights that can be utilized to improve them. Big data may also be utilized in a therapeutic setting to track the impact of specialist drugs, such as the cost of oncolytic, depending on the individual as well as malignancy (genetic) features. It will contribute to the advancement of targeted therapy because it will give important information about how much different treatment regimens will cost [28].

11.2 The Source and Type of Big Data, and Their Concern

Care delivery actions generate huge volumes of data daily due to interactions between people and organizations within the medical system. Demographic and medical data sources include laboratory and radiographic findings (both verbatim and as a digital file), programmed throughput from patient testing equipment, enrollment, and financial records. Consider the following clinical and demographic data sources [29]. Unstructured and structured data can be found on the internet.

Structured data is defined as information that could be arranged into spreadsheets and are accessible and frequently obtained from a set of pre-determined answer possibilities. When dealing with an EHR, a clinician can select values from a drop-down menu or be presented with a list of diagnostic or billing codes to choose from. Patient or doctor responses in free-text fields, narrative notes, handwritten or scanned papers, pictures, and other unstructured data formats account for up to 80% of all healthcare data. As the concept of big data gained traction across industries, several underlying concepts arose, including an ever-expanding list of characteristics that can be used to identify exceedingly large datasets [30]. What began as a simple list of three big data "Vs" volume, velocity, and variety has evolved into a list of ten criteria to examine when analyzing a source of information about any industry that demonstrates characteristics such as truthfulness, variation, authenticity, vulnerability, instability, transparency, and value. Additional components of health relational databases should be researched to aid in quality improvement efforts. Due to errors or fragmentation caused by patient migration across many institutions, missing data might result in gaps and, in certain cases, erroneous interpretations of service provision frequency and clinical outcomes; thus, EHRs. Discrimination and inaccurate data could also contribute to incorrect conclusions. For instance, each investigation found that service users who starved to death in the nearest hospital and yet will not survive long enough to be admitted to an inpatient unit had less diagnostic and side effect documentary evidence in their patient history as a result of the brief encounter [31]. This finding was made while examining mortality associated with community-acquired pneumonia using EHR-derived data. Because of the systematic lack of such data, the analytic algorithms assigned these patients a lower acuity score, but a physician's manual assessment would have swiftly placed them among the most critical. Interoperability issues commonly impede data aggregation across systems, resulting in different EHR systems being unable to share or use data. Despite how difficult it may seem, this is a goal that can be reached. All it takes is one phone call to show that exceptional interoperability can be achieved [29–33].

11.2.1 The challenge of big data

Interoperability issues, on the other hand, are not limited to the capacity to transmit data from one device to another. Additionally, similar contents ought to be sufficiently interpretable. The widespread acceptance of EHRs by medical practices, laboratories, and hospital systems have resulted in the widespread utilization of standardized terminology to document clinical manifestations, symptoms, diagnostic tests, test results, and procedures. As a result, many businesses and their EHR systems get entrenched in the usage of the local language, creating barriers to "semantic interoperability." While this may not be an issue if the EHR is simply used to refer to a patient's condition rather than a specific patient, performance evaluation aims to collect and evaluate health information at the community level, and research is hampered by the difficulty of correctly mapping comparable phrases or codes together [33]. The terms sonar, radiology, ultrasonography, and imaging tests, for example, all seem to be rationally connected to a real practitioner, even though they are not unless explicitly taught otherwise. There are currently two possible solutions to this problem: upgrading both EHR systems and physician behavior to use standard terminology consistently and creating artificial intelligence tools to recognize and precisely map phrases [34–37].

11.3 Big Data Sources and Platforms

11.3.1 The national population-based cancer database

One of the foremost well-known population-based cancer datasets is the National Cancer Institute's investigating, medicine, and end results (SEER) program and, thus, the Centers for Illness Management and Prevention's NPCR. SEER now collects and disseminates tumor prevalence and mortality data from community-based confirmed cases, which cover around one-third of the population [38]. SEER collects information on patients' demography (age, sexual identity, race, ethnicity, and place of birth); melanoma aspects (tumor cell sorts, biochemical and genetic characteristics, as well as many screening tools and genomic details on tumors); illness phases; therapy details (surgery, radioactivity, drug treatment, secretion therapy, and monoclonal antibodies); and health experience (surgery, radiation, therapy, secretion medical aid, immunotherapy, vital standing, and cause of death). The Centers for Disease Control and Prevention manage the National Preventive Cardiology Registry, which is established by Congress in 1992. It collects data on cancer incidence, early treatment methods, and outcomes. NPCR and SEER work along to gather knowledge on 97% of the North American population. With such near-universal coverage, researchers could study the cancer burden and highlight the importance of cancer interference and management activities at the national, state, and native levels. Since 1990, the American College of Surgeons and the American Cancer Society have partnered on the National Cancer Information (NCDB) [39]. The NCDB today has over 1500 institutions licensed by the Commission on Cancer, which account for almost 70%

of newly diagnosed cancer cases in the United States, as well as over 34 million historical records. While SEER, NPCR, and NCDB have data on the vast majority of cancer patients in the United States, they do not cover every patient's entire treatment history. There are no statistics on recurrence or longitudinal follow-up data. Furthermore, the most commonly stated goal is general survival; however, no further information is available. There is no record of any second-line or salvage therapy after the original course of treatment. Inhabitant tumor databases (SEER and NPCR) collect treatment data at a high level, but descriptions of the criteria, prescription brands, and quantities are not collected. Due to these limitations, conducting research based solely on current national demographic figures is difficult [40]. Despite these constraints, SEER is taking numerous initiatives to improve the quality of its existing data. The American Society of Clinical Oncology launched the CancerLinQ project with the purpose of gathering and analyzing the details collected out of each cancer sufferer in the U.S. CancerLinQ collects data from EHRs directly, eliminating the need to switch data sources. CancerLinQ uses cloud-based algorithms to analyze and convert the datasets. Due to the absence of a standardized data layout in actual medical environments or EHR platforms, a range of computer technologies is being employed to convert complex data into analyzable structured information [41]. On the other hand, human data abstraction and/or natural language processing will be required for unstructured clinical notes. SEER and CancerLinQ cooperated to make it easier for people to share cancer information. This would improve patient treatment while also raising cancer awareness in the country [38–45].

11.3.2 Commercial and private cancer databases

Many industrial large info enterprises have evolved to collect and combine actual data such as patient history, medical reports, and billing records in order to provide cancer care stakeholders with real-time feedback on treatments and outcomes. Iron Health, as an associate example, created the medication cloud for this purpose. Flatiron's network includes more than one-quarter thousand tumor health centers and 1.5 million active patients [46], all of whom are joined through a cloud-based EHR platform to form a unified system. Their associate in Nursingalytical tools is employed to boost associate EMR system, analyze the value of particular sufferer care, gather quality indicators, and see realizable clinical trial candidates. Flatiron's methodology is utilized in conjunction with clinical biological science info from Foundation Medicine by the FDA to see the importance of "actual proof." A pair of such private-sector health maintenance organizations' exploitation

and databases are Kaiser Permanente clinical analysis networks and Truven Health's MarketScan. These sorts of large info are distinct from the quality giant information bases created exclusively to assemble information for clinical trials or analysis objectives [47]. There are some ways to use the above-mentioned info sources to induce unstructured info and provide an amount of your time analysis and feedback, which could facilitate improved cancer medical aid delivery [46–49].

11.3.3 Cancer biological science and various "Omic" databases

The NCI's center for cancer genomics established the Cancer info workplace in 2018. The project's objective is to supply laptop tools for analyzing and mixing info from cancer biological science labs and patients. These techniques would probably be applied to elementary scientific queries like cancer genetic predisposition, status, and response to treatment. The cancer Genome Atlas (TCGA), which was maintained by the NCI and so the NHOAI, was one in each of the first large-scale cooperative genomic datasets. As of March 2019 [50], the TCGA had sequenced and molecularly profiled tumors from over 33,000 patients. UN agency had been diagnosed with around 70 different types of cancer, beat 22,000 genes, and characteristic over 3,140,000 alterations [51, 52]. Notably, academics from all around the world have profited from the TCGA's intensive publicly out molecular identification, which has been cited in over 5000 studies. The National Cancer Institute's Clinical Proteomic Growth Analysis pool could also be a collaboration of institutions and researchers that uses "pan" and "omic" analyses to investigate the molecular foundations of cancer. Genetic science info generated by Clinical Proteomic Growth Analysis pool analysis is kept publicly out that is accessible to researchers worldwide. To boot, the National Cancer Institute's Therapeutically applicable research to generate effective treatments (TARGET) program collaborates with a variety of clinical trial cooperative groups and consortia to assemble clinical info and tissue samples for the aim of generating, analyzing, and coding genomic info [53]. TARGET investigates genomes and transcriptomes through the utilization of a "multiomic" approach that comes with a style of sequencing and array-based technologies. Attributable to the advancement to facilitate the development of genomic sequence therapeutic targets, a majority of educational medical centers had also established one's committee polymer sequence alignment strategies (e.g., Memorial Sloan Kettering-Integrated Mutagenesis Recognition of Inequitable Melanoma Objectives) or rely on commercially available frames such as those from base treatments. Although knowledge was first gathered

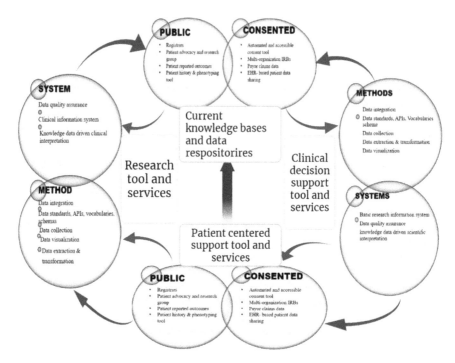

Figure 11.1   Big data collaborate with oncology to predict the best treatment for patient.

in silos, the American Association for Cancer Research is attempting to combine it through the biological science-based pathological process information sharing [54]. The data generated by these efforts have benefited researchers in gaining associated improved information on the genetic makeup of various cancers and seeking out therapeutic targets. In some ways, we have barely scraped the surface, and there is still a mountain of data to assemble and analyze. As a result, academics have increased their efforts to discover new ways to increase the accessibility of this information, as well as the tools required to view it [50–55]. Figure 11.1 shows the collaboration of big data with oncology.

11.4 Data Collection in Big Data for Oncology Treatment

In terms of digital imaging of the treatment delivery process, radiation oncology is one of the most modern medical fields. We have a computerized record of each patient's radiation medicine and know exactly where it is going in the body. Current linear accelerators with on-board imaging capabilities can assist researchers better comprehend the relevance of typical life variations in the victim's framework in terms of radioactive delivery position [56]. Our understanding of the impact of radiation on human function will increase as automated anatomy segmentation and tools for analyzing the topological effect change in the dose of ionizing radiation within a certain anatomical structure [57]. As a result of the highly automated nature of radiation therapy, the treatment delivery process may be measured and evaluated. Radiation medicine needs to pay more attention to the increase of digital data collection of critical patient states and, perhaps more importantly, oncologic outcomes of interest. Pathology, genetics, and, more recently, radiomics (six to nine) have all contributed to our understanding of the ailments we treat. Furthermore, genetics provides information on the radiation susceptibility of normal anatomy [56–61].

Improving knowledge integration can greatly improve the flexibility to refine groups of patients that are otherwise "similar." The cost of genetic sequencing has declined significantly in recent years, and it appears that it will remain low for the foreseeable future. The most challenging knowledge to collect is outcome knowledge, which should be obtained throughout the review process to ensure the accuracy of knowledge. Otherwise, the time it takes to gather knowledge after patient examination erodes accuracy and introduces bias. This issue is not confined to the large knowledge enterprise; it is going to even be encountered in early-stage clinical investigations. Each patient visit provides an opportunity to learn more about treatment side effects, symptom management, illness response, and quality of life. The challenge is to get this information while not distracted by the patient–physician connection or a frantic clinical procedure [60]. As a result, establishing the right technology to support these goals while maintaining existing clinical practices and documentation standards could be a major challenge. This technique of text-based record-keeping and dictation is shy for capturing data. Consequently, the observer may interpret these findings as crucial; thus, various documentations opt to report just the relevant positive findings. Analytical algorithms, on the other hand, need important negatives to produce context for relevant positive findings [61]. Therefore, optimal knowledge capture strategies should include these issues, with the optimum knowledge capture strategy favoring the use of organized knowledge objects. While fast processing and abstraction are beneficial, the ability to evaluate the knowledge/information integrity required for many huge data projects' insight finding goals may be limited. It is vital to develop more tailored user interfaces for physicians and nurses that allow for coordinated data collection while keeping practitioners present for patients.

These user interfaces can be changed to fit different types of encounters; also, they have displays that show how the patient is doing and allow for quick changes [62].

Along with medical evaluations, physician consequences enable the collection of some more frequent data on a person's well-being. Validated Quality of Life devices can be electronically collected to retain consistent and global assessments of individual health. Portable gadgets have the potential to do more continuous patient evaluations on a daily (or even minute) basis. These devices are essential for recording real-time patient encounters. To drive a culture transformation, we must demonstrate the value of the learning health system in supporting and improving patient care [63]. Data collection should be emphasized as a paradigm for better serving the particular patient. We must be cautious not to discourage its use by imposing rigorous practice monitoring, which may be perceived as punitive while not influencing patient care. If quality metrics are to be used at all, they must be linked to relevant therapeutic objectives.

11.4.1 Data management and aggregation in big data

Over the last two decades, system-to-system syntactic compatibility has improved substantially. DICOM RT has allowed methods to communicate about therapy, freely. Additionally, HL7 enables the linking of test results, patient schedules, and demographic data. On the other hand, semantic interoperability is still in its infancy. Ontologies such as SNOWMED-CT are gaining popularity, and their adoption will allow systems and organizations to interchange vast amounts of data while maintaining semantic integrity. Ontologies for radiation oncology have been improved further, and we will be able to correlate data with meaning even more in the future.

Although ontologies give a uniform terminology, the underlying concept is difficult to communicate to all clinicians. These ontologies need to be set up in such a way that clinicians can utilize them without having to worry about how they operate while caring for patients [64, 65].

When it comes to data aggregation, the most often used method is to employ a data warehouse, which was created for business analytics. Radiation oncology is unaffected by these fundamental business analytics tools [66]. These off-the-shelf solutions are incapable of handling our data due to its complexity and the number of variables in our queries. The architectures of medical data warehouses range from classic relational databases to more contemporary no-SQL versions that put a larger emphasis on documents. To do research, we need a model capable of storing all patient data. Analytical tools must be used by medical researchers and clinicians, and they must help to make pragmatic controlled trials possible [66–70].

All types of radiation oncology providers may participate in big central data repositories such as the NROR, QOPI, and ASCO CancerLinQ. The ultimate goal is to centralize all data to facilitate quality comparisons and validation assessments across institutions. The issue is that, historically, these models limited data collection to what was deemed "acceptable" for each patient. As a result, data extraction through healthcare workflows into such centralized information sources requires approval from the participating institution, takes more time and effort to maintain interfaces with electronic medical records, and limits access to the participating institution's data. This has made exporting data from clinical systems to these central data repositories very costly.

Both Oncospace and MAASTRO/EuroCAT25 use federated data architectures, which allows each institution to keep track of its own data, which offers several advantages. Each institution retains total control over data access and analysis. This expands your options for connecting to internal electronic medical records. Flexible criteria for data gathering offer more accurate analysis and decision support with less worry about data consistency. This method protects patient privacy because the institution has complete control over the data and makes sure that it never leaves the institution.

11.4.2 The data sources for big data in medicine

There are several massive data sources accessible, each with its unique set of properties and dimensions. The findings of clinical trials conducted on cancer patients are the most evident. Often utilized for therapeutic purposes, computerized patient files include a high number of data points or subjects [61]. These files contain a wealth of information about patients and their tumors and treatments, as well as demographic information such as the patient's gender, age, family history, and comorbidities. They additionally contain tomography knowledge (such as magnetic resonance imaging, CT, PET, and US) and solid and liquid tissue-based analysis (such as histopathological identification, DNA/RNA sequencing blood analyses, and whole order BAM files), which can be used to perform whole-genome sequencing. *In vitro* data, on the other hand, may be informative and important [62]. The computational analysis of massive datasets is another source of big data. Indirect and calculated data are incorporated in the processed information, such as radiomic and digital picture analysis, as well as genetic expression and mutation research. Massive computer files, particularly structured data, are a rapidly rising source of machine-learning-processed data. Another major source of information is data derived from patients' self-reported outcomes and experiences, commonly referred to as "big data." This data is collected using software on computers and mobile devices and may be given by their healthcare professionals (eHealth, telemedicine, etc.) or collected by the patients themselves. A fourth source is the published literature (IBM project). Given that over 1 million biomedical publications are published each year, it is difficult for any clinician to study even a fraction of the available material. However, one critical component of cancer is the volume of data acquired on each patient. Despite the relatively small patient population in oncology, the sector creates and maintains a vast number of observables (thousands to millions) [64]. Patients with rare conditions, such as head and neck cancer, confront an even bigger disparity between the quality of their data and the size of their cohorts. Recent breakthroughs in machine learning and neural networks may be highly beneficial if examples are available. For example, hundreds to millions of instances must be employed to create algorithms for object recognition in photos. We will need a bigger sample size to use this data to build personalized medicines. Some important parts of cancer research are important for all types of cancer, but especially for head and neck oncology [75, 76].

11.5 Therapy Plan for Cancer

11.5.1 Big data will aid in the development of novel cancer therapies

Consistency in data collection is critical, given the abundance of sources of data. In the long term, this standardization will result in more consistent and full datasets, which will connect them to other data sources easier. Consistency is required for data integration to make sense of and produce value from data. When assessing the quality of care delivered after various surgical operations, patients' tumor stages, adjuvant treatments, co-morbidity, and other aspects must be considered. In Netherlands, there are a lot of national databases that collect medical, pathogenic, hereditary, chromosomal, and PROM/PREM data from people [75].

The Dutch Head and Neck Audit (DHNA) started collecting clinical data in 2014 and are now part of the Dutch Institute for Clinical Auditing (DICA), which has subsequently established subgroups for numerous medical disorders (cancer and non-cancer). Oncological tissue-based data has been collected meticulously and is now accessible synoptically for over 20 unique tumor types throughout the country. The organization of radiological data remains an open question. The NET-QUBIC organization has developed a nationwide website for reporting PREMs and PROMs for HNC and other forms of cancer. In Netherlands, the majority of patient-derived data sources have been integrated into HNC [76]. The Cancer Imaging Archive (TCIA) and the Head and Neck Squamous Cell Carcinoma (HNSCC) Collection are two instances of similar global efforts, which collect this data as soon as possible, preferably on a patient-by-patient basis. In Netherlands, the Dutch Head and Neck Society runs eight hospitals in collaboration with six approved partners (NWHHT). The head and neck cancer community has a lot of potential in terms of pooling resources and coordinating actions to universally complete existing databases [76]. There is an enormous possibility to link these data-sets throughout the country and develop algorithms for analyzing data from several locations simultaneously in head and neck cancer [77, 78].

11.5.2 A cancer treatment plan that is tailored to each patient

Personalized medicine is based on actionable insights derived from big data, which are used to translate current knowledge and data into improved treatment results. New technologies, including sequencing and imaging, are generating terabytes of data that are becoming more available to the scientific community. In terms of volume, radiomic and digital image analyses account for the great majority of data output, rather than direct patient-related information. Head and neck tumors provide a unique diagnostic and therapeutic challenge due to their intricate architecture and unpredictability. Radiomic may be able to assist in addressing some of these concerns. Radiomic is a non-invasive and low-cost approach for collecting medical imaging features. Imaging characteristics may be utilized to quantify the phenotypic characteristics of a whole tumor, which is the foundation for the radiomic notion. It is possible to identify differences in expected/predicted survival rates between groups of patients and treatment outcome(s) prediction using radiomics, which enables prognostic and reliable machine learning methods, to aid in the selection of the best possible treatment for patients with head and neck cancer. Medical and radiation oncologists may be able to (de)increase the doses of systemic treatment or irradiation administered to cancer patients in various circumstances [71].

11.6 Big Data Powers the Design of "Smart" Cell Therapies for Cancer

By merging machine intelligence with biological innovation, scientists can create "living medicines" that can target malignancies with pinpoint accuracy.

The goal of oncology research is to find drugs that can destroy cancer cells while sparing healthy tissue. Researchers at Princeton and UC San Francisco (UCSF) have different ideas about how to solve this problem by using "smart cell therapies," or biological drugs that are only activated by specific protein combinations found in cancer cells. Scientists from Princeton University and the University of California, San Francisco (UCSF) have different ideas on how to overcome this issue [72].

Researchers at the UCSF Cell Design Initiative and National Cancer Institute-Sponsored Centre for Synthetic Immunology have been studying the biological elements of this wide approach for many years. By merging cutting-edge therapeutic cell engineering with cutting-edge computational approaches, the most recent study brings a significant new dimension to this endeavor. Princeton's Lewis-Sigler Institute for Integrative Genomics and the Flatiron Institute of the Simons Foundation collaborated with Lim's team, as did the computer scientist Olga G. Troyanskaya, Ph.D. The researchers examined massive datasets, including hundreds of proteins identified in both cancerous and healthy cells, using machine learning technologies. They narrowed down the millions of possible protein combinations to those that may be utilized to target cancer cells while leaving healthy cells alone [73]. Using this computationally produced protein data, Lim and his colleagues were able to develop cancer treatments that were successful and precisely targeted [74].

According to the UCSF professor and head of the cellular and molecular pharmacology, Dr. Lim, most cancer therapies, including chimeric antigen receptor (CAR) T cells, are now focused on "stopping this" or "killing this," as opposed to "treating this," as in the past. A therapeutic cell's choices should be more nuanced and sophisticated, according to the researchers.

CAR T cells have received a lot of interest in recent years as a prospective cancer treatment; however, solid tumors have not responded well to this method. CAR T cell therapy is genetically modified immune system cells taken from a patient's blood to develop a particular receptor that recognizes a specific marker or antigen on cancer cells. Cancers of the blood, such as leukemia and lymphoma, have shown great promise when treated with CAR T cells, but the treatment has not yet been shown successful against solid tumors like breast, lung, or liver cancer. Since cells in solid tumors commonly interchange antigens with normal cells, CAR T cells may have unexpected consequences by attacking healthy organs. CAR T cells are often limited in their ability to combat solid tumors because of the hostile milieu they face [74]. When it comes to evaluating their surroundings and making decisions, cells are analogous to molecular computers, according to Lim. Solid tumors are more difficult to treat than blood cancers; so "you have to construct a more comprehensive strategy," he says. The Cell Systems team, which comprised a graduate student in Troyanskaya's Princeton lab and a clinical fellow in Lim's lab, examined at almost 2300 genes in normal and malignant cells to see which antigens could help separate one from the other. Researchers employed machine learning methods to discover prospective hits and identify clusters of antigens. Antigen combinations based on gene expression data may help T cells recognize cancer while disregarding healthy tissue, according to Lim, Troyanskaya, and colleagues. The Boolean AND, OR, or NOT operators may be used to differentiate tumor cells from normal tissue, for example, using markers "A" or "B," but not "C," where "C" is an antigen that is detected exclusively in normal tissue [74].

11.6.1 Instructions are inserted into the cells

SynNotch, a molecular sensor that enables synthetic biologists to fine-tune cell programming, was utilized to encode these instructions into T cells. It was developed at the Lim lab in 2016 and is a receptor that can be designed to identify a broad variety of target antigens. If synNotch's output response can be controlled, the cell may be able to perform a variety of things in response to the discovery of an antigen [75].

They were able to utilize synNotch as a tool to educate T lymphocytes to attack kidney cancer cells that produce a particular combination of antigens known as CD70 and AXL, demonstrating the potential importance of the data they had collected. Using a designed synNotch AND logic gate, T cells were able to target cancer cells and not healthy ones. Tumor-specific T cells were able to target just cancer cells even though CD70 and AXL are present in both healthy immunological and lung cells [75].

In the last five years, Troyanskaya said, "Cancer big data analysis and cell engineering have both expanded, but these accomplishments have not been brought together." Thanks to the computational abilities of therapeutic cells and machine learning methodologies, it is now possible to act on the growing amount of comprehensive genetic and proteomic data regarding cancer [76].

The discoveries of Jasper Williams, a former UCSF graduate student, are taken into account when numerous synNotch receptors are daisy-chained to construct a variety of sophisticated cancer detection circuits. As a result of synNotch's ability to "plug and play" the expression of particular genes, these components may be connected in many ways to form circuits with a wide range of Boolean functions, allowing for accurate detection and a variety of reactions when sick cells are found.

Lim described his research as "basically a cell engineering manual" to develop therapeutic T cells that can detect nearly any combination of antigen sequences on a cancer cell; it teaches us how to produce multiple types of therapeutic T cells.

Antigen A may be recognized by a synNotch receptor and then utilized to generate a secondary synNotch that recognizes antigen B, triggering the production of a CAR that recognizes antigen C. As a result, for a T cell to be deadly, all three antigens must be present. When a T cell recognizes an antigen that is present in normal tissues but not in the tumor, it can be trained to self-destruct. No harm will be done to normal cells, and unwanted repercussions will be avoided [77].

Lim and colleagues demonstrate that by utilizing synNotch arrangements of this type, they can selectively kill cells that exhibit different combinatorial indicators of melanoma and breast cancer. T cells coupled with synNotch were also tested in mice with two distinct malignancies, each with a unique antigen combination, and they quickly and properly detected the tumors expected [80]. In addition to pancreatic cancer, the researchers are looking into how similar circuits could be used in CAR T cells to treat aggressive brain malignancies like glioblastoma, which is nearly always fatal when treated conventionally. "It is not just a one-shot wonder bullet that you are looking for," Lilim saw that Lim was using every piece of data he had. This might open the way for the generation of more intelligent cells that can make use of the computational complexity of biology, which could have a big effect on cancer therapy [82–84].

11.7 Future and Challenges of Big Data in Oncology

11.7.1 Challenges

Methods for large knowledge administration and assessment are constantly changing for period knowledge monitoring, collection, compilation, statistics (including cubic centimeter plus data modeling), and visual image technologies which will aid within the integration of EMRs into care. For example, in the United States, the adoption rate of state-tested and authorized EHR applications in the healthcare industry is nearing completion [85]. However, the provision of many government-certified EHR programs with its clinical nomenclature, technical necessities, and purposeful capabilities has hampered knowledge interchange and sharing. However, we can confidently conclude that the healthcare industry has reached the "post-EMR" phase of deployment. The first objective now is to draw unfair insights from the massive amounts of data generated by EMRs. We will go over a few of these challenges in further depth in this section.

11.8 Perspectives for the Future

In the present era of technologies and digitalization, numerous medical and research techniques, including genome sequencing, digital biosensors, and digital assistants, create a massive volume of information. Therefore, one must be aware of or evaluate the benefits which can be obtained by using such information. Examination of these kinds of information will provide more perspectives on methodological, conceptual, therapeutic, as well as other sorts of transforming healthcare [86].

Working with advanced analytics is difficult with many relational database management systems, portable statistical, and visualization programs [87]. Several assessments seem impossible to finish in a reasonable timeframe. ML, information retrieval, and computation evaluation are all names used to represent certain "computation" processes [88]. The two principal components for reinforcement methods and logistic regression exhibit a wide range of aims they are attempting to achieve. Several computer models that have already gained broad appeal include neural networks, support vector machines, Bayesian network modeling, choice branches as well as randomized, and cluster approaches [89].

As the usage of digitalization grows, more processing effort will be required for multivariable calculus [90]. To accommodate this form of research, the infrastructure must also be extensible or even contain the statistical methods along with facts. This is necessary for understanding the additional issues presented by many records, as well as how to properly utilize them. Throughout the next decade, the overall space of CER would shift, and the effects might still be seen among allied subspecialties investigation. Major University commitments involving new clusters of attendees (beneficiaries, practitioners, victims, or even corporate entities) would then arise to construct information of such area. Additionally, they will construct uniqueness designed to sustain sophisticated analytics strategies to evaluate meta-analyses and thus to acknowledge person variability throughout terms of diagnosis curriculum and drug sensitivity. Some comparisons are frequently included throughout, including both terms of information references as well as the utilization of widely diverse data sources. Multiple examples of such efforts are the Observational Medical Outcomes Partnership [91] and eMERGE [92]. A study of lithium medication efficacy conducted through the PatientsLikeMe® platform (MA, USA) [93] demonstrates patient involvement in efficacy research. CER research will become significantly more multipolar, encompassing both government and academia, as well as the biopharmaceutical industry. Researchers categorized the major conceptual framework of investigation in health informatics based on the relevant evaluation. Considering the variances from every study's methods, we found that the future perspectives were often comparable. We categorized the data into three different categories: "technological" observations (succeeding digitalization innovations to host a variety and perhaps forecast technological advancements mostly in the surrounding region), "organizational" viewpoints (possible enhancements to hospital methods in order of confidentiality, mentoring, service delivery, cost-cutting, and so on), and "investigation" outlooks (future study objectives that all mentioned strategies would then be bridging). Technology will play the most important contribution to the BDA's development in health since the volume of health data will expand at a rapid pace around the world, increasing the demand for information technology infrastructure [94].

Many researchers indicated that their primary goal was to improve the technological technique presented or evaluated in their study so that more advanced versions of their approaches will be accessible in the future. Scientists determined features of their methods that may be enhanced with expert input, and they expect more innovative strategies for mining of large datasets, as well as new platforms or procedures, to optimize data value. They also anticipate that technology will expand the capacity of systems and increase the accuracy of health data, allowing for better risk adjustment. Advanced new broadcast technology has the potential to improve, communicate, and provide cost-effective therapies to patients [95], thereby minimizing resource misallocation (e.g., tracking patients across service sites, aggregating more data, providing another more detailed perspective of statistics, accessibility of the patient history to every physician, and so forth) [96]. Programs that permit or enhance clinical experts' judging abilities, particularly when diagnosing complex diseases or pathologies, appear to be in high demand in medical services. Finally, numerous researchers saw room for change in their computational method in terms of offering new modalities to provide more adequate outcomes [97].

Among the "organizational viewpoints," future advancements in the field of private information secrecy, as a result of the emergence of technologies that will standardize and secure the process of combining anonymous aid data from aid organizations, are the most important. Aid companies and researchers are ready to employ these datasets for extra worth creation if privacy and cybersecurity are ensured. Another future approach is for aid organizations to train and educate all physicians as well as the general public about the next redeveloping era or the period in which people play an active role in the creation and management of health and medication data using digital technology. Environmental factors are expected to be incorporated into the data analysis method and judgment calls to observe natural dangers, like gas outpouring. On the other hand, in order to reduce our negative impact on the environment, we can sanction the real-time implementation of water closet light sources, reduce unnecessary journeys, prevent patient falls, and use other alternative methods. Because of the need for brand new patient treatment protocols and patient–clinician relationships, victimization of artistic and computer-aided identification systems, like questioning ancient practices, can lead to new social group norms. Given the value of knowledge analysis in aid, additional investments in IT infrastructure and individuals with appropriate interest/expertise from aid firms or governments for aid monitoring, such as national machine-driven bio-surveillance systems, are expected. Scientists additionally indicated that understanding the extent to that style selections are created on things like drug prescription patterns, which might impact value outcomes, would be useful. In the formation of broad collaborations between makers, payers, providers, and regulators within the healthcare system, very little attention is currently being paid to organizational future views, to say nothing of true worth that massive knowledge can provide in medicine.

The main focus of the "research views" is researchers' need for a large number of investigations to support their theory, as well as their conviction that their approach may "provide value" in alternative help applications. Incorporating period sensing element knowledge from patients' devices into the BDA method, for instance, will enhance clinical call support for a polygenic disorder, similarly to alternative sophisticated medical diseases like Alzheimer's and psychopathy. Alternative researchers envision a future in which massive data is used to guide time clinical decision-making based on individual patient characteristics, and in which large teams of patient data are pooled from multiple institutions so that everyone, including associated physicians, will notice "people accept them" to aid in clinical decision-making during the particular duration. As a result, many writers argue that their knowledge analytic methodologies ought to be valid with patient demographic options from totally different areas or nations to broaden the reach of their model and provide world assumptions. Finally, there is proof that the authors' projected approach can be utilized in alternative scientific fields in the future. For example, knowledge-based innovations in a variety of large knowledge fields, such as precision medicine, nutrigenomics, vaccinomics, personalized medicine, protocols in order, and other possible suspects, are expected to be centered on the interpretation of post-genomics data using specific algorithms.

11.9 Conclusion

Cancer is a leading cause of death in both genders. Patients with tumors have a variety of options. Big data approaches could be employed in tumor research and in improving cancer therapy and delivery decisions. The term "big data" is vague, but it often refers to large datasets that require human analysis. Huge datasets typically are both organized (test results or CPT codes) and unstructured (doctor's notes). There is a large amount of patient-level data, which may help overcome the clinical–translational research gap. However, processing such large amounts of data requires a high level of knowledge. Radiation oncology data includes diagnostic, therapeutic, and symptomatic management. The purpose is to influence treatment decisions using diagnostic and prognostic criteria. A project's biological impact should be considered in every case. The requirement for proof is increasing in the field of cancer treatment development. Big data analytics can help solve several problems in cancer therapy and medication research. Data from routine treatment and real-world settings can be used to learn from each patient. It may also help design and refine disease preventive measures. The National Cancer Institute's SEER program, and hence the Centers for Disease Control and Prevention's NPCR, can now follow tumor frequency and mortality rates in communities, thanks to this new data collection.

11.10 Acknowledgment

We are thankful to our co-author to be a part of this chapter.

11.11 Funding

No funding source

11.12 Conflict of Interest

There is no conflict of interest.

References

[1] Cox, M., & Ellsworth, D. (1997, October). Application-controlled demand paging for out-of-core visualization. In *Proceedings. Visualization'97 (Cat. No. 97CB36155)* (pp. 235–244). IEEE.

[2] Bryant, R., Katz, R. H., & Lazowska, E. D. (2008). Big-data computing: creating revolutionary breakthroughs in commerce, science and society.

[3] Press G. Big Data Definitions: What's yours? 2014. URL http://www. forbes. Com/sites/gilpress/2014/09/03/12-big-data-definitions-whats-yours's. 12.

[4] W.H. Organization, World Cancer Report 2020, 2020

[5] Berger, M. L., & Doban, V. (2014). Big data, advanced analytics and the future of comparative effectiveness research. *Journal of comparative effectiveness research*, *3*(2), 167–176.

[6] Krulwich, R. (2012). Which is greater, the number of sand grains on earth or stars in the sky. *Krulwich Wonders.: NPR*.

[7] Rashbass, J., & Peake, M. (2014). The evolution of cancer registration. *European Journal of Cancer Care*, *23*(6), 757–759.

[8] Hanahan D, Weinberg RA. Hallmarks of cancer: the next generation. cell. 2011 Mar 4;144(5):646–74.

[9] Bill-Axelson, Anna; Holmberg, Lars; Ruutu, Mirja; Garmo, Hans; Stark, Jennifer R.; Busch, Christer; Nordling, Stig; Häggman, Michael; Andersson, Swen-Olof; Bratell, Stefan; Spångberg, Anders; Palmgren, Juni; Steineck, Gunnar; Adami, Hans-Olov; Johansson, Jan-Erik (2011). *Radical Prostatectomy versus Watchful Waiting in Early Prostate Cancer. New England Journal of Medicine, 364(18), 1708–1717.* doi:10.1056/ Nejmoa1011967

[10] Evans, J., Ziebland, S., & Pettitt, A. R. (2012). Incurable, invisible and inconclusive: watchful waiting for chronic lymphocytic leukaemia and implications for doctor–patient communication. *European journal of cancer care*, *21*(1), 67–77.

[11] Tsai, C. J., Riaz, N., & Gomez, S. L. (2019, October). Big data in cancer research: real-world resources for precision oncology to improve cancer care delivery. In *Seminars in radiation oncology* (Vol. 29, No. 4, pp. 306–310). WB Saunders.

[12] Kahl BS. With a Little Help from My Friends. Clinical Advances in Hematology and Oncology. 2018 Dec 1; 16(12).

[13] Dewdney, S. B., & Lachance, J. (2017). Electronic records, registries, and the development of "Big Data": crowd-sourcing quality toward knowledge. *Frontiers in oncology*, *6*, 268.

[14] Hilal, T., Sonbol, M. B., & Prasad, V. (2019). Analysis of control arm quality in randomized clinical trials leading to anticancer drug approval by the US Food and Drug Administration. *JAMA oncology*, *5*(6), 887–892.

[15] Nazha, B., Mishra, M., Pentz, R., & Owonikoko, T. K. (2019). Enrollment of racial minorities in clinical trials: old problem assumes new urgency in the age of immunotherapy. *American Society of Clinical Oncology Educational Book*, *39*, 3–10.

[16] Taofeek K. Owonikoko. American Society of Clinical Oncology Educational Book 39:3–10, 2019

[17] Tsai, C. J., Riaz, N., & Gomez, S. L. (2019, October). Big data in cancer research: real-world resources for precision oncology to improve cancer care delivery. In *Seminars in radiation oncology* (Vol. 29, No. 4, pp. 306–310). WB Saunders.

[18] Shaikh, A. R., Butte, A. J., Schully, S. D., Dalton, W. S., Khoury, M. J., & Hesse, B. W. (2014). Collaborative biomedicine in the age of big data: the case of cancer. *Journal of medical Internet research*, *16*(4), e2496.

[19] Roman-Belmonte JM, De la Use the "Insert Citation" button to add citations to this document.

[20] Roman-Belmonte, J. M., De la Corte-Rodriguez, H., & Rodriguez-Merchan, E. C. (2018). How blockchain technology can change medicine. *Postgraduate medicine*, *130*(4), 420–427.

[21] Bourne, Philip E (2014). *What <i>Big Data</i> means to me. Journal of the American Medical Informatics Association, 21(2), 194–194*.

[22] McNutt, T. R., Moore, K. L., & Quon, H. (2016). Needs and challenges for big data in radiation oncology. *International Journal of Radiation Oncology, Biology, Physics*, *95*(3), 909–915.

[23] Wu, B., McNutt, T., Zahurak, M., Simari, P., Pang, D., Taylor, R., & Sanguineti, G. (2012). Fully automated simultaneous integrated boosted–intensity modulated radiation therapy treatment planning is feasible for head-and-neck cancer: a prospective clinical study. *International Journal of Radiation Oncology* Biology* Physics*, *84*(5), e647–e653.

[24] Petit, S. F., Wu, B., Kazhdan, M., Dekker, A., Simari, P., Kumar, R., ... & McNutt, T. (2012). Increased organ sparing using shape-based treatment plan optimization for intensity modulated radiation therapy of pancreatic adenocarcinoma. *Radiotherapy and Oncology*, *102*(1), 38–44.

[25] Moore, J. A., Evans, K., Yang, W., Herman, J., & McNutt, T. (2014, March). Automatic treatment planning implementation using a database of previously treated patients. In *Journal of Physics: Conference Series* (Vol. 489, No. 1, p. 012054). IOP Publishing.

[26] Moore, K. L., Brame, R. S., Low, D. A., & Mutic, S. (2011). Experience-based quality control of clinical intensity-modulated radiotherapy planning. *International Journal of Radiation Oncology* Biology* Physics*, *81*(2), 545–551.

[27] Meropol, N. J. (2018). Opportunities for using big data to advance cancer care. *Clinical Advances in Hematology & Oncology: H&O*, *16*(12), 807–809.

[28] Willems, S. M., Abeln, S., Feenstra, K. A., de Bree, R., van der Poel, E. F., de Jong, R. J. B., ... & van den Brekel, M. W. (2019). The potential use of big data in oncology. *Oral Oncology*, *98*, 8–12.

[29] Healthcare Information and Management Systems Society. Blending structured and unstructured data to develop healthcare insights. Updated 2016. http://www.himss.org/library/ blending-structured-and-unstructured-data-develop-healthcareinsights. Accessed December 19, 2017.

[30] Fessele, K. L. (2018, May). The rise of big data in oncology. In *Seminars in Oncology Nursing* (Vol. 34, No. 2, pp. 168–176). WB Saunders.

[31] Hripcsak, G., Knirsch, C., Zhou, L., Wilcox, A., & Melton, G. B. (2011). Bias associated with mining electronic health records. *Journal of biomedical discovery and collaboration*, *6*, 48.

[32] Geraci, A. (1991). *IEEE standard computer dictionary: Compilation of IEEE standard computer glossaries*. IEEE Press.

[33] Heubusch, K. (2006). Interoperability: what it means, why it matters. *Journal of AHIMA*, *77*(1), 26–30.

[34] Dixon, B. E., Vreeman, D. J., & Grannis, S. J. (2014). The long road to semantic interoperability in support of public health: experiences from two states. *Journal of biomedical informatics*, *49*, 3–8.

[35] Moreno-Conde, A., Moner, D., Cruz, W. D. D., Santos, M. R., Maldonado, J. A., Robles, M., & Kalra, D. (2015). Clinical information modeling processes for semantic interoperability of electronic health records: systematic review and inductive analysis. *Journal of the American medical informatics association*, *22*(4), 925–934.

[36] Putta A, Ramireddy RD. An adaptive machine learning approach for semantic analysis to extract medical knowledge. Int J Adv Res Sci Technol. 2014; 3:140–144.

[37] Kim, H., El-Kareh, R., Goel, A., Vineet, F. N. U., & Chapman, W. W. (2012). An approach to improve LOINC mapping through augmentation of local test names. *Journal of biomedical informatics*, *45*(4), 651–657.

[38] Surveillance, Epidemiology, and End Results (SEER) Program. 2019. https://seer.cancer.gov/ztml(Accessed 1 May 2019).

[39] National Program of Cancer Registries (NPCR). 2019. https://www.cdc.gov/cancer/npcr/about.htm.

[40] Garfinkel, L. (1993). The National Cancer Data Base: a cancer treatment resource. *CA: A Cancer Journal for Clinicians*, *43*(2), 69–70.

[41] Murphy GP. The National Cancer Data Base. CA Cancer J Clin. 1991 Jan-Feb;41(1):5–6. doi: 10.3322/canjclin.41.1.5. PMID: 1898636.

[42] National Cancer Database (NCDB). 2019. https://www.facs.org/quality programs/cancer/ncdb (Accessed 1 May 2019).

[43] ASCO forges ahead with CancerLinQ. Cancer Discov2014; 4: OF4.

[44] Sledge Jr, G. W., Miller, R. S., & Hauser, R. (2013). CancerLinQ and the future of cancer care. *American Society of Clinical Oncology Educational Book, 33*(1), 430–434.

[45] Miller, R. S., & Wong, J. L. (2018). Using oncology real-world evidence for quality improvement and discovery: the case for ASCO's CancerLinQ. *Future Oncology, 14*(1), 5–8.

[46] Flatiron Health. 2019. https://flatiron.com/oncology/ (Accessed 1 May 2019).

[47] Singal, G., Miller, P. G., Agarwala, V., Li, G., Kaushik, G., Backenroth, D., ... & Miller, V. A. (2019). Association of patient characteristics and tumor genomics with clinical outcomes among patients with non–small cell lung cancer using a clinicogenomic database. *Jama, 321*(14), 1391–1399.

[48] Khozin, S., Abernethy, A. P., Nussbaum, N. C., Zhi, J., Curtis, M. D., Tucker, M., ... & Pazdur, R. (2018). Characteristics of real-world metastatic non-small cell lung cancer patients treated with nivolumab and pembrolizumab during the year following approval. *The Oncologist, 23*(3), 328–336.

[49] Kulaylat, A. S., Schaefer, E. W., Messaris, E., & Hollenbeak, C. S. (2019). Truven health analytics MarketScan databases for clinical research in colon and rectal surgery. *Clinics in colon and rectal surgery, 32*(01), 054–060.

[50] Ruppin E. Cancer Data Science Laboratory. 2019. https://ccr.cancer. Gov/cancer-data-science-laboratory (Accessed 1 May 2021).

[51] Wang, Z., Jensen, M. A., & Zenklusen, J. C. (2016). A practical guide to the cancer genome atlas (TCGA). In *Statistical Genomics* (pp. 111–141). Humana Press, New York, NY.

[52] Edwards, N. J., Oberti, M., Thangudu, R. R., Cai, S., McGarvey, P. B., Jacob, S., ... & Ketchum, K. A. (2015). The CPTAC data portal: A resource for cancer proteomics research. *Journal of proteome research, 14*(6), 2707–2713.

[53] Rudnick, P. A., Markey, S. P., Roth, J., Mirokhin, Y., Yan, X., Tchekhovskoi, D. V., ... & Stein, S. E. (2016). A description of the clinical proteomic tumor analysis consortium (CPTAC) common data analysis pipeline. *Journal of proteome research, 15*(3), 1023–1032.

[54] TARGET: Therapeutically Applicable Research To Generate Effective Treatments. 2019.

[55] Cheng, D. T., Mitchell, T. N., Zehir, A., Shah, R. H., Benayed, R., Syed, A., ... & Berger, M. F. (2015). Memorial Sloan Kettering-Integrated

Mutation Profiling of Actionable Cancer Targets (MSK-IMPACT): A hybridization capture-based next-generation sequencing clinical assay for solid tumor molecular oncology. *The Journal of molecular diagnostics*, *17*(3), 251–264.

[56] El Naqa, I., Suneja, G., Lindsay, P. E., Hope, A. J., Alaly, J. R., Vicic, M., ... & Deasy, J. O. (2006). Dose response explorer: an integrated open-source tool for exploring and modelling radiotherapy dose–volume outcome relationships. *Physics in Medicine & Biology*, *51*(22), 5719.

[57] NAqA, I. E., Deasy, J. O., Mu, Y., Huang, E., Hope, A. J., Lindsay, P. E., ... & Bradley, J. D. (2010). Datamining approaches for modeling tumor control probability. *Acta Oncologica*, *49*(8), 1363–1373.

[58] Lambin, P., Rios-Velazquez, E., Leijenaar, R., Carvalho, S., Van Stiphout, R. G., Granton, P., ... & Aerts, H. J. (2012). Radiomics: extracting more information from medical images using advanced feature analysis. *European journal of cancer*, *48*(4), 441–446.

[59] Parmar, C., Leijenaar, R. T., Grossmann, P., Rios Velazquez, E., Bussink, J., Rietveld, D., ... & Aerts, H. J. (2015). Radiomic feature clusters and prognostic signatures specific for lung and head & neck cancer. *Scientific reports*, *5*(1), 1–10.

[60] Rosenstein BS, West CM, Bentzen SM, Alsner J, Andreassen CN, Azria D, Barnett GC, Baumann M, Burnet N, Chang-Claude J, Chuang EY, Coles CE, Dekker A, De Ruyck K, De Ruysscher D, Drumea K, Dunning AM, Easton D, Eeles R, Fachal L, Gutiérrez-Enríquez S, Haustermans K, Henríquez-Hernández LA, Imai T, Jones GD, Kerns SL, Liao Z, Onel K, Ostrer H, Parliament M, Pharoah PD, Rebbeck TR, Talbot CJ, Thierens H, Vega A, Witte JS, Wong P, Zenhausern F; Radiogenomics Consortium,"Radiogenomics: radiobiology enters the era of Big Data and team science," Int J Radiat Oncol Biol Phys. 2014 Jul 15;89(4):709–13

[61] Barnett GC, Thompson D, Fachal L, Kerns S, Talbot C, Elliott RM, Dorling L, Coles CE, Dearnaley DP, Rosenstein BS, Vega A, Symonds P, Yarnold J, Baynes C, Michailidou K, Dennis J, Tyrer JP, Wilkinson JS, Gómez-Caamaño A, Tanteles GA, Platte R, Mayes R, Conroy D, Maranian M, Luccarini C, Gulliford SL, Sydes MR, Hall E, Haviland J, Misra V, Titley J, Bentzen SM, Pharoah PD, Burnet NG, Dunning AM, West CM," A genome wide association study (GWAS) providing evidence of an association between common genetic variants and late radiotherapy toxicity," Radiother Oncol. 2014 May;111(2):178–85

[62] W Y Yang, J Moore, H Quon, K Evans, A Sharabi, J Herman, A Hacker-Prietz and T McNutt, "Browser Based Platform in Maintaining Clinical

Activities – Use of The iPads in Head and Neck Clinics," Phys.: Conf. Ser. 489 012095 doi:10.1088/1742–6596/489/1/012095 Int't Conf. on Computers in Radiotherapy, Melbourne AUS 2013

[63] Dekker, A. "Radiation Oncology Ontology White Paper,", https://devwiki.maastro.nl/display/ROO/White+Paper, http://bioportal.bioontology.org/ontologies/ROO

[64] Ruch, P., Gobeill, J., Lovis, C., & Geissbühler, A. (2008, October). Automatic medical encoding with SNOMED categories. In *BMC medical informatics and decision making* (Vol. 8, No. 1, pp. 1–8). BioMed Central.

[65] Ibrahim, A., Bucur, A., Perez-Rey, D., Alonso, E., de Hoog, M., Dekker, A., & Marshall, M. S. (2015). Case study for integration of an oncology clinical site in a semantic interoperability solution based on HL7 v3 and SNOMED-CT: data transformation needs. *AMIA Summits on Translational Science Proceedings*, *2015*, 71.

[66] Marungo, F., Robertson, S., Quon, H., Rhee, J., Paisley, H., Taylor, R. H., & McNutt, T. (2015, January). Creating a data science platform for developing complication risk models for personalized treatment planning in radiation oncology. In *2015 48th Hawaii International Conference on System Sciences* (pp. 3132–3140). IEEE.

[67] Robertson, S. P., Quon, H., Kiess, A. P., Moore, J. A., Yang, W., Cheng, Z., ... & McNutt, T. R. (2015). A data-mining framework for large scale analysis of dose-outcome relationships in a database of irradiated head and neck cancer patients. *Medical physics*, *42*(7), 4329–4337.

[68] Roelofs, E., Persoon, L., Nijsten, S., Wiessler, W., Dekker, A., & Lambin, P. (2013). Benefits of a clinical data warehouse with data mining tools to collect data for a radiotherapy trial. *Radiotherapy and Oncology*, *108*(1), 174–179.

[69] Godwin, M., Ruhland, L., Casson, I., MacDonald, S., Delva, D., Birtwhistle, R., ... & Seguin, R. (2003). Pragmatic controlled clinical trials in primary care: the struggle between external and internal validity. *BMC medical research methodology*, *3*(1), 1–7.

[70] Lurie, J. D., & Morgan, T. S. (2013). Pros and cons of pragmatic clinical trials. *Journal of comparative effectiveness research*, *2*(1), 53–58.

[71] Palta, J. R., Efstathiou, J. A., Bekelman, J. E., Mutic, S., Bogardus, C. R., McNutt, T. R., ... & Rose, C. M. (2012). Developing a national radiation oncology registry: From acorns to oaks. *Practical radiation oncology*, *2*(1), 10–17.

[72] Jason A. Efstathiou, MD, DPhil-, Deborah S. Nassif, PhD, Todd R. McNutt, PhD, C. Bob Bogardus, MD, Walter Bosch, DSc, Jeffrey Carlin,

Ronald C. Chen, MD, MPH, Henry Chou, PhD, Dave Eggert, MS, Benedick A. Fraass, PhD, Joel Goldwein, MD, Karen E. Hoffman, MD, Ken Hotz, PhD, Margie Hunt, MS, Marc Kessler, PhD, Colleen A.F. Lawton, MD, Charles Mayo, PhD, Jeff M. Michalski, MD, Sasa Mutic, PhD, Louis Potters, MD, Christopher M. Rose, MD, Howard M. Sandler, MD, Gregory Sharp, PhD, Wolfgang Tomé, PhD, Phuoc T. Tran, MD, PhD, Terry Wall, MD, JD, Anthony L. Zietman, MD, Peter E. Gabriel, MD and Justin E. Bekelman, MD, "Practice-Based Evidence to Evidence-Based Practice: Building the National Radiation Oncology Registry", J. of Oncology Practice, May 2013 vol. 9 no. 3 e90–e95

[73] Guo, Y., Jiang, W., Lakshminarayanan, P., Han, P., Cheng, Z., Bowers, M., Hui, X., Shpitser, I., Siddiqui, S., Taylor, R. H., Quon, H., & McNutt, T. (2019). Spatial Radiation Dose Influence on Xerostomia Recovery and Its Comparison to Acute Incidence in Patients With Head and Neck Cancer. *Advances in radiation oncology*, *5*(2), 221–230. https://doi.org/10.1016/j.adro.2019.08.009

[74] Moore, K. L., Kagadis, G. C., McNutt, T. R., Moiseenko, V., & Mutic, S. (2014). Vision 20/20: Automation and advanced computing in clinical radiation oncology. *Medical physics*, *41*(1), 010901.

[75] Zhang, C., Bijlard, J., Staiger, C., Scollen, S., van Enckevort, D., Hoogstrate, Y., ... & Abeln, S. (2017). Systematically linking tranSMART, Galaxy and EGA for reusing human translational research data. *F1000Research*, *6*.

[76] Willems, S. M., Abeln, S., Feenstra, K. A., de Bree, R., van der Poel, E. F., de Jong, R. J. B., ... & van den Brekel, M. W. (2019). The potential use of big data in oncology. *Oral Oncology*, *98*, 8–12.

[77] Grossberg, A. J., Mohamed, A. S., Elhalawani, H., Bennett, W. C., Smith, K. E., Nolan, T. S., ... & Fuller, C. D. (2018). Imaging and clinical data archive for head and neck squamous cell carcinoma patients treated with radiotherapy. *Scientific data*, *5*(1), 1–10.

[78] Prior, F., Smith, K., Sharma, A., Kirby, J., Tarbox, L., Clark, K., ... & Freymann, J. (2017). The public cancer radiology imaging collections of The Cancer Imaging Archive. *Scientific data*, *4*(1), 1–7.

[79] Govers, T. M., Rovers, M. M., Brands, M. T., Dronkers, E. A., de Jong, R. J. B., Merkx, M. A., ... & Grutters, J. P. (2018). Integrated prediction and decision models are valuable in informing personalized decision making. *Journal of Clinical Epidemiology*, *104*, 73–83.

[80] Wong, A. J., Kanwar, A., Mohamed, A. S., & Fuller, C. D. (2016). Radiomics in head and neck cancer: from exploration to application. *Translational cancer research*, *5*(4), 371.

[81] Parmar, C., Grossmann, P., Rietveld, D., Rietbergen, M. M., Lambin, P., & Aerts, H. J. (2015). Radiomic machine-learning classifiers for prognostic biomarkers of head and neck cancer. *Frontiers in oncology*, 5, 272.

[82] Dannenfelser, R., Allen, G. M., VanderSluis, B., Koegel, A. K., Levinson, S., Stark, S. R., ... & Lim, W. A. (2020). Discriminatory power of combinatorial antigen recognition in cancer T cell therapies. *Cell systems*, 11(3), 215–228.

[83] Williams, J. Z., Allen, G. M., Shah, D., Sterin, I. S., Kim, K. H., Garcia, V. P., ... & Lim, W. A. (2020). Precise T cell recognition programs designed by transcriptionally linking multiple receptors. *Science*, 370(6520), 1099–1104.

[84] https://www.ucsf.edu/news/2020/11/419121/big-data-powers-design-smart-cell-therapies-cancer

[85] Reisman, M. (2017). EHRs: the challenge of making electronic data usable and interoperable. *Pharmacy and Therapeutics*, 42(9), 572.

[86] Dash, S., Shakyawar, S.K., Sharma, M. *et al.* Big data in healthcare: management, analysis and future prospects. *J Big Data* 6, 54 (2019). https://doi.org/10.1186/s40537–019-0217–0

[87] Jensen, C. S., Pedersen, T. B., & Thomsen, C. (2010). Multidimensional databases and data warehousing. *Synthesis Lectures on Data Management*, 2(1), 1–111.

[88] Provost, F., & Fawcett, T. (2013). *Data Science for Business: What you need to know about data mining and data-analytic thinking*. " O'Reilly Media, Inc.".

[89] Wikipedia: machine learning. http://en.wikipedia.org/wiki/Machine_learning

[90] Thomson, R., & Sordo, M. (2002). Open Clinical: Knowledge Management for Medical Care. *Introduction to Neural Networks in Healthcare, Harvard*.

[91] Ryan, P. B., Stang, P. E., Overhage, J. M., Suchard, M. A., Hartzema, A. G., DuMouchel, W., ... & Madigan, D. (2013). A comparison of the empirical performance of methods for a risk identification system. *Drug safety*, 36(1), 143–158.

[92] McCarty, C. A., Chisholm, R. L., Chute, C. G., Kullo, I. J., Jarvik, G. P., Larson, E. B., ... & Wolf, W. A. (2011). The eMERGE Network: a consortium of biorepositories linked to electronic medical records data for conducting genomic studies. *BMC medical genomics*, 4(1), 1–11.

[93] Wicks, P., Vaughan, T. E., Massagli, M. P., & Heywood, J. (2011). Accelerated clinical discovery using self-reported patient data collected

online and a patient-matching algorithm. *Nature biotechnology*, *29*(5), 411–414.

[94] Abbas, A., Bilal, K., Zhang, L., & Khan, S. U. (2015). A cloud based health insurance plan recommendation system: A user centered approach. *Future Generation Computer Systems*, *43*, 99–109.

[95] Lupton, D. (2014). The commodification of patient opinion: the digital patient experience economy in the age of big data. *Sociology of health & illness*, *36*(6), 856–869.

[96] Barkley, R., Greenapple, R., & Whang, J. (2014). Actionable data analytics in oncology: Are we there yet?. *Journal of Oncology Practice*, *10*(2), 93–96.

[97] Lopez-Martinez, F., Schwarcz, A., Núñez-Valdez, E. R., & Garcia-Diaz, V. (2018). Machine learning classification analysis for a hypertensive population as a function of several risk factors. *Expert Systems with Applications*, *110*, 206–215.

12

Targeted Drug Delivery in Cancer Tissues by Utilizing Big Data Analytics: Promising Approach of AI

Brojendra Nath Saren[#], Vikram Prajapat[#],
Subham Appasaheb Awaghad[#], Indrani Maji, Mayur Aalhate,
Srushti Mahajan, Jitender Madan, and Pankaj Kumar Singh[*]

Department of Pharmaceutics, National Institute of Pharmaceutical
Education and Research (NIPER), India
 # Authors Contributed Equally
*Corresponding Author
Pankaj Kumar Singh M. Pharm, Ph.D.
Assistant Professor, Department of Pharmaceutics, National Institute of
Pharmaceutical Education and Research (NIPER), India
Email: pankajksingh3@gmail.com

Abstract

Cancer is considered as one of the fastest evolving, ever-changing complex disease. According to the WHO, cancer is the leading cause of death worldwide, accounting for nearly 10 million deaths in 2020. The global burden of cancer is continuously growing and creating tremendous emotional, physical, and financial strain on individuals, families, communities, and health systems. The fight against cancer has made huge progress over the last 30 years with a great improvement in the survival rate, but the general cure is still elusive. Targeting cancer depends on the proper understanding of cancer biology by applying different "omics" approaches like genomics, proteomics, and transcriptomics, which are considered as the predictive analytical tool for different cancers. Approaches for targeted delivery of therapeutics in cancer typically involve the systemic and localized administration of therapeutics or drug entrapped nanocarriers. Several steps are involved in designing a targeted drug delivery system, such as identifying suitable small molecular therapeutics and selecting different biomarkers, a ligand for targeting, optimization, and evaluation of the formulation. This is the era where most of the research activities are based on AI (artificial intelligence) or information technology, as they are less time-consuming and cost-effective. Despite such advancements, the challenges that oncologists face are managing huge data coming from different high throughput sources like computer-aided drug discovery (CADD), molecular biology, ADMET profiling, imaging, and pathological studies *in vitro* experiments, and statistical analysis. To mitigate this problem, "big data analytics" comes into the picture. "Big data analytics" plays an important role in integrating and interpreting the massive amount of data scattered around the world of cancer research. In a data-rich field like oncology, interpretation, storage, standardization, and sharing of data are also important, for example, different databases available in the public domain like Drug Bank which is a database composed of detailed information of different approved, investigational, and withdrawn drugs, The Cancer Genome Atlas (TCGA) which is a database that includes detailed information regarding the cancer patients, PubChem which is a chemical compound database, and Protein Data Bank which is crystal structure database for different proteins as well as ligands. In short "big data analytics" fuels cancer research by offering quality and precise data, thus intern accelerating the decision making, risk stratification, and prevention program. This book chapter deals with the emerging advances in "big data analytics" concerning targeted drug delivery toward cancer and its utility in the screening of drug molecules, selection of target, and ADMET profiling; further, the current challenges, as well as future applications of "big data analytics" in oncology, are also enlightened.

12.1 Introduction

Cancer is one of the severe pathological conditions in which abnormal cells grow out of control and, by invading the nearby tissues, ultimately causes disturbances in the normal health condition. The therapy in cancer patient is still a challenging and uphill task due to the lack of ability to distinguish between normal cells from cancer cells [1, 2]. Cancer is a serious global public health concern, with millions of new cases and deaths, which ultimately motivates the researcher to explore the new therapeutic options to cater the unmet clinical needs in the cancer therapy [3].

Till date, more than 200 types (or subtypes) of cancers have been identified based on the shape, location, and metastatic behavior of the tumor cells [4]. Due to the changes in the genomic behavior among the cases and the development of drug resistance, it is very difficult to justify the generalized treatment toward metastatic tumor. According to the investigation, cancer biologists identified that the changes in the genome-like altered gene expression of certain proteins, mutations (somatic or genetic), copy number aberrations (CNAs), and changes in epigenetic patterns are the reasons behind the huge dynamic behavior in the cancer treatment [5]. This variations intern causes vulnerability in the translation of successful treatment; that is why the customized and precision medicine came into the picture [6]. Precision medicines are developed on the basis of proper understanding of a patient's genome structure and function, which gives them direction to target either specific receptor or molecular pathways for the betterment of therapeutic efficacy and safety.

Conventionally, the role of cytotoxic therapies predominated in the treatment of cancer and is the mainstay in cancer therapeutics for several decades. Most of the existing anticancer drugs are unable to show specific action toward malignant cells and, thus, generate severe adverse reactions, which may prompt the cessation of treatment in certain patients. Another major issue with all chemotherapies is the emergence of chemoresistance as a result of the continued use of cytotoxic medicines [7].

The failure rate of drug products in oncology suffers a greater setback. Despite the various challenges in creating clinically effective antineoplastic drugs, pharmaceutical companies are still interested to pursue novel opportunities for anticancer drug candidates due to their high returns on investment [8]. Figure 12.1 depicts global expenditure on oncology medicines for example, global sales of oncology drugs reached $164 billion in 2020 and are estimated to be around $269 billion by 2025 [9].

To date, many anticancer drugs are being used to ultimately kill cancer cells, thus increasing overall survival rates and patient quality of

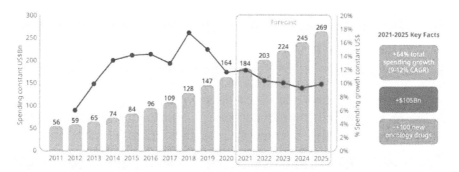

Figure 12.1 Global spending on oncology medicines [10].

life. Traditionally discovered anticancer cytotoxics act by blocking single essential function and killing the dividing cancer cells, categorized as alkylating agents (e.g., Cyclophosphamide, Thiotepa, Carmustine, Dacarbazine, and others), forming altered DNA complexes (e.g., Cisplatin, Carboplatin, etc.), antimetabolites (e.g., Azathioprine, Methotrexate, 5-flurouracil, etc.), as well as those altering microtubule formation and chromosome topology (e.g., Paclitaxel and Irinotecan) [11, 12]. Further, newer cytotoxic agents explored in the journey that are used to target dividing tumor cells in new ways or targeting over-expressed proteins in tumors include kinase inhibitors (e.g., Barasertib, Alisertib, Danusertib, etc.), cyclin-dependent kinase inhibitors (e.g., Abemaciclib, Palbociclib, etc.), modifying chromatin topology (histone deacetylase inhibitors), as well as proteasome inhibitors (Bortezomib) [13–16]. Since time, the oncological drug research has progressed from hormonal therapy (e.g., Tamoxifen), immunomodulators (e.g., nivolumab), to targeted therapy (e.g., Bevacizumab). Now, the focus shifts toward targeted drug delivery in cancer tissues directly along with the immune modulators which aids in the patient's immune system to defeat in the fight against cancer. In a real sense, the path of a novel molecular dosage form or innovator product is not simple and needs years of incumbent scientific research; thus, most of the pharmaceutical industries are facing the major issues of low productivity and return on investment in R&D [17]. At the same time, the large capital investment required to bring a new drug to market, estimated at around USD 2.6 billion, is constantly accompanied by high complexity and high failure rates, amidst intense scrutiny from regulatory bodies and the general public concerned about the safety of drug products [18].

The emergence of translational research in drug discovery is a pioneering step toward the development of novel therapeutics, which has resulted in the emergence of new key market players, such as academic institutions, biotech firms, large pharmaceutical organizations, and the National Institutes of Health (NIH), all focusing on developing the expertise required to generate new therapies by linking basic drug discovery directly to unmet clinical therapeutic needs [19]. However, there are still increasing concerns over traditional drug discovery programs resulting in inefficiency and failure to provide pharmaco-technological breakthroughs that have transformed other scientific industries. With the introduction of translation research and the current buzzwords "artificial intelligence," "machine learning," and "big data analytics," a changing paradigm shift in the drug discovery and development is observed; intending to drive healthcare forward through clinical breakthroughs in the treatment of cancer diseases [20].

12.2 Tools and Techniques for Targeted Drug Discovery and Delivery

Over the past three decades, drug discovery and drug delivery in various diseases, particularly in oncology that too site-specific targeted drug delivery in cancer tissues utilizing the Internet of Things (IoT), artificial intelligence-machine learning (AI-ML) clubbed with big data analytics (BDA), have revolutionized the multiple domains in pharmaceutical and healthcare industry. However, big data is being explored in multiple sectors including healthcare [21, 22], pharma sector [23–25], biotechnological sector [26], genomics [27], agriculture [28], business management [29], and many others. BDA changes the scenario of the drug discovery and drug formulation development and continuously leads toward development of bio-pharmaceutically viable, stable, safe, and efficacious product (Figure 12.2). The usage of BDA and AI has also led to reduction in the overall drug development costs as well as timelines for the launching of a pharmaceutical drug product [30].

Targeted drug delivery of chemotherapeutic agents in cancer tissues is dependent on understanding of the cancer cell biology, the role of cell cycle regulators, or tumor microenvironment (TME) along with exploring different predictive analytical tools, for example, "omics" such as genomics, proteomics, and transcriptomics. The research in the pharmaceutical domain has witnessed the translation of several new chemical entities (NCEs) as well as drug repositioning from "omics" data to securing FDA approval for clinical practice. The research in the targeted drug delivery in cancer tissues has benefitted from the implementation of many such innovative approaches and technologies right from the drug discovery phase to utilizing next-generation sequencing coupled with RNAi interference and, more recently, the CRISPR technology. These approaches are found to be important for understanding the biological system, cancer cell growth, and physiology and identification of drug targets, while exploring targeted drug delivery within cancerous cells, thus attaining the therapeutic efficacy and excluding the chances of adverse effects [31]. In that case, the usage of IoT, AI, and BDA could be harnessed to fulfill the unmet clinical needs, especially in oncology. There is a huge diversity as well as complexity in data generated during extensive drug discovery programs. The oncology sector apart from biopharmaceuticals has witnessed a steep rise in getting innovative drug approval from regulatory agencies in the last few decades [32]. Recent drug discovery research has focused on understanding the tumor microenvironment, identifying key molecular events in the pathophysiology of specific cancer diseases, and investigating abnormalities at the genomic level, leading to the development of targeted therapeutics that selectively kill tumor cells resulting in a higher safety and efficacy profile [33]. Traditionally, the development of antineoplastic agents involved screening of lead molecules in *in vitro* studies on cultured cancer lines, followed by *in-vivo* testing in human cancer xenografts growing in immunodeficient mice, assessing pharmacokinetics and toxicological profile, and thereon testing their therapeutic potential in a human clinical trial. Following this pipeline strategy, many successful anticancer therapeutics were made available in clinical practice and still making their existence in the clinical domain [34].

For instance, Imatinib (a tyrosine kinase inhibitor) was the first targeted therapeutic agent developed, which acts by targeting specific oncogenic signaling cascade (constitutively including active Bcr-Abl, c-KIT, and PDGFR) used for the treatment of chronic myeloid leukemia (CML) or gastrointestinal stromal tumor (GIST) patients [35, 36].

Even though traditional rational drug design has resulted in many blockbuster successes in the development of anticancer therapeutics, there is still a high rate of attrition, and selected lead molecules in the preclinical phase are still found to be failed to exert therapeutic effects in phase I/II clinical trials [37]. The primary underlying hypothesis is that malignant cells are constantly monitored by the immune system and that immunosuppression (active inhibition of immune responses) and immunoselection (recognition of suitable immunogenic determinants by immune detectors) allow them to grow into clinically viable tumors [38, 39].

As a result, researchers from various backgrounds are collaborating to develop safe and therapeutically effective diagnostic tools and targeted drug delivery systems, in shorter time frames. Nanomedicines, nanoneedles, pulsed laser surgery, injectable, genetic testing tools, nanotube-based biosensing devices, and integration of intelligent/smart biomaterials for site-specific tumor eradication, all belong to hot topics recently [40].

In the industry, the data revolution is both thrilling and worrisome because of the enormous volume of unstructured data generated and because they are difficult to synchronize structurally. Currently, big data companies are collaborating with major pharmaceutical companies, biotech companies, and academic research institutions to derive newer findings at every stage of the drug discovery process, from target identification to lead discovery, molecular design, precision medicine, and clinical development. The computational R&D platform aims at understanding the intricacies of human disease and increases the chances of developing a viable treatment [41, 42].

The drug discovery program in the oncology sector remains a highly challenging endeavor due to tumor heterogeneity and genetic complexes of many tumors and, thus, creates a formidable challenge in the development

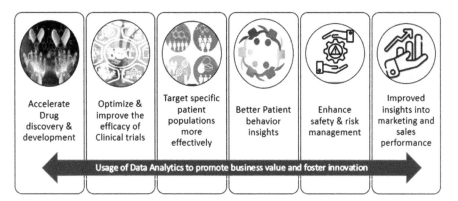

| Accelerate Drug discovery & development | Optimize & improve the efficacy of Clinical trials | Target specific patient populations more effectively | Better Patient behavior insights | Enhance safety & risk management | Improved insights into marketing and sales performance |

Usage of Data Analytics to promote business value and foster innovation

Figure 12.2 "Big data analytics" to accelerate business value and foster innovation.

of targeted drug delivery [43]. Big data analytics in combination with cloud services and knowledge graphs can be used to drive to a particular conclusive decision. A knowledge graph basically integrates multiple data sources, allows users to gather informative insights from those sources, and ultimately makes the information available for convenient questioning [44].

We are familiar with the complexity of big data (the five Vs: volume, velocity, variety, veracity, and value), and it is an important attribute since the generation of zettabyte (10^{21}) and yottabyte levels (10^{24}) of data is expected in early drug discovery programs such as chemistry, toxicological screenings, pharmacological studies, bioassays, and mechanistic mapping of cancer disease [45]. The diversity generated in data obtained through different drug discovery disciplines illustrates the complexity of the analytics required to fully unlock the potential insights in real-time applications.

Big data is increasingly being used to generate novel leads for the development of new chemical entities (NCEs) to address unmet clinical needs in oncology (Figure 12.3). In addition, big data plays an important role in generating conclusive evidence of the mechanism of drug resistance possessed by cancer therapeutics, which is a key challenge to tackle among oncology clinicians and drug discoverers [46].

The next-generation genome sequencing (NGS) approach often provides numerous candidate targets in mechanism-based cancer drug development, which must be systematically evaluated along with the progression from hit identification to lead optimization. The integrated approach is needed to prioritize and validate the selected target site(s) using big data both in chemistry and biology. All of this adds to the difficulty of obtaining focused knowledge for the selection of animal models, experimental methods, druggability prediction, and biomarker discovery [47].

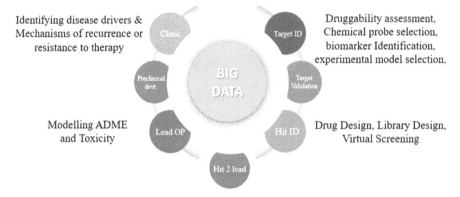

Figure 12.3 Applications of BDA and AI-ML throughout the phase of drug discovery.

A fast-paced optimization process of lead molecules and clinical candidates highly depends on the application of tools like AI, and various computational techniques. In a drug discovery cycle where past data and insightful expertise generated are critical assets, during the clinical development of the drug, the candidate will give feedback into future target identification attempts.

12.2.1 Data sources available for drug discovery

Many publicly supported organizations and institutes operate as data custodians and have allowed open access to enormous data being generated available to the scientific community. Examples include the European Bioinformatics Institute (EMBL-EBI) https://www.ebi.ac.uk/, The National Center for Biotechnology Information (NCBI) https://www.ncbi.nlm.nih.gov/, Swiss Institute of Bioinformatics (SIB) https://www.sib.swiss/, American Type Culture Collection (ATCC) genome portal https://genomes.atcc.org/, Addgene https://www.addgene.org/, Broad Institute https://www.broadinstitute.org/, BGI Group https://www.bgi.com/global/home, The Cambridge Crystallographic Data Center https://www.ccdc.cam.ac.uk/, and many others. Drug developers and oncologists recognize the value of big data in cancer biology by utilizing complex data to achieve breakthroughs in the cancer drug discovery and development.

12.2.2 Anticancer drug target discovery and validation

Earlier, drug discovery was mainly done by detecting phenotypic changes and afterward investigating the mechanism of action of the drug. However, the drug's safety and efficacy are more important than its pharmacological mechanism for successful regulatory approval [48]. For example, in the Human Genome Project (HGP), a number of potential therapeutic targets has been discovered with possibly hundreds of mutations, which can help to set a plethora of cancer targets in each type of tumor for their specific treatment [49]. Anticancer drugs include small molecular therapeutics and biologicals (including peptides, nucleic acid derivatives, and vaccines) that interact with either druggable or undruggable target moieties. The majority of clinically authorized therapies interact with protein targets, except in a few situations where they might be dubious targets (e.g., gaseous anesthetics), lipidic targets (e.g., amphotericin B), free-radical targets (e.g., antioxidants), or undistinguishable targets (e.g., loop diuretics) [50, 51]. Lazo *et al.* have explored the undruggable proteinaceous targets lacking enzymatically active sites, such as protein phosphotransferases, transcription factors (MYC-MAX, p53, and so on), RAS oncoprotein, and so on and, thus, need to be reinvestigated as potential drivers of cancer disease [51]. Among the approximately 30,000 genes discovered in humans, around 6000–8000 sites are regarded to represent promising therapeutic targets. However, only roughly 400 encoded proteins are effective in drug development so far [52].

Cancer is the most complicated disorder and involves several possible molecular targets; thus, identification of drug–target interactions is a crucial step in drug discovery program. High-throughput screening and other biological procedures are considered as the most typical approaches utilized in drug discovery, but they are expensive, time-consuming, and complex to execute. Nowadays, models for predicting drug–protein interactions utilizing network-based and machine learning approaches have been created using a variety of computational tools [53]. The discovery of various anticancer drugs along with specific therapeutic target sites (both essential and non-essential) still remains a mystery [43]. Multiple target proteins can also be found for a single therapeutic agent. Sildenafil, for example, was created to treat angina pectoris but was subsequently repurposed to treat erectile dysfunctioning also [54]. The science of molecular oncology is booming, from oncogenes and tumor suppressor genes to metastasis regulators. Cancer pharmacology has been changed by advances in the genetics and immunology. There is still a lot of work to be done to find novel targets [51]. The screening of nearly 200,000 genomes from patients with a variety of disorders using AI becomes possible, and, as a result, the identification of a disease target as well as prospective medication candidates become easy to find out. Nowadays, established drug companies are also applying AI-ML in their multi-omics operations, like Roche and Genentech, which are recently partnered with Stanford

University to employ AI to discover novel ways that a drug can target cancer targets [55].

The presence of multi-omics databases has made targeted drug delivery in cancer conceivable on a large scale. Using computational techniques like The Cancer Genome Atlas (TCGA) [56], The Human Protein Atlas [57], International Cancer Genome Consortium (ICGC) [58], and many cancer-related proteins can be recognized as therapeutic targets. The collective use of these large databases helps in the identification of appropriate targets, selective biomarkers involved in the oncologic pathway, and creating a rational hypothesis behind the mechanistic pathway. Computer-aided drug design (CADD) is a more rational, efficient, and economical strategy for drug discovery in this virtual era. The molecular docking is usually utilized to find out the best ligand binding site in the promising therapeutic target molecule, and based on the binding energy of docked hit molecules, the lead molecules are selected. The successful development of viracept (an HIV protease inhibitor) in the United States in 1997 was the first structure-based CADD concept of its kind, laying the groundwork for CADD to become an important tool in drug discovery initiatives [59].

12.3 Sources of Big Data in Drug Discovery and Delivery

In oncology, drug discovery is triggered by newer approaches like gene sequencing, high throughput screening, and *in silico* study while the chemical library and protein library have helped in the design hit to lead molecule. Big data generated across different pharmaceutical, biotechnological, medical, and genetic sectors caused complexity in drug discovery. Recently, big data analytics are being employed in multiple domains allowing the academic section to analyze a massive number of diverse datasets. Some resources of big data have been explained in this section and summarized in Table 12.1, which are utilized in the discovery of drugs for cancer.

12.3.1 COSMIC-3D

COSMIC-3D datasets include mutations involved in cancer, protein structure, human genetic, and proteomics data. Handling genomic profiles of humans specifically allow more precision in drug discovery. Thus, in order to achieve better precision, this database combined the COSMIC database related to somatic mutations in cancer of the human body and the 3D structure of human proteomic data for mapping, translating, and coordinating the position in protein structure [65].

Table 12.1 Some important resources for different databases.

Resource databases	Description	Reference	URL
Drugbank	Since 2006, this database has been available for free and contains detailed drug information such as mechanisms, drug interactions, and drug targets, as well as transcriptomic, metabolomic, and proteomic data	[60]	https://www.drugbank.com/
The Cancer Genome Atlas (TCGA)	The program started in 2006, containing over 2.5 petabytes of genomic, epigenomic, transcriptomic, and proteomic data	[56]	https://portal.gdc.cancer.gov/
International Cancer Genome Consortium (ICGC)	Comprehensive genomic resource in oncology, representing data from 86 cancer projects, representing around overall 81 million somatic mutations and data-insights obtained from over 22,000 donor samples	[61]	https://dcc.icgc.org/
The 100,000 Genome Project	Genomic database covering 100,000 genomes obtained from around 97,000 patients, including both rare disease and cancer patients	[62]	https://www.genomicsengland.co.uk/
Genomics of Drug Sensitivity (GDSC)	Unique database facilitating cancer drug discovery by identifying molecular biomarkers and resources on therapeutic sensitivity in cancer cell lines	[63]	https://www.cancerrxgene.org/
Genomic Data Commons (GDC)	GDC is a repository and cancer knowledge base for cancer research community	[64]	https://gdc.cancer.gov/

12.3.2 The cancer genomic atlas

The Cancer Genomic Atlas (TCGA) is a genomic algorithm that generates a database of cancer-causing genetic mutations. The researchers can utilize this database to discover the most effective compound that will be suitable for the treatment of cancer. TCGA BRCA RNA-sequence data expression was studied by Gruener *et al.*, and it has been found that AZD-1775 is the most effective compared to AZD7762 and leptomycin B specifically inhibiting the G2/M checkpoint [66]. ComBat software plays an important role in integrating already filtered common genes between Cancer Cell Line Encyclopedia (CCLE) and TCGA expression datasets. TCGA was utilized in the study for breast cancer patients who have similar characteristics called a cohort. The imputation-based analysis also uncovered the mutation status of 13 genes, allowing for the discovery of a possible biomarker for the screening. Recently, Loxo Oncology developed LOXO-101 by utilizing publicly available TCGA data, which is found to be effective in metastatic cancer [67]. By analyzing large-scale cancer databases, CiDD helps to uncover prospective medications for *in vitro* tests. Scientists have used datasets of mutation and RNA-sequencing from the TCGA colon for the treatment of BRAF V600E mutations to discover the lead compound [68]. According to the findings of this investigation, the TCGA-derived classifier displayed lesser sensitivity but greater specificity (62%) resulting in competition with the PETACC3-derived signature, and it is more specific in distinction between BRAF wild-type and BRAF mutant cell lines [69].

12.3.3 Gene expression omnibus (GEO)

Gene Expression Omnibus archive database serves as a public library for people all around the globe. It gathers and shares high-throughput gene expression data as well as information from other functional genomic studies for free. This database is also supported by the NCBI for profiling RNA methylation. Next-generation sequencing is used for drug discovery as well as for the identification of the target. Several software are available for analyzing and labeling datasets of GEO, which include GEO2R, ScanGEO, and GEOracle query. Researchers studied gene expression profile, GSE108524, to identify genes associated with the disease called acoustic neuromas and possible therapeutic medications by utilizing GEO database [70]. While screening 542 differentially expressed genes by the GEO2R tool, 12 genes were found to be promising for the target of disease. Recent reports suggested that under some circumstances, miRNAs can act as either oncogenes or tumor suppressors [71, 72]. Scientists have discovered the genes that code for miR-452-5p, a protein that has previously been linked to prostate cancer [73].

12.3.4 Human protein atlas

The Human Protein Atlas is another technique for examining and mapping protein localization, expression in human tissues and cells in addition to antibody-based imaging. This is a free resource that allows scientists to investigate various human proteomes. While we are focusing on cancer here, the human protein atlas is already being used on COVID-19, tuberculosis, cardiovascular disease, and other disorders. Kaplan *et al.* implement a Kaplan–Meier plotter to discover the gene of interest that seems to be involved in pancreatic cancer so that it may be detected early [74]. The model included characteristics like gender/race and compared the outcomes to the typical clinical information from Human Protein Atlas databases to acquire a clearer picture of the targeted gene for validation. The assessment of mRNA expression of the RAS-related protein, Rab-1A in tumor and normal pancreatic tissue revealed that pancreatic cancer patients with reduced Rab1A gene expression have a higher survival rate with a p-value of 0.048 [74]. CanSAR is another largest scientific database in the world for drug discovery in the field of oncology from fundamental to clinical research. This database contains 562,375 proteome sequences along with data on protein–protein interactions from other databases such as MSigDB and TRRUST and over 300,000 medication combinations derived from various cancer cell line models [75].

12.4 Big Data Analytics

The word "data" is derived from a Latin word and it is a classical plural form of the word "datum" which means "things given." Data refers to the massive volume of information. Currently, we are living in an information age where vast amount of information is easily available for us. This information is crucial to the development of infrastructure and future creativity in medicine and healthcare. Big data is fundamentally a massive volume of data related to a particular topic. Big data analytics acquire and analyze this vast quantity of data for future applications. In the medical field, big data includes two types of information: 1) disease-related and 2) data related to the patients [76]. The big data allows understanding of mechanisms of disease, treatment plans, recovery rates, future medicine inventions, and time-saving treatment strategies. There was a time when gathering and storing huge amounts of data for medical use was very costly, complex, and even time-consuming. But, with growing technologies, it has become easier to collect and store such data which can be utilized to provide a better care. In the case of cancer, the main purpose of big data analytics is to predict and provide solutions to a problem before it is too late by using the data-driven findings, assessing different methods to provide faster treatment, and involving the patients more in their health by providing them essential tools to do so (e.g., smartwatches tracking blood pressure).

Several genetic mutations are linked to cancer [77, 78]. Cancer may grow and propagate as a result of these mutations. Using various tools and technologies to handle data, linked to DNA sequencing can aid in the early detection of genetic alterations. Big data is gathered from several sources for the treatment of cancer along with medical records in electronic form, pharmacological data, environmental variables, and dietary patterns [79]. In big data analytics, the machine learning technique is being utilized to handle complicated data. Among the few are MapReduce framework for execution of clustering algorithm [80], MapReduce-based hierarchical clustering algorithm [81], density-based spatial clustering of applications with noise (DBSCAN) [82], P3C algorithm [83], MapReduce-based subspace-clustering algorithm [84], an automatic segmentation method [85], Bland–Altman analysis, and concordance correlation coefficients (CCC) [86].

Manogaran *et al.* were particularly interested in variations in the number of copies of DNA in the genome to discover a genetic abnormality and make an early cancer diagnosis. A Bayesian hidden Markov model (HMM) is utilized for cancer modeling and change-point detection in sequences of the genome by applying Gaussian mixture (GM) clustering methodology [84]. The MATLAB cghcbs algorithm failed to detect the second mutation cluster region and split the first MCR into two halves. The CBS technique fails to find any outliers in sample PA's Chromosome 12. C.Dan.G. With 80.22% accuracy and 12.22% error, the Bayesian hidden Markov model (HMM) with GM clustering-based change detection methodology outperforms the pruned exact linear time (PELT), binary segmentation, and segment neighborhood method [84]. Data analysis is also important for the detection of genetic disorders. The reported HMM with Gaussian clustering change has shown effectiveness in marking changes in DNA sequencing; thus, pre-indication of cancer can be known. The HDFS-based phenotype method uses national language processing (NLP) to examine large amounts of data such as DNA sequencing, imaging, ultrasound analysis, and age-related data among other things and production values that aid in cancer diagnosis [87].

12.5 Future Prospects of "Big Data Analytics"

Receptors are complex protein structures comprising folded, winded twisted amino acid chains. These different structures are results of differential amino acid sequences and folding patterns uncommon to each other. In the last two to three decades, substantial advancements in the crystallographic method have resulted in a tremendous rise in the establishment of structural data for receptors proteins and enzymes. Nowadays, biological data is available in the public domain. Atlas is a database that provides information about cancer-related genomes [88] Figure 12.4 depicts future prospects of big data analytics (BDA).

Omics science is a combined term used for genomics, transcriptomics, proteomics, and epigenomics. It includes the study of cancer pathophysiology. High throughput sequencing or next-generation sequencing is recent technology being used in the sequencing of DNA and RNA, which is faster and more economical as compared to older sequencing technologies. The generated data is tremendous and is effective only when analyzed and studied in-depth. It is challenging to explore this whole data appropriately without proper tools, ultimately resulting in the slowing of process of genome study. It further requires critical thinking, a scientific approach, and a complex computation method to extract knowledge from this data. Already existing data is enormous and there is necessity to evaluate it with different approaches. Better predictive analysis of omics data and getting the best conclusions as a new chemical entity is possible when there will be good collaboration among various organizations working on cancer research as well as sharing of data.

Figure 12.4 Future prospects of big data analytics (BDA).

Currently, omics data is easily assessable to scientists and is freely available on different platforms like The Cancer Genome Atlas (TCGA) data portal and International Agency for Research on Cancer (IARC) (TP 53 gene) data. ASCO, and CancerLinQ allow patients and physicians to share their findings and access the data. The APOLLO platform intends to build a well-coordinated network for generating long-term informative data on detailed medical records as well as data from their tissue specimens [89].

AI is taking part in pharmacotherapy and more particularly in oncology. AI is a simulation of the human intellectual process by machines, which learns by experience or by monitoring the activities and does not require a specific program to work. AI is widely utilized in medicine and is employed in diagnostics, especially to identify minor differences from reference, to detect progression of the disease and to imaging of disease. IBM Watson for oncology is a project started in 2011 and had promised to convert big data into personalized medicines. IBM Watson gathers information from oncologists, medical publications, manuals, medical evidence, and data from cancer patients. Currently, it provides some AI-enabled tools for medicines to doctors and institutions along with few mobile-based applications to consumers for deciding the best-suited treatment on a particular cancer. Other AI platforms focused on the implementation of AI in healthcare are Microsoft Hanover and Google Deep Mind. The use of AI in medication is being debated till now, as AI had little success in medicine and data analysis in properly understanding cancer pathology [90]. Regulators have only allowed a handful of AI-based technologies for usage in hospitals and doctors' offices. Those ground-breaking products generally work in the visual arena, analyzing pictures such as X-rays and retina scans with computer vision. We need to comb through all of the available cancer data to find the unambiguous combinatorial signature of cancer.

More and more data are being generated in the field of oncology, and, therefore, it will be a good fit for predictive analytics. In cancer therapy, predictive analytics is used to better estimate the average lifespan, critical care use, adverse effects, genomics, and genomic risk. Predictive analytics is emerging as a future tool in clinical practice for pathological interpretation, drug development, and population health management. The research, technical, and regulatory barriers that are preventing analytics from being used in oncology must be addressed by developers and policymakers [91]. Big data and predictive analytics tools will encourage clinicians to use this technology in routine patient care activities and decision-making in choosing treatment strategies. Along with decision support, predictive analytics will be utilized in genomic risk stratification [91]. Clinical studies collect a massive quantity of possibly relevant information on multiple patients, even those who are getting routine therapies. Different stakeholders have advocated for data-sharing initiatives on a national and international level to create detailed data related to clinical studies accessible for researchers [92]. Scientists may utilize shared data to build a more inclusive network, solve research-related problems faster than they would ever do with traditional randomized clinical trials, reduce redundant work, and maximize productivity. Surrogate endpoint verification, advancement of strong predictive or accurate designs, the patient choice for phase II randomized trials, and recognition of subgroups of patients for newer therapeutic strategies could help in industrial research and accuracy in innovation and provide a guide in upcoming development [92].

The National Cancer Database (NCDB) is a worldwide open-access database used openly in oncology. It is used to collect approximately 70% of all newly reported melanoma diagnoses in the United States each year. The American College of Surgeons (ACoS) Commission on Cancer (CoC) has recognized the NCDB as effective clinical monitoring and quality enhancement technique for cancer programs. In addition, the NCDB collects data on operative treatment such as radiotherapy, immunotherapy, chemotherapy, and hormonal therapy [93]. The NCDB is a huge, robust database that offers a variety of clinical and industrial research-related data. Importantly, the NCDB can be used to benchmark hospitals on performance measures and act as a catalyst for hospital-level quality improvement programs. Cancer registration activities are costly for hospitals, but the benefits are demonstrated by the NCDB's extensive efforts to feed important information back to participating institutions. The NCDB is the only cancer registry that provides hospitals with feedback data [93]. Because of the comparatively large number of cases in adult group compared to the pediatrics patient population, big data has been actively applied to adult oncology. Even though the application of big data in pediatrics is limited, the National Cancer Institute has led efforts to encourage joint usage and sharing of big data across oncology subspecialties. Other existing data-sharing programs, such as pediatric cancer data, are supported by sub-discipline interest organizations, national health agencies, and university consortia. In Cancer Research Data Commons (CRDC), the ecosystem is included. The ecosystem's purpose is to construct data common nodes and sources where data storage and computing infrastructure as well as services, tools, and applications can be located. Therapeutically Applicable Research to Generate Effective Treatments (TARGET), Proteomics Data Commons (PDC), Imaging Data Commons, Genomic Data Commons (GDC), Integrated Canine Data Commons, and the Human Tumor Atlas Network are all part of the ecosystem [94]. Social media will increase communication between patients, providers, and communities, e.g., patients with similar conditions and providers with similar specialties. This will not only work to globalize and democratize healthcare, but it is also a potentially important source of big data [95].

Cloud-hosted software as service (SaaS) solutions reduce the obstacles while entering into the big data domain. With the aid of MapReduce-based solutions and thousands of machines, Google and Amazon, like a huge firm, process and analyze large amounts of complex data in the terabytes. MapReduce algorithms break down complex challenges to small tasks that may then be shared over a multitude of devices for analysis and the findings compiled into the best response. Tableau is another option where a visualization-focused cloud-based service can be obtained [95]. Accessible Hadoop is a high-performance, flexible, and typically low-cost solution for handling huge datasets used by many enterprises. To implement Hadoop systems successfully by utilizing an open platform, one needs training, professional expertise, and support. Hadoop has been monetized by companies like Greenplum, Oracle, and other businesses like British Airways and Expedia [95]. SaaS is a key tool for making big data results more accessible to the general public. Healthcare organizations that manage subsets of data can disclose access through SaaS-based solutions, which eliminate some of the aggregation and integration problems. Additional analytics-related services, both basic and advanced, can be incorporated into the total offering [95].

The quality of medical records has an impact on the accuracy of real-world data. There is a requirement of clinicians to enter the structured data into the medical record. As a result, new approaches to increase the completeness and accuracy of data entered into the electronic health record that is consistent across platforms and does not disrupt physician workflow are needed [96]. In ordinary practice, real-world evidence can be used to conduct evaluations of the efficacy of diagnostic instruments and treatment strategies. Real-world evidence may be used to predict outcomes such as short-term survival or the chance of hospitalization, identify the likelihood of benefit and adverse events with certain medicines, and choose the treatments that are most likely to benefit individual patients [96].

Larger the data is, better will be the conclusion with minimal error and more precision. Clinicians planning large-scale observational studies should engage population scientists to ensure that patient identification, endpoint selection, and analytical method are all thoroughly scrutinized. When every aspect of patient treatment and tumor characteristics are recorded, large-scale observational studies will become even more beneficial for clinical conditions [97].

12.6 Challenges in the Field of "Big Data Analytics"

Parikh *et al.* have described future challenges in front of big data analysis as (1) challenges related to collection of exhaustive data and clinical end points from patient population, (2) unavailability of anticipated validation process for predictive tools in big data analytics, and (3) risk of automating bias in observational dataset and risk of spontaneous bias in observed dataset.

Due to emergency hospitalization and severe chemotherapies, collecting multiple samples from patients may be difficult, resulting in patient non-compliance. Furthermore, collection of samples from critically ill patient seems to be unethical and has social constraints. Overall, this makes collection of real-time data difficult and expensive for research organization [91]. Several studies have demonstrated the benefits of big data, but the problems of incorporating big data into everyday healthcare have received significantly less attention. The benefits of big data analytics technologies have been proven in a different field of healthcare, extending from medical imaging technology and chronic disease treatment to public health services for precise and personalized medicine. This strategy has the potential to increase the efficiency of hospital-patient care, lower management costs, and accelerate disease identification. Apart from all of the benefits, data access, collection biases, and data hoarder trust in big data analytics systems have hampered their adoption in everyday healthcare. While communication with both the provider and the end user is crucial during this process, it is often just half the battle. It takes a long time and a lot of effort to collect high-quality data in order to construct unbiased analytical tools [98]. Patients are also concerned about data privacy and security when developing algorithms [98].

Troyanskaya described that the field of cancer big data analysis and cell engineering have grown popularity in recent years. Researchers are analyzing a massive database of thousands of proteins in cancer and normal cells using machine learning and artificial intelligence. This has led to millions of alternative protein combinations that may be utilized to precisely target only cancer cells while leaving healthy cells alone [99, 100]. Big data and predictive analytics may face significant challenges in the daily healthcare system, including data collection, potential algorithm evaluation, and bias prevention [91]. Big data has similar issues when it comes to integrating into other businesses as it does in healthcare. In some fields, the major difficulty has shifted from data collection to knowing how to properly interpret and exploit the data. To become efficient to store the datasets, advances in computational biology are required. Apart from the difficulty of organizing and keeping this data, the equipment required to do so is incredibly expensive and takes up lots of space [3, 101].

The main difficulties in big data arise due to the decentralized and faded data among the healthcare provider including health insurance companies, hospitals, laboratories, auxiliary venders, and regulatory agencies. The obsolete model, in which all data were stored into a warehouse, is being replaced by online big data architecture. Data federation will emerge as a solution in which the big data architecture is built on a collection of nodes both inside and outside the company and accessed through a layer that combines data and analytics [102]. Big data solution architectures must be adaptable enough to handle not only the addition of new sources but also the growth of schemas and structures for transferring and storing data. Metadata and semantic layers that accurately evaluate the large data and provide suggestions and direction, including appropriate uses of the data, are required to guarantee that analytics are meaningful, accurate, and suitable. This growth of strictness will help to improve data effectiveness in the long run [102].

The goal of data-driven science and discovery is to uncover the insights that lead to valuable inventions. Without a thorough understanding of the data and the domain, it is easy to fall into the trap of simplistic and erroneous correlation, resulting in the development of false discoveries. It is critical to completely know and appreciate the domain in which one is working as well as to structure all observations and ideas inside that domain effectively [103].

In the current state of the art, the implication of novel algorithms to address the four Vs (volume, value, velocity, variety, and veracity) of big data with consideration of previous advances and identification of potential bottlenecks, challenges, and pitfalls creates a different impact. Any advancements in scalable algorithms should be linked to architecture, systems, and new database constructs. The schema-free environment of new data types and the predominance of unstructured data led to a shift toward technologies like NoSQL and Hadoop. The construction of data storage and the computational fabric is a collaborative work usually done by both an algorithmic researcher and a system/database researcher with proper integration of machine learning and data mining algorithms [103].

12.7 Conclusion

The 21st century is the era of newly invented technologies generating a large volume of datasets. Data may be social, economic, and related to healthcare; this large volume of dataset cannot be easily analyzed by using traditional computing tools. This exponentially increasing data is called big data. Big data analytics is the process of extracting meaningful insights from raw data using advanced analytical techniques (BDA).

Big data is characteristically described by fundamentally five V's – volume, value, velocity, variety, and veracity. Big data is not only relevant to oncology and genomics, but it applies to online social platforms like Google, Facebook, different E-commerce companies, and the insurance sector too. The velocity concept of big data explains that the data generation is a high-speed process in which the data is computed at a faster rate. As the worldwide cases of cancer are increasing on every passing day, the huge data generation in genomics and cancer is taking place. Advanced technologies are being employed in the field of genomics and cancer diagnosis to enable patient-centric treatment. In the upcoming 10–15 years, personalized medication, more particularly in chronic diseases, will be of prime importance. More patient-centric diagnosis and treatment will automatically lead to an increased need for superior data generation and analysis techniques. The term variability of big data represents a variation in data from subject-to-subject and on a case-to-case basis. The variability in data makes it necessary to increase the numbers of subjects to be studied. The data may vary for different types of disease, genetics of individuals, different time points, as well as different places. The value or importance of big data can change the prospects of the whole healthcare sector. The current conventional disease-centric aspects will be converting into patient-centric diagnosis and personalized medication very soon. The research and development field are more likely to benefit from big data. The dominating phase of "genome-wide association studies," or GWAS, is gradually shifting to a phase of "data wide association studies," or DWAS, with a focus on big data.

Reuse of already generated data is of great importance for new research; the new conclusion may be derived from existing data, and, hence, accessibility, security, and privacy of this data are the main concerns of discussion. Findable, accessible, interoperable, and reusable (FAIR) data principles were first published in 2014 and recognized by the G20 (2016) and G7 (2017). Findability (F) of data relies on a persistent identifier in the place. Accessibility (A) defines rules and regulations for accessing data and licenses for data, by keeping a vision on data privacy. Interoperability (I) describes ready to be exchanged, interpreted, and combining with other means of datasets by humans as well as large computer networks. Reusability (R) refers to future usage of digital assets in a well-defined way, allowing for data integration to be made compatible with data resources for reaping fruitful drug development in oncology sector.

Owing to sophisticated research technologies, the amount of data or metadata collected during biomedical research has increased dramatically. The accessibility of these metadata, visualization, and deriving potential insights in exploring human biology of disease has excited the research community about the potential of speeding up progress toward precision medicine-tailoring preventative measures, diagnosis, and therapeutic interventions depending on the molecular characteristics of a diseased patient. For uncommon malignancies, such as those that affect children, the return on investment from pooling and sharing research data is extremely high. However, sorting through huge volumes of "big data" to find answers to the complex biological concerns that will introduce precision medicine into clinical practice remains a challenging endeavor. This difficulty is exemplified in oncology, where a large portion of the data will originate from clinical studies of cancer patients. Pharmacovigilance (PV) is an important part of clinical research, which detects, assesses, and understands the adverse effects associated with new medications and generates a relatively large extent of data. PV is extensively used in clinical trials of all medications, including anticancer drugs. The usage of generated data is improving public health and safety, as well as the decision-making process for medical practitioners. AB Cube Safety Easy, Oracle Argus Safety, and AIR Sg are some private platforms engaged in the generation and analysis of data regarding the safety of new drug applications.

12.8 Acknowledgment

We are thankful to our co-authors for their contribution in this chapter.

12.9 Funding

None.

12.10 Conflict of Interest

There is no conflict of interest.

References

[1] Bertram, J.S.J.M.a.o.m., The molecular biology of cancer. 2000. 21(6): p. 167–223.
[2] Alberts, B., et al., Cancer as a microevolutionary process, in Molecular Biology of the Cell. 4th edition. 2002, Garland Science.

[3] Sung, H., et al., Global cancer statistics 2020: GLOBOCAN estimates of incidence and mortality worldwide for 36 cancers in 185 countries. 2021. 71(3): p. 209–249.

[4] UK, C.R., Types of cancer. 2020: Cancer Research.

[5] Pereira, B., et al., The somatic mutation profiles of 2,433 breast cancers refine their genomic and transcriptomic landscapes. 2016. 7(1): p. 1–16.

[6] Sapiezynski, J., et al., Precision targeted therapy of ovarian cancer. 2016. 243: p. 250–268.

[7] Benstead-Hume, G., S.K. Wooller, and F.M.J.E.o.o.d.d. Pearl, 'Big data'approaches for novel anti-cancer drug discovery. 2017. 12(6): p. 599–609.

[8] Kola, I. and J. Landis, Can the pharmaceutical industry reduce attrition rates? Nature Reviews Drug Discovery, 2004. 3(8): p. 711–716.

[9] Report, I.I. Global Oncology Trends 2021, Outlook to 2025. 2021 [cited 2022 February 08]; Available from: https://www.iqvia.com/insights/the-iqvia-institute/reports/global-oncology-trends-2021.

[10] Report, I.I. Global Oncology Trends 2021, Outlook to 2025. 2021 [cited 2022 February 08]; Available from: https://www.iqvia.com/insights/the-iqvia-institute/reports/global-oncology-trends-2021.

[11] Faivre, S., S. Djelloul, and E. Raymond. New paradigms in anticancer therapy: targeting multiple signaling pathways with kinase inhibitors. in Seminars in oncology. 2006. Elsevier.

[12] Rixe, O. and T.J.C.c.r. Fojo. Is cell death a critical end point for anticancer therapies or is cytostasis sufficient? 2007 [cited 13 24]; 7280–7287].

[13] Chohan, T.A., et al., An insight into the emerging role of cyclin-dependent kinase inhibitors as potential therapeutic agents for the treatment of advanced cancers. Biomedicine & Pharmacotherapy, 2018. 107: p. 1326–1341.

[14] Bavetsias, V. and S. Linardopoulos, Aurora Kinase Inhibitors: Current Status and Outlook. 2015. 5.

[15] Lents, N.H. and J.J. Baldassare, Cyclins and Cyclin-Dependent Kinases, in Encyclopedia of Cell Biology, R.A. Bradshaw and P.D. Stahl, Editors. 2016, Academic Press: Waltham. p. 423–431.

[16] Xu, W.S., R.B. Parmigiani, and P.A. Marks, Histone deacetylase inhibitors: molecular mechanisms of action. Oncogene, 2007. 26(37): p. 5541–5552.

[17] Gaudelet, T., et al., Utilizing graph machine learning within drug discovery and development. 2021. 22(6): p. bbab159.

[18] Romasanta, A.K.S., P. van der Sijde, and J.J.S. van Muijlwijk-Koezen, Innovation in pharmaceutical R&D: mapping the research landscape. 2020. 125(3): p. 1801–1832.

[19] Fishburn, C.S.J.D.d.t., Translational research: the changing landscape of drug discovery. 2013. 18(9–10): p. 487–494.

[20] Stahel, R., et al., Optimising translational oncology in clinical practice: Strategies to accelerate progress in drug development. 2015. 41(2): p. 129–135.

[21] Asri, H., et al. Big data in healthcare: Challenges and opportunities. in 2015 International Conference on Cloud Technologies and Applications (CloudTech). 2015. IEEE.

[22] Shilo, S., H. Rossman, and E.J.N.m. Segal, Axes of a revolution: challenges and promises of big data in healthcare. 2020. 26(1): p. 29–38.

[23] Tormay, P.J.P.m., Big data in pharmaceutical R&D: creating a sustainable R&D engine. 2015. 29(2): p. 87–92.

[24] Cattell, J., et al., How big data can revolutionize pharmaceutical R&D. 2013. 7.

[25] Izmaylov, A., A. Evstratov, and E. Heidelberg. Big Data Applications in the Pharmaceutical Industry. in International Scientific Conference "Digital Transformation of the Economy: Challenges, Trends, New Opportunities". 2021. Springer.

[26] Verma, A.J.C., How Big Data is Anchoring Biotechnological and Medical Industries to Solve Newer Obstacles? 2020. 2(4): p. 25–27.

[27] Bhardwaj, R., A. Sethi, and R. Nambiar. Big data in genomics: An overview. in 2014 IEEE International Conference on Big Data (Big Data). 2014. IEEE.

[28] Roy, S., et al. IoT, big data science & analytics, cloud computing and mobile app based hybrid system for smart agriculture. in 2017 8th Annual Industrial Automation and Electromechanical Engineering Conference (IEMECON). 2017. IEEE.

[29] Darvazeh, S.S., I.R. Vanani, and F.M.J.N.T.i.t.U.o.A.I.f.t.I. Musolu, Big data analytics and its applications in supply chain management. 2020: p. 175.

[30] Taglang, G. and D.B.J.G.o. Jackson, Use of "big data" in drug discovery and clinical trials. 2016. 141(1): p. 17–23.

[31] Workman, P., A.A. Antolin, and B.J.E.o.o.d.d. Al-Lazikani, Transforming cancer drug discovery with Big Data and AI. 2019. 14(11): p. 1089–1095.

[32] Batta, A., et al., Trends in FDA drug approvals over last 2 decades: An observational study. 2020. 9(1): p. 105.

[33] Friedman, J.R., et al., Chapter Six - Capsaicinoids enhance chemosensitivity to chemotherapeutic drugs, in Advances in Cancer Research, K.D. Tew and P.B. Fisher, Editors. 2019, Academic Press. p. 263–298.

[34] Zitvogel, L., et al., Mechanism of action of conventional and targeted anticancer therapies: reinstating immunosurveillance. 2013. 39(1): p. 74–88.

[35] Iqbal, N. and N. Iqbal, Imatinib: a breakthrough of targeted therapy in cancer. Chemotherapy research and practice, 2014. 2014: p. 357027–357027.

[36] Druker, B.J., et al., Five-year follow-up of patients receiving imatinib for chronic myeloid leukemia. 2006. 355(23): p. 2408–2417.

[37] Ocana, A., et al., Preclinical development of molecular-targeted agents for cancer. 2011. 8(4): p. 200–209.

[38] Kim, R., M. Emi, and K. Tanabe, Cancer immunoediting from immune surveillance to immune escape. Immunology, 2007. 121(1): p. 1–14.

[39] Beatty, G.L. and W.L. Gladney, Immune escape mechanisms as a guide for cancer immunotherapy. Clinical cancer research : an official journal of the American Association for Cancer Research, 2015. 21(4): p. 687–692.

[40] Biswas, A.K., et al., Nanotechnology based approaches in cancer therapeutics. 2014. 5(4): p. 043001.

[41] Groves, P., et al., The'big data'revolution in healthcare: Accelerating value and innovation. 2016.

[42] Blackburn, M., et al., Big Data and the Future of R&D Management: The rise of big data and big data analytics will have significant implications for R&D and innovation management in the next decade. 2017. 60(5): p. 43–51.

[43] Kamb, A., S. Wee, and C.J.N.r.d.d. Lengauer, Why is cancer drug discovery so difficult? 2007. 6(2): p. 115–120.

[44] Brown, N., et al., Big data in drug discovery. 2018. 57: p. 277–356.

[45] Raghupathi, W., V.J.H.i.s. Raghupathi, and systems, Big data analytics in healthcare: promise and potential. 2014. 2(1): p. 1–10.

[46] Al-Lazikani, B., U. Banerji, and P.J.N.b. Workman, Combinatorial drug therapy for cancer in the post-genomic era. 2012. 30(7): p. 679–692.

[47] Workman, P. and B.J.N.r.D.d. Al-Lazikani, Drugging cancer genomes. 2013. 12(12): p. 889–890.

[48] Moffat, J.G., J. Rudolph, and D.J.N.r.D.d. Bailey, Phenotypic screening in cancer drug discovery—past, present and future. 2014. 13(8): p. 588–602.

[49] Overington, J.P., B. Al-Lazikani, and A.L.J.N.r.D.d. Hopkins, How many drug targets are there? 2006. 5(12): p. 993–996.

[50] Dang, C.V., et al., Drugging the'undruggable'cancer targets. 2017. 17(8): p. 502–508.

[51] Lazo, J.S., E.R.J.A.r.o.p. Sharlow, and toxicology, Drugging undruggable molecular cancer targets. 2016. 56: p. 23–40.

[52] Drews, J.J.s., Drug discovery: a historical perspective. 2000. 287(5460): p. 1960–1964.

[53] Chen, X., et al., Drug–target interaction prediction: databases, web servers and computational models. 2016. 17(4): p. 696–712.

[54] Ghofrani, H.A., I.H. Osterloh, and F.J.N.r.D.d. Grimminger, Sildenafil: from angina to erectile dysfunction to pulmonary hypertension and beyond. 2006. 5(8): p. 689–702.

[55] Roche. Harnessing the power of AI. 2022; Available from: https://www.roche.com/partnering/harnessing-the-power-of-ai.htm.

[56] Hutter, C. and J.C.J.C. Zenklusen, The cancer genome atlas: creating lasting value beyond its data. 2018. 173(2): p. 283–285.

[57] Uhlén, M., et al., A human protein atlas for normal and cancer tissues based on antibody proteomics. 2005. 4(12): p. 1920–1932.

[58] Zhang, J., et al., International Cancer Genome Consortium Data Portal—a one-stop shop for cancer genomics data. 2011. 2011.

[59] Kaldor, S.W., et al., Viracept (nelfinavir mesylate, AG1343): A potent, orally bioavailable inhibitor of HIV-1 protease. 1997. 40(24): p. 3979–3985.

[60] Wishart, D.S., et al., DrugBank 5.0: a major update to the DrugBank database for 2018. 2018. 46(D1): p. D1074–D1082.

[61] Zhang, J., et al., The International Cancer Genome Consortium Data Portal. Nature Biotechnology, 2019. 37(4): p. 367–369.

[62] England, G. 2022.

[63] Yang, W., et al., Genomics of Drug Sensitivity in Cancer (GDSC): A resource for therapeutic biomarker discovery in cancer cells. Nucleic Acids Research, 2012. 41(D1): p. D955–D961.

[64] Institute, N.C. 2022.

[65] Tate, J.G., et al., COSMIC: the Catalogue Of Somatic Mutations In Cancer. Nucleic acids research, 2019. 47(D1): p. D941–D947.

[66] Gruener, R.F., et al., Facilitating Drug Discovery in Breast Cancer by Virtually Screening Patients Using In Vitro Drug Response Modeling. Cancers, 2021. 13(4): p. 885.

[67] Institute, N.C., TCGA Data Leveraged for Developing FDA-Designated Breakthrough Therapy. 2016.

[68] Zaman, A., W. Wu, and T.G. Bivona, Targeting Oncogenic BRAF: Past, Present, and Future. Cancers, 2019. 11(8): p. 1197.

[69] San Lucas, F.A., et al., Cancer in silico drug discovery: a systems biology tool for identifying candidate drugs to target specific

molecular tumor subtypes. Molecular cancer therapeutics, 2014. 13(12): p. 3230–3240.

[70] Huang, Q., et al., Gene Expression, Network Analysis, and Drug Discovery of Neurofibromatosis Type 2-Associated Vestibular Schwannomas Based on Bioinformatics Analysis. Journal of oncology, 2020. 2020: p. 5976465–5976465.

[71] Shenouda, S.K. and S.K. Alahari, MicroRNA function in cancer: onco-gene or a tumor suppressor? Cancer Metastasis Rev, 2009. 28(3–4): p. 369–78.

[72] Zhang, B., et al., microRNAs as oncogenes and tumor suppressors. Developmental Biology, 2007. 302(1): p. 1–12.

[73] Strand, S.H., T.F. Orntoft, and K.D. Sorensen, Prognostic DNA Methylation Markers for Prostate Cancer. International Journal of Molecular Sciences, 2014. 15(9): p. 16544–16576.

[74] Zwyea, S., L. Naji, and S. Almansouri. Kaplan-Meier plotter data analy-sis model in early prognosis of pancreatic cancer. in Journal of Physics: Conference Series. 2021. IOP Publishing.

[75] Mitsopoulos, C., et al., canSAR: update to the cancer translational research and drug discovery knowledgebase. Nucleic Acids Research, 2021. 49(D1): p. D1074–D1082.

[76] Raghupathi, W. and V. Raghupathi, Big data analytics in healthcare: promise and potential. Health information science and systems, 2014. 2: p. 3–3.

[77] Mohamed, A., et al., The state of the art and taxonomy of big data analyt-ics: view from new big data framework. Artificial Intelligence Review, 2020. 53(2): p. 989–1037.

[78] Beckmann, J.S., X. Estivill, and S.E. Antonarakis, Copy number vari-ants and genetic traits: closer to the resolution of phenotypic to geno-typic variability. Nature Reviews Genetics, 2007. 8(8): p. 639–646.

[79] Zarrei, M., et al., A copy number variation map of the human genome. Nature reviews genetics, 2015. 16(3): p. 172–183.

[80] Bates, D.W., et al., Big data in health care: using analytics to identify and manage high-risk and high-cost patients. Health affairs, 2014. 33(7): p. 1123–1131.

[81] Zhao, W., H. Ma, and Q. He. Parallel k-means clustering based on mapreduce. in IEEE international conference on cloud computing. 2009. Springer.

[82] Sun, T., et al. An efficient hierarchical clustering method for large data-sets with map-reduce. in 2009 International conference on parallel and distributed computing, applications and technologies. 2009. IEEE.

[83] Alghzzy, M.A., Density-based spatial clustering of applications with noises for DNA methylation data. 2017: Northern Illinois University.

[84] Moise, G., J. Sander, and M. Ester. P3C: A robust projected clustering algorithm. in Sixth international conference on data mining (ICDM'06). 2006. IEEE.

[85] Manogaran, G., et al., Machine learning based big data processing framework for cancer diagnosis using hidden Markov model and GM clustering. Wireless personal communications, 2018. 102(3): p. 2099–2116.

[86] Jung, J.W., et al., Application of an automatic segmentation method for evaluating cardiac structure doses received by breast radiotherapy patients. Physics and imaging in radiation oncology, 2021. 19: p. 138–144.

[87] Nishino, M., et al., CT tumor volume measurement in advanced non-small-cell lung cancer: performance characteristics of an emerging clinical tool. Academic radiology, 2011. 18(1): p. 54–62.

[88] Sivakumar, K., N. Nithya, and O. Revathy, Phenotype Algorithm based Big Data Analytics for Cancer Diagnose. Journal of Medical Systems, 2019. 43.

[89] Liu, B., et al., Artificial intelligence and big data facilitated targeted drug discovery. Stroke and vascular neurology, 2019. 4(4).

[90] Barbosa, C.D., Challenges with big data in oncology. J Orthop Oncol, 2016. 2(112): p. 2.

[91] Wilson, B. and G. Km, Artificial intelligence and related technologies enabled nanomedicine for advanced cancer treatment. 2020, Future Medicine. p. 433–435.

[92] Parikh, R.B., et al., Using big data and predictive analytics to determine patient risk in oncology. American Society of Clinical Oncology Educational Book, 2019. 39: p. e53–e58.

[93] Green, A.K., et al., The project data sphere initiative: accelerating cancer research by sharing data. The oncologist, 2015. 20(5): p. 464.

[94] Bilimoria, K.Y., et al., The National Cancer Data Base: a powerful initiative to improve cancer care in the United States. Annals of surgical oncology, 2008. 15(3): p. 683–690.

[95] Major, A., S.M. Cox, and S.L. Volchenboum, Using big data in pediatric oncology: Current applications and future directions. Seminars in Oncology, 2020. 47(1): p. 56–64.

[96] Hamilton, B., Big data is the future of healthcare. Cognizant 20–20 insights, 2012.

[97] Meropol, N.J., Opportunities for using big data to advance cancer care. Clinical advances in hematology & oncology: H&O, 2018. 16(12): p. 807–809.

[98] Tsai, C.J., N. Riaz, and S.L. Gomez, Big Data in Cancer Research: Real-World Resources for Precision Oncology to Improve Cancer Care Delivery. Semin Radiat Oncol, 2019. 29(4): p. 306–310.

[99] Wang, Y., et al., An integrated big data analytics-enabled transformation model: Application to health care. Information & Management, 2018. **55**(1): p. 64–79.

[100] Siegel, R.L., et al., Cancer statistics, 2021. 2021. 71(1): p. 7–33.

[101] Troyanskaya, O.G. Health IT Analytics.

[102] Tabata, R.C., The Future Challenges Of Big Data In Healthcare. 2021: Forbes.

[103] Tabata, R.C., The Future Challenges Of Big Data In Healthcare. 2021.

[104] Zhou, Z.-H., et al., Big Data Opportunities and Challenges: Discussions from Data Analytics Perspectives [Discussion Forum]. IEEE Computational Intelligence Magazine, 2014. 9: p. 62–74.

Index

365

About the Authors

Sonali Vyas is currently serving as an Academician and Researcher for more than a decade. Currently, she is working as an Assistant Professor (Selection Grade) with the School of Computer Science, University of Petroleum and Energy Studies, Uttarakhand, India. Her research interests include healthcare informatics, blockchain, database virtualization, data mining, and big data analytics. She has authored an ample number of research papers, articles, and chapters in refereed journals/conference proceedings and books. She authored a book on *Smart Health Systems* (Springer). She is also an Editor for *Pervasive Computing: A Networking Perspective and Future Directions* (Springer Nature) *and Smart Farming Technologies for Sustainable Agricultural Development* (IGI Global). She acted as a Guest Editor in a special issue of *"Machine Learning and Software Systems"* in *Journal of Statistics & Management Systems (JSMS)* (Thomson Reuters). She has also authored three patents in the area of smart and sustainable system. She has also acted as a Resource Person in AICTE-ISTE Faculty Refresher Course on "Embedded Systems, IoT, Pervasive Computing" and delivered many talks in reputed international and national conferences. She is also a member of Editorial Board and Reviewer Board of many referred national and international journals. She has also been a member of Organizing Committee, National Advisory Board, and Technical Program Committee at many international and national conferences. She has also Chaired Sessions in various reputed international and national conferences. She has been awarded the "National Distinguished Educator Award 2021" and "International Young Researcher Award 2021," instituted by the "International Institute of organized Research (I2OR)" which is a registered MSME, Government of India and Green ThinkerZ. She was also awarded the "Women Researcher Award 2021," by VDGOOD Professional Association in International Conference on Award Winners in Engineering, Science and Medicine. She was also awarded the "Best Academician of the Year Award (Female)" in "Global Education and Corporate Leadership Awards (GECL-2018)." She is a professional member of IEEE, ACM-India, CSI, IFERP, IAENG, ISOC, SCRS, and IJERT.

Deepshikha Bhargava has rich experience of around 22+ years as an Academician. She is currently working as a Professor with the School of Computing, DIT University, Dehradun, India. She has authored 16 books and 14 book chapters, edited 02 books, and published 60+ research papers in journals and conference proceedings. She has recently been nominated as a member in Project Review Steering Group (PRSG), Ministry of Electronics & IT (MeitY), Government of India. She has also served as a Visiting Fellow with Université des Mascareignes (UDM), Ministry of Education and Human Resources, Tertiary Education and Scientific Research, Mauritius. She has been nominated by MeitY, Government of India to visit Drone Application and Research Center (DARC), Information Technology Development Agency (ITDA), Department of Information Technology (Government of Uttarakhand). She is also empaneled in PaperVest Press Scientific Advisors, PaperVest University Publisher of Centro Universitário Facvest – UNIFACVEST, Brazil. Recently, she has been included as Reviewer for 2022 NSF Graduate Research Fellowship Program (GRFP) by National Science Foundation (NSF), USA. Prof. Bhargava also received the following awards: "Active Participation Woman Award," "Best Faculty of the year," under subcategory "Authoring Books on Contemporary Subjects," to name a few. She was also awarded by MHRD, Government of India in 1992 for academic excellence. Overall, 04 Ph.D. students completed under her guidance. Her research thrust areas are artificial intelligence, soft computing, bio-inspired computation, and healthcare informatics.

Samiya Khan is currently a Research Fellow with the University of Wolverhampton, Wolverhampton, United Kingdom. She is an alumna of University of Delhi, New Delhi, India. She received the Ph.D. degree in computer science from Jamia Millia Islamia, New Delhi, India. She has contributed several research papers and her publications extend across journal articles, conference papers, book chapters, magazine articles, and edited books in high impact publications of international repute. Dr. Khan's research interests include the multi-faceted use of artificial intelligence and edge computing for IoT applications.

Yogita Kumari has done her post-graduation under the specialization quality assurance from Lovely Professional University. Last year, she was nominated for the best researcher award 2021. Her area of research is targeted drug delivery for the treatment of various diseases like diabetes, ulcerative colitis, and Alzheimer's. She also has published quality research and review articles in highly reputed journals and also has presented research papers in national and international conferences. She is currently working as an Assistant Professor with the School of Pharmacy, Arka Jain University, Jharkhand, India.

Khushboo Raj is currently working toward the Ph.D. degree with Guru Ghasidas Vishwavidyalaya, Bilaspur (C.G), India. She has done her M.Pharm. with specialization in pharmaceutical chemistry from Guru Ghasidas Vishwavidyalaya, Bilaspur (C.G). She has done her B.Pharm. from Galgotias University, Greater Noida. She has also qualified GPAT in the year 2018. Her area of research interest includes "synthesis of novel compounds and biological evaluation; isolation & structure elucidation for targeted anti-cancer drugs." She has published quality research and review articles in highly reputed journals and also has presented research papers in national and international conferences. This book chapter is a very important contribution to her interest in different facets of research area and future endeavors.

Dilipkumar Pal (born in W.B., India), Ph.D., M.Pharm., Chartered Chemist, Postdoc (Australia), is an Associate Professor with the Department of Pharmaceutical Sciences, Guru Ghasidas Vishwavidyalaya (A Central University), Bilaspur (C.G.), India. He received the master's and Ph.D. degrees from Jadavpur University, Kolkata and performed postdoctoral research as "Endeavor

Post-Doctoral Research Fellow" in the University of Sydney, Australia. His areas of research interest include "Synthesis of novel compounds and biological evaluation; isolation, structure elucidation, and biological evaluation of indigenous plants." He has published 182 full research papers in peer-reviewed reputed national and international scientific journals, having good impact factor and contributed 118 abstracts in different national and international conferences. He has written one book, 68 book chapters, and edited 08 books published by reputed international publishers. His research publications have acquired a highly remarkable cited record in Scopus and Google Scholar (H-Index: 49; i-10-index: 110, total citations 7452 till date). He was granted one patent also. Dr. Pal, in his 22 years' research-oriented teaching profession, received 13 prestigious national and international professional awards also. He had received 03 research projects. He had guided 07 Ph.D. and 40 master's students for their dissertation thesis. He is the Reviewer and Editorial Board Member of 30 and 29 scientific journals, respectively. He is the member and life member of 15 professional organizations.

Ankita Moharana has completed post-graduation specialized in pharmaceutics from the School of Pharmaceutical Sciences, Shiksha 'O' Anusandhan University, Bhubaneswar, Odisha, India. Her area of research is formulation development and bioavailability enhancement. She has presented research and review papers in various seminar and conferences. She is currently working as an Assistant Professor with the School of Pharmacy, Arka Jain University, Jamshedpur, Jharkhand, India.

Thota Ramathulasi works as a Full-Time Research Scholar with the School of Computer Science and Engineering, Vellore Institute of Technology (V.I.T University), Vellore, Tamil Nadu, India. She has ten years of teaching experience. She received the M.Tech. degree in computer science and engineering from JNTUA, Anantapuram, Andhra Pradesh, India. Her research interests include service-oriented computing, software engineering, recommendation system, and data mining.

Rajasekhara Babu works as a Professor with the School of Computer Science and Engineering, Vellore Institute of Technology (V.I.T University), Vellore, Tamil Nadu, India. He had his primary education in Anantapur District, Andhra Pradesh. He received the B.Tech. degree in electronics and communication engineering from Sri Venkateswara University, Tirupathi, Andhra Pradesh, India, and the M.Tech. degree in computer science and engineering from R.E.C. Calicut (presently known as N.I.T. Calicut), Calicut, Kerala, India. He received the Ph.D. degree from V.I.T. University. He has published more than 100 publications in peer-reviewed journals and national/international conferences and has authored a couple of books in computer architecture, compiler design, and grid computing. He produced around 10 Ph.D. students from V.I.T. University. His area of research includes Internet of Things (IoT) and high-performance computing (HPC).

Mohamed Yousuff works as a Full-Time Research Scholar with the School of Computer Science and Engineering, Vellore Institute of Technology (V.I.T University), Vellore, Tamil Nadu, India. He has nine years of teaching experience. He received the bachelor's degree from Anna University, Chennai, India, and the master's degree from the B.S. Abdur Rahman Crescent Institute of Science and Technology, Chennai, India. His areas of interest include artificial intelligence, machine learning, deep learning, data analytics, and data visualization.

Akanksha Sharma is currently an Assistant Professor with the Monad College of Pharmacy, Monad University, Hapur. She received the B.Pharm. degree from the Kashi Institute of Pharmacy, Varanasi, in 2018. She received the M.Pharm. degree (Pharmaceutics) from Galgotias University in 2020. She has also published research and review papers mostly in scopus indexed journal. She is a good motivator and provides guidance to students for being a good pharmacy professional, researcher, and to be more human.

Ashish Verma is currently a student at School of Pharmacy, Monad University, Hapur. He received the B.Pharm. degree from the Kashi Institute of Pharmacy, Varanasi, in 2018. He always does work with students at ground level to understand along with utilization of different basic concepts of pharmacy in real life.

Dr. Rishabha Malviya is presently working as an Associate Professor with Galgotias University in the Department of Pharmacy, School of Medical and Allied Sciences Greater Noida, UP, India. He received the Ph.D. degree from Galgotias University, India. He received the M.Pharm. degree in pharmaceutics (2008–2010) and the B.Pharm. degree (2004–2008) from U.P Technical University, Lucknow, India. He has published more than 115 articles in different international and national journals. He has published two books and four book chapters and also has some patents. He has guided 23 M.Pharm. students for their respective project work. He was qualified in Gate with 94.36 percentile during 2008–2009 and also received Faculty Appreciation Award from Meerut Institute of Engineering and Technology and Appreciation Letter from Rexcin Pharmaceuticals Private Limited. He is serving as Editorial Board Member and Reviewer of various international and national journals. He is also serving as consultant/advisor to industry for process optimization and formulation. development.

Deepika Bairagee, B.Pharm., M.Pharm. (Quality Assurance), is an Assistant Professor with the Oriental College of Pharmacy and Research, Oriental University, Indore, India. She has five years of teaching experience and two years of research experience. She has spoken at more than 20 national and international conferences and seminars, presenting over 20 research articles. She has over 20 publications in international and national journals. She is the author of over 18 books. She has over 50 abstracts that have been published at national and international conferences. Young Researcher, Young Achievers, and Excellent Researcher were among the honors bestowed upon her. Proteomics and metabolomics are two areas of research that she is currently interested in.

Mrs. Vandana Tyagi has more than 5year of experience in teaching domain. Mrs. Vandana Tyagi did her graduation from Mahatma Jyotiba Phule University, Bareilly in 2014 and her post graduation in Pharmaceutics in 2016 from Galgotia University, Greater Noida attaining Gold Medal. She is currently working in Starex University, Gurugram, Haryana and pursuing PhD (Pharmaceutics (part time) from Amity University,Noida. She has worked as Assistant professor in Ramgopal College of pharmacy, RNRM college of pharmacy & RAS college of pharmacy, Gurgaon. She has published 3 review article and 1 research paper, presented over 4 abstracts in various national seminars

Dr. Amrish Chandra is an established Academician, a Research scientist, and a Registered Patent agent with Indian Patent Office. He has extensive experience of 19 years and is associated with Amity Noida from 2011 and is currently serving as Acting Director AIPH.

Dr Chandra did his Master's in pharmacy from Birla Institute of Technology, Mesra in 2002, a GATE scorer with 92.32 percentile, he did his Ph.D. in Transdermal Drug Delivery System in Pharmaceutical Sciences. His research profile includes more than 100 publications with more than 1250 citations, h-index of 17 & i-10 index of 26, his publications have a cumulative impact factor of 70+, has supervised 8 PhD and 35 PG students & has a DST sponsored Indo-Hungary Project with cumulative total of 1.75 Crore & an Italian consultancy project. He has written one book with Springer & 3 book chapters.

Dr Amrish Chandra has been dedicated to patent drafting and prosecution. He has a vast array of experience with all aspects of the Indian patent process & PCT filing, including assessing patentability and performing freedom-to-operate searches; drafting applications; reporting and responding to Office Actions; preparing infringement and validity opinions, and counselling researchers regarding portfolio management strategies. He has also advised researchers extensively on and assisted with the preparation of industrial design & copyright applications. He actively supports more than 250 researchers every year in filing patent, design & copyright.

Dr. Chandra has delivered numerous talks and influenced more than 10,000 audience till date.

His passion to help academicians bring their research to the forefront is what drives him to this platform today.

Dr. Neelam Dhankhar is a distinguished professor of Pharmaceutics, Principal in School of Pharmaceutical Sciences, Starex University, Gurugram. She has served as head of Institution and Associate Professor for curriculum and teaching. She also works for training and placement cell of University. She is having PhD. in Pharmaceutical Sciences. She also works in various pharmacy institutions.

Professor Neelam, Research, teaching and administrative interests are like a supply chain for her institutions and university where she works. She has done a Certified Course in IPR from WIPO, Geneva, Switzerland and M. Pharmacy project from Ranbaxy Research Laboratories, PDR Department. She has published more than 20 research and review papers in various International Journals, presented over 8 abstracts in various national seminars. She has attended many conferences like IPC & IPGA, various FDP and done numerous poster presentations. She is also the Editor and Reviewer of various journals of pharmacy. Her research activities include Nano medicine and Innovative technologies in Drug delivery and analysis, Pre-formulation; formulation development; physicochemical assessment; stability assessment and prediction; in-vitro performance tests, transport; pharmacokinetics and bio-distribution and in vitro-in vivo correlation.

Dr. Bhavna Tyagi has more than 5 year of experience in Academic Writer/ Medical writer. Mrs. Bhavna Tyagi did her BDS from IDST, Meerut in 2013 and her post MDS in Oral Pathology and Microbiology in 2017 from Sharda University, Greater Noida. She has published 3 review article and 1 research paper, presented over 3 poster presentations in various national seminars and attended 8 conferences.

Neeraj Kumar completed his B. Pharm from College of Pharmacy Jhansi and M.Pharm from Advance Institute of Biotech & Paramedical Sciences Kanpur and I presently working as an Assistant Professor, Department of Pharmacy at Dr. M.C. Saxena College of Pharmacy. I have teaching experience of more than 2 years and I taught a diverse set of subjects like Pharmaceutics, Herbal Drug Technology, Pharmaceutical Engineering, and Novel Drug Delivery systems. I have published different research papers and participated in many conferences of National and International. One of the most significant impacts on my interest in technology has been this book chapter along with different faces of the research area and future ventures.

Amrita Shukla M.Pharm (Pharmacology) from Integral University, Lucknow and I am presently working as Assistant Professor, at Dr. M.C. Saxena College of Pharmacy, Lucknow. I have teaching experience for more than 5 years and I have taught various subjects like Human Anatomy and Physiology, Pathophysiology, Pharmacology, and Pharmacy Practice. I have been actively participating in various research activities including research papers, seminars, webinars and conferences at the National and International levels. I have also worked in the area of Intellectual Property rights at Patent Facilitating Centre, New Delhi.

Shahid Raza graduated from Dr. M.C. Saxena College of Pharmacy, Lucknow in B Pharmacy I am currently working as Lecturer, at Dr. M.C. Saxena College of Pharmacy, Lucknow. I have teaching experience of 18 months and I have taught various subjects like Pharmaceutical Chemistry, Food Microbiology and Drug Store and Business Management. I have actively participated in various activities like seminars, webinars, conferences and workshops at the National and International levels which is ISO certified.

Simran Ludhiani has completed her M.Pharm (Pharmaceutics) from Shri Govindram Sakseria Institute Of Technology And Science in 2019. She has worked as a medical writer in APCER Life Sciences from Aug 2021 to Aug 2022. Currently, she is working IQVIA as an associate medical writer since Aug 2022. She has exposure in CSR regulatory writing. She has worked on the development of a liquisolid system of a poorly water-soluble drug, furosemide by mixed solvency concept under the mentorship of Dr. R.K. Maheshwari. She has worked on other projects with mixed solvency concepts. She has been a part of many academic and research-based activities. This book chapter is an effort the extension of knowledge in the area of research.

Sudhanshu Mishra completed my M. Pharm (Pharmaceutics) from Rajiv Gandhi Proudyogiki Vishwavidyalaya and currently working as a Teaching faculty in the Department of Pharmaceutical Science & Technology, Madan Mohan Malaviya University of Technology, Gorakhpur. He worked on the herbal topical formulation for the treatment of arthritis during my M. Pharm research work and develop an interest in herbal formulation-related research work. Meanwhile, he is working on various literature, like writing review articles on the different novel approaches and technology for targeting chronic diseases. He is also participating in various academic activities like international seminars, conferences, workshops, and oral presentations. This book chapter is one of the important contributions to his interest in technology and the future research area.

Prof. Sonali Sundram completed B. Pharm & M. Pharm (pharmacology) from AKTU, Lucknow. She has worked as research scientist in project of ICMR in King George's Medical University, Lucknow after that she has joined BBDNIIT and currently she is working in Galgotias university, Greater Noida. Her PhD (Pharmacy) work was in the area of Neurodegeneration and Nanoformulation. Her area of interest is neurodegeneration, clinical research, artificial intelligence. She has Edited 4 books (Wiley, CRC Press/Taylor and Francis, River Publisher) She has attended as well organized more than 15 national and international seminar/conferences/workshop. She has more than 8 patents national and international in her credit.

Shilpa Rawat has completed her B. Pharmacy from HNBU Garhwal University, Uttarkhand and M. Pharma (pharmaceutics) From Galgotias University, Greater Noida, India. Her area of interest is in the area of Nano-formulation, targeted drug delivery, artificial intelligence, big data, Deep learning and Machine learning. She has also published 2 chapters in the field of Bioinformatics and deep learning with the prestigious CRC press publisher, Wiley publisher. She has attended 3 international and 5 national seminars/conferences/workshops. Her strength is research skill, thinking innovation, leadership qualities, decision making and positive thinking.

Shilpa Singh has completed her B.Pharmacy from Akshar preet institute of pharmacy, affiliated by Gujarat Technological University, Gujarat and M.Pharm (Pharmaceutics) From Galgotias University, Greater Noida, India. Her area of interest is in the area of nano-formulation, targeted drug delivery, cloud computing, artificial intelligence, Deep learning and machine learning. She has also published 2 chapters in the field of cloud computing and deep learning with the prestigious CRC press publisher, Wiley publisher. She has attended 3 international and 5 national seminars/conferences/workshops. Her strength is research skill, leadership qualities, innovative thinking, decision making, quality of leadership and positive thinking.

Shweta Kumari is currently working at T.John College of Pharmacy, Bangalore as an Assistant Professor in Pharmaceutics. She received her undergraduate degree (B.Pharma) from Sham Higginbottom University of agriculture, technology & Sciences Allahabad. and M.Pharma from Guru Ghasidas Vishwavidyalaya Bilaspur. She has been awarded the gold medal in the undergraduate program. She has qualified GATE & GPAT with AIR 2170 & 1372 respectively. She has published one research article and 3 book chapters in a highly indexed journal. She also filed an Indian patent. Her research interest is in nanoparticle, microsphere, and targeted drug release. And her area of subject is novel drug delivery, biopharmaceutics, etc.

Pawan Upadhyay is currently working as an assistant professor at the Maharishi University of Information Technology, Lucknow in the School of Pharmaceutical sciences (Department of Pharmaceutics). He has one year of teaching and research experience. He received his bachelor's degree in Pharmacy from Raja Balwant Singh Engineering Technical Campus (Bichpuri Agra Uttar Pradesh) affiliated to Dr. A.P.J. Abdul Kalam Technical University Lucknow. He completed their Master's degree in Pharmacy from the Arya College of Pharmacy kookus,

Jaipur affiliated to Rajasthan University of Health Sciences Jaipur. He had work experience on the matrix tablet during his masters. He had published two review papers in a different journal. He qualified in GRADUATE PHARMACY aptitude test in the year 2019 and the NIPER exam in 2019. He also attended many webinars, conferences, and training.

Akhalesh Kumar, Obtained His B. Pharm And M. Pharm Degree From Dr. A.P.J. Abdul Kalam Technical University, Lucknow, Uttar Pradesh. He Has Worked As An Assistant Professor at Maharishi University Of Information Technology, Lucknow. He Has One Year Of Teaching And Research Experience. His Research Interests are Organics Synthesis, Phytochemical Analysis, and Biological Evaluation of Anticancer and Anticonvulsant Activity. He Has Published Many Reviews and Research Articles in both National and International Peer-Reviewed Journals. He Presented Two Papers In International Conferences.

Ruchi Singh, pursuing PhD (Pharmacology) from Banasthali Vidyapith, Newai, Rajasthan. I have approx 15 research and review articles in highly indexed journal with high impact factors. I have also presented research papers in national and international conferences. Presently I am working as Assistant Professor at Narayan Institute of Pharmacy in Gopal Narayan Singh University, Jamuhar, Bihar (India).

Dr Smriti Ojha is an Experienced Professor with a demonstrated history of working in the pharmaceutical education industry. She earned a Doctorate in Pharmaceutical sciences from Dr. A. P. J. Abdul Kalam Technical University, Lucknow. She is actively engaged in the research area of Drug Delivery, Pharmaceutics, Pharmaceutical Research, and nanotechnology. Currently, she is working at Madan Mohan University of Technology, Gorakhpur, Uttar Pradesh, India. This book chapter is a very important contribution to her interest in different facets of the research area and future endeavors.

Poojashree Verma, Associate professor Poojashree Verma pursuing a Ph.D. Pharmaceutics (novel drug delivery system)) from oriental university, Indore Madhya Pradesh. She is over 8-year teaching and research experience. She has been working on a project on women hormonal related disorders from novel drug delivery systems and related to various research work in formulation development of transdermal drug delivery system and also biotechnology and gene therapy, currently she is working as an Associate professor in a Pacific University, Udaipur, India. She published 10 research paper in novel drug delivery systems, herbal drug delivery systems and formulation technology in various reputed national and international journals.

Dr. Neelam Jain, Prof. (Dr.) Neelam Jain holds a Doctorate in Pharmaceutics (specialization in Nanomedicine) from the Institute of Foreign Trade and Management, Moradabad, Uttar Pradesh, India. She has over 11 years of research and teaching experience. She bagged Gold medal by Dr. A.P.J. Abdul Kalam for securing first position in B. Pharm from Uttar Pradesh Technical University, Lucknow, Uttar Pradesh and fellowships for her excellent contributions in her fields. She has been working on an array of projects relating to nanomedicine, gene therapy, nanotechnology, and pharmacology. Her research vicinity focuses on novel formulation and clinical trials etc. She uses state-of-the-art technology, for systematic evaluation of the efficiency of novel polymeric nanoparticles encapsulated with biologically active agents. Currently, she is Associate Professor and HOD in the Faculty of Pharmacy, Oriental University, Indore, India. She published extensively more than 50 research/review papers in nanomedicine, drug delivery, and formulation technology in peer-reviewed reputed journals and books.

Porf. Dr. Neetesh Kumar Jain having total 14 years of experience in Academic and Research. Currently working as Professor & Dean in Faculty of Pharmacy, Oriental University, Indore and Director, Drug & Disease Information Center, Indore. He has strong academic and research background, as he has 09 Patents, published more than 146 Research papers in National and International journals and presented more than 186 Research papers in National and International conferences. In addition to that, He also received many prestigious awards i.e. Eminent Professor Award, Inspirational Associate Professor Award, Visionary Principal Award, Inspirational Leadership Award, Young Scientist Award, Young Talent Award, Award for Excellence in Research, Outstanding Research in Pharmacy in Central India" for efforts made by him to bridging the gap between Indian & global standards and recently he got "Academician of the Year-2018" by EET CRS -Academic Brilliance Awards-18, Excellent Teacher Award etc. for his research contribution in the field of Medical and Pharmaceutical sciences. He is approved Ph.D supervisor in Oriental University, Indore and Awarded 04-Ph.D, Submitted-03 and having 8 Ph.D research scholars under his guidance.

He has International exposure as Plenary Speaker in 22nd International Conference on Research in Life-Sciences & Healthcare (ICRLSH), 18-19 October, Kasetsart University, Chatuchak Bangkok-Thailand for delivering "Plenary Address" and to preside over the Conference as Conference Chair and Discussant.

Dr. Sumeet Dwivedi has completed his B.Pharm from SCOPE, Indore, M.Pharm with Honors & Gold Medal in Pharmacognosy & Phytochemistry from Vinayaka Mission's University, Salem, TN and Ph.D. from Suresh Gyan Vihar University, Jaipur, RJ, India. He has awarded with Fellowship FLSL in 2013 from Pavan Education society Gujrat for his outstanding performance in research on herbal medicine and herbal formulations. Presently Dr. Dwivedi is working as Professor & Principal, Faculty of Pharmacy (UIP), Oriental University, Indore, M.P. He has his credit of more than 280 research/review papers in various national and international Journal of repute, has more than 100 abstract in various Seminar & Conferences, published 30 books with 8 Indian Patent and 1 International Patent and serving as Executive Editor, International Journal of Pharmacy and Life Sciences and Managing Editor, International Journal of Drug Discovery and Herbal Research, Editor, Life Science Leaflets, Reviewer of many Journals and Editorial Board Member of several International Journal. He has been awarded as Young Scientist award in 2018, Young Researcher award in 2018, Inspirational Teacher award in 2019 & Young Achiever award in 2020 by various organizations. He has also received letter of appreciation by DHR Industrial, India for adopting his research work formula and methodology for formulation of herbal oil and shampoo in Nov, 2018. He has delivered more than 20 lectures in various conferences & Seminar on herbal medicine, traditional claims, folk lore uses of herbs & Herbal formulations. Dr. Dwivedi has successfully organized 6 Seminar funded by MPCST & AERB; 1 Workshop on Medcinal plants and completed one major research project funded by Omar Al-Mukhtar University, Al- Bayda, Libya . He has guided more than 25 Postgraduate students, awarded one Ph.D, submitted 2 Ph.D and presently 8 students are working under him for their Ph.D. He has keen interest in Herbal Medicine, Documentation and Validation of herbal dosage form. He is involved in research on biological & pharmacological screening, formulation, standardization of herbals, cultivation practices of medicinal plants and biotechnological approaches of herbs.

Swati Verma has completed B. Pharm from KIET (AKTU), Ghaziabad & M. Pharm (pharmaceutical chemistry) from Jaipur. She has joined BBDNIIT as Assistant Professor and currently she is working in Galgotias university, Greater Noida. Her area of interest is computer aided drug design (CADD), peptide chemistry, analytical chemistry, medicinal chemistry, artificial intelligence, neurodegeneration, and gene therapy. She has attended and well organized more than 15 national and international seminar/conferences/workshop.

Akanksha Pandey, the author in this book, has done her schooling from a convent school in Jhansi. Later on, she did her graduation in pharmacy (B. Pharm) from S. R. Group of Institutions, Jhansi. Currently, she is pursuing her masters in pharmacy with specialization in pharmaceutics from Galgotias University, Greater Noida. Her area of interest is in the area of nano-formulation, targeted drug delivery, artificial intelligence, big data and machine learning. At present, she has several chapters in different books. She has attended 2 international and 5 national Seminars/Conferences/Workshops.

Brojendra Nath Saren completed his B. Pharm (2016-2020) from Bengal College of Pharmaceutical Sciences & Research, West Bengal, India, and received his M.S. in Pharmaceutics (2020-2022) from the National Institute of Pharmaceutical Education and Research, Hyderabad. He has co-authored a review article on "Applications of lipid-engineered nano-platforms in the delivery of various cancer therapeutics to surmount breast cancer" in the Journal of Controlled Release. Currently, he is pursuing doctoral degree in the area of Pharmaceutical Nanotechnology at National Institute of Pharmaceutical Education and Research, SAS Nagar.

Vikram Prajapat has completed graduation in Pharmacy (2016-2020) from Govt. College of Pharmacy, Karad, Maharashtra, India, and received his M.S. in Pharmaceutics (2020-2022) from The National Institute of Pharmaceutical Education and Research, Hyderabad, India. His research interests include the application of nanotechnology in drug delivery. He has developed an understanding of the QbD approach, Strategy Planning, DoE and risk assessment to achieve the target quality product profile. Currently, he is working as an analyst in Commercial Analytics and Solutions team at IQVIA India.

Shubham Appasaheb Awaghad has completed his M.S. (Pharm) from National Institute of Pharmaceutical Education and Research, Hyderabad. He has been a research intern in Lupin Pharmaceuticals, Pune where he has worked on formulation development of oral solid dosage forms. He is currently working as a formulation scientist in Jodas Expoim, Hyderabad. His area of focus include development of complex injectable, biosilimilars, stability enhancement of highly sensitive formulations, impurity profiling of complex injectable and quality by design.

Indrani Maji, is the PhD candidate in Department of Pharmaceutics, NIPER Hyderabad. She has completed her MS Pharm Pharmaceutics from Department of Pharmaceutics NIPER Mohali. Her research interests include Novel drug delivery system and Nano medicine. Currently, she has 3 peer reviewed publications in journals and 2 book chapters.

Mayur Aalhate is a Doctoral Researcher at National Institute of Pharmaceutical Education and Research, Hyderabad. He has earned his M.S. (Pharm.) from National Institute of Pharmaceutical Education and Research, S.A.S. Nagar (Mohali), Punjab. He served as a formulation scientist, Dr. Reddy's Laboratories, Hyderabad (2018-2021) in the development, scale-up, and technology transfer of complex Inhalation formulations. He is well versed with product life cycle, regulatory requirements and quality guidelines for

USA and China market. He has co-authored 2 review publications. His work focuses on cancer nano medicine and targeted therapeutics.

Srushti Mahajan is a Doctoral Researcher at National Institute of Pharmaceutical Education and Research, Hyderabad. She has earned her M.S. (Pharm.) from National Institute of Pharmaceutical Education and Research, S.A.S. Nagar (Mohali), Punjab. She has worked as a formulation scientist, Dr. Reddy's Laboratories, Hyderabad (2018-2020) in the formulation development and scale of oral solid dosage forms. Her area of interest is design thinking, process capability analysis, risk assessment and sustainable product development. She has co-authored 3 review publications and 8 book chapters. She is currently working on formulation development of various targeted cancer therapeutics.

Dr. Pankaj Kumar Singh, currently working as an Assistant Professor in the Department of Pharmaceutics at NIPER Hyderabad, India. He received a Ph.D. degree from CSIR-CDRI, Lucknow, India. His research of interest includes the development of targeted drug delivery nano-medicines including micelles, nanoparticles, & liposomes for Cancer therapy. He has published more than 55 quality research papers along with reviews in journals of high repute and has an h-index of 17, i10 index of 22, 662 citations with cumulative impact factor 285. He has edited 4 books, co-authored 18 book chapters, filed 6 patents with 1 patent granted. He has also bagged several accolades like the "CSIR-Research Associate" award, "CSIR-Senior Research Fellowship" award, Travel grant award, financial Assistance, and Incentive award for research paper publication.

About the Editors

Rishabha Malviya completed B. Pharmacy from Uttar Pradesh Technical University and M. Pharmacy (Pharmaceutics) from Gautam Buddha Technical University, Lucknow Uttar Pradesh. His PhD (Pharmacy) work was in the area of Novel formulation development techniques. He has 11 years of research experience and presently working as Associate Professor in the Department of Pharmacy, School of Medical and Allied Sciences, Galgotias University since past 8 years. His area of interest includes formulation optimization, nanoformulation, targeted drug delivery, localized drug delivery and characterization of natural polymers as pharmaceutical excipients. He has authored more than 150 research/review papers for national/international journals of repute. He has 51 patents (12 grants, 38 published, 1 filed) and publications in reputed National and International journals with total of 91 cumulative impact factor. He has also received an Outstanding Reviewer award from Elsevier. He has Edited 13 books (Wiley, CRC Press/Taylor and Francis, Springer Nature, Apple Academic Press, River Publisher, Lambert and OMICS publication) and authored 15 book chapters. His name has included in word's top 2% scientist list for the year 2020 by Elsevier BV and Stanford University. He is Reviewer/Editor/Editorial board member of more than 50 national and international journals of repute. He has invited as author for "Atlas of Science" and pharma magazine dealing with industry (B2B) "Ingredient south Asia Magazines".

Naveen Chilamkurti is currently working as Professor and Head of Cybersecurity Discipline Director, La Trobe Cybersecurity Research Hub, La Trobe University, Melbourne, Australia. He has completed master's and Ph.D. in computer science from La Trobe University. His passion is teaching and adapts different design thinking principles while delivering his lectures. He has published more than 10 books and book chapter on various technologies. He has published more than 250 review and research article in national and international journal.

Sonali Sundram completed B.Pharm and M.Pharm (pharmacology) from AKTU, Lucknow, Uttar Pradesh, India. She has worked as a Research Scientist in the project of ICMR in King George's Medical University, Lucknow, Uttar Pradesh, India. After that, she has joined BBDNIIT and currently she is working with Galgotias University, Greater Noida, India. Her Ph.D. (pharmacy) work was in the area of neurodegeneration and nanoformulation. Her areas of interest are neurodegeneration, clinical research, and artificial intelligence. She has attended more than 15 well-organized national and international seminar/conferences/workshops. She has more than 8 patents in her credit.

Rajesh Kumar Dhanaraj is currently an Associate Professor with the School of Computing Science and Engineering, Galgotias University, Greater Noida, India. He holds a Ph.D. degree in information and communication engineering from Anna University, Chennai, India. He has contributed 20+ books on various technologies and 35+ articles and papers in various refereed journals and international conferences and contributed chapters to the books. His research interests include machine learning, cyber–physical systems, and wireless sensor networks. He is an Expert Advisory Panel Member of Texas Instruments, Inc., Dallas, TX, USA

Balamurugan Balusamy has served up to the position of Associate Professor in his stint of 14 years of experience with VIT University, Vellore, Tamil Nadu, India. He had completed his bachelor's, master's, and Ph.D. degrees from top premier institutions in India. His passion is teaching and adapts different design thinking principles while delivering his lectures. He has published 30+ books on various technologies and visited 15+ countries for his technical course. He has several top-notch conferences in his resume and has published over 150 quality journals, conferences, and book chapters combined. He serves in the advisory committee for several startups and forums and does consultancy work for industry on industrial IoT. He has given more than 175 talks in various events and symposium. He is currently working as a Professor with Galgotias University, Greater Noida, India and teaches students and does research on blockchain and IoT.

Milton Keynes UK
Ingram Content Group UK Ltd.
UKHW051535141024
449569UK00001B/37